HISTORY OF PHYSICS

SELECTED REPRINTS

EDITED BY

STEPHEN G. BRUSH

PUBLISHED BY

AMERICAN ASSOCIATION OF PHYSICS TEACHERS

History of Physics
©1988 American Association of Physics Teachers

Published by:
American Association of Physics Teachers
Publications Department
5112 Berwyn Road
College Park, MD 20740, U.S.A.

Cover Art: The diagram of Maxwell's representation (corrected slightly) of an electromagnetic ether in and surrounding a current-carrying conductor appears on page 102 in "The Mutual Embrace of Electricity and Magnetism" by M. Norton Wise. The cover rendition of this diagram was designed by Rebecca Heller Rados.

ISBN: #0-917853-29-6

Contents

RESOURCE LETTER

Roger H. Stuewer, *Editor*

School of Physics and Astronomy, 116 Church Street
University of Minnesota, Minneapolis, Minnesota 55455

This is one of a series of Resource Letters on different topics intended to guide college physicists, astronomers, and other scientists to some of the literature and other teaching aids that may help improve course contents in specified fields. No Resource Letter is meant to be exhaustive and complete; in time there may be more than one letter on some of the main subjects of interest. Comments on these materials as well as suggestions for future topics will be welcomed. Please send such communications to Professor Roger H. Stuewer, Editor, AAPT Resource Letters, School of Physics and Astronomy, 116 Church Street SE, University of Minnesota, Minneapolis, MN 55455.

Resource letter HP-1: History of physics

Stephen G. Brush

Department of History and Institute for Physical Science & Technology, University of Maryland, College Park, Maryland 20742

(Received 14 November 1986; accepted for publication 19 November 1986)

This Resource Letter provides an introduction to the literature on the history of physics. Because of the overwhelming quantity of publications, the emphasis is on works covering broad topics and historical periods rather than on monographs dealing with special subjects or individual physicists, and on recent reference works that facilitate retrieval of more specialized information. The letter E after an item indicates elementary level or material of general interest to persons becoming informed in the field. The letter I, for intermediate level, indicates material of somewhat more specialized nature; and the letter A indicates rather specialized or advanced material. An asterisk (*) indicates those articles to be included in an accompanying Reprint Book.

Table of Contents

I. INTRODUCTION

Physics is a small science, if judged by number of researchers, teachers, and students. It is somewhat larger if judged by the current rate of article publication and amount of funding (see recent issues of *Science Indicators* published by the National Science Board). But physics is biggest of all for its history: The number of works published in recent years about physics from 1800 to 1914, for example, is several times as great in relation to the number of physicists in that period as the number of works on any other science. (This assertion can be checked by glancing at recent issues of the annual Isis Critical Bibliography published by the History of Science Society.)

Estimating the audience for history of physics is more difficult. One indication is the sales of a recent, fairly technical biography of Albert Einstein by Abraham Pais (Ref. 207): 60 000 copies in the first three years after publication. Thomas Kuhn's *The Structure of Scientific Revolutions* (Ref. 29), a very popular work that makes considerable use of examples from the history of physical science, sold 250 000 copies in the dozen years following its publication in 1962. [I thank Professor Pais for information about the sales of his book. The figures for Kuhn's book are taken from an article by John D. Heyl, "Kuhn, Rostow, and Palmer: The problem of purposeful change in the 60s," Historian **44**, 299–313 (1982).] These figures are, of course, small compared to the initial print orders for today's fiction bestsellers, but they are quite respectable for scholarly or scientific books. Another gauge is the number of physicists who are interested enough to keep up with reports of current research and other historical activities: The Division of History of Physics of the American Physical Society now has more than 1700 members and that figure does not include members of the American Association of Physics Teachers who are interested in history but do not belong to APS. Historical sessions at APS and joint APS/AAPT meetings have consistently attracted large crowds, sometimes more than any other subfield of physics.

The history of physics has been used in a variety of ways. Tales of yesterday's historic deeds have traditionally been used to inspire physics students. Policymakers invoke earlier experience in justifying support for basic research; the recent proposal to consolidate several federal agencies in a single Department of Science and Technology has been criticized on the grounds that past attempts at centralized control of research (e.g., in 19th century France) stifled the freedom to seek unorthodox solutions to new problems. Colleagues in other sciences look to the history of physics for ideas about what makes a science successful. Teachers have employed the historical approach to make science more interesting to nonscience majors, portraying research as a human and socially involved activity. Finally, in a highly developed discipline like physics, whose professionals and graduate students tend to spend most of their time solving ever more complicated technical problems, history gives an opportunity to discuss old, but still important, philosophical issues such as whether the world exists independently of the observer.

The purpose of this Resource Letter is to help physics teachers find books and articles on the history of physics, and to guide librarians in developing collections to support education and research. It is impossible to list more than a small selection from the huge literature on this subject, so preference has been given to items most likely to be useful to someone not already familiar with it. Thus we favor recent publications on broad topics, together with reference works that allow one to retrieve more specialized monographs and articles, editions of collected works, biographies, and publications in foreign languages. Anthologies of original sources are cited here only if they contain new translations of articles, since such translations are often difficult to locate.

Readers who want an elementary introduction to the subject are advised to scan selected chapters in general works on the history of science as indicated in Sec. IV (Refs. 2, 4, 5, 9) and then look at a thoughtful physicist's account of the development of modern physics (Ref. 176). The articles marked by asterisks, which will appear in a separate Reprint Book, give a sampling of what historians of science have written on a few topics of general interest. (The Reprint Book does not include articles from *Physics Today* reprinted in Ref. 177, which may be considered a companion volume emphasizing 20th century physics.) Three surveys of historical research (Refs. 188, 212, 213) offer recommendations on where to go for more detailed studies.

An important resource is the Center for History of Physics at the American Institute of Physics, 335 East 45 Street, New York, NY 10017. Information about historical materials of all kinds can be obtained from the Center's Manager, Spencer Weart, or Associate Manager, Joan Warnow. They publish a Newsletter that includes reports on current activities, recent publications, and archival collections. Members of the American Physical Society may join the Division of History of Physics and receive its History of Physics Newsletter at no additional cost; others may subscribe to this Newsletter for a small fee. (Write to Professor Albert Wattenberg, Physics Department, University of Illinois, Urbana, IL 61801.)

II. JOURNALS

Articles on the history of physics appear regularly in the following journals. A more complete list of journals that publish articles on the history of science, together with information about societies and graduate programs, may be found in the Isis Guide to the History of Science (published about once every three years by the History of Science Society). Isis publishes each year a "Critical Bibliography" of publications in the history of science.

Historical Studies in the Physical and Biological Sciences
Archive for History of Exact Sciences
Isis
Osiris
British Journal for the History of Science
Centaurus
Physics Today
American Journal of Physics
Physics Teacher
Soviet Physics Uspekhi
Studies in History and Philosophy of Science
Archives Internationales d'Histoire des Sciences
History of Science

III. CONFERENCE PROCEEDINGS

The only continuing series of conference proceedings that regularly includes several articles on history of physics is the **International Congress of the History of Science,** which meets every three or four years. The most recent Congress, number XVII, was held in Berkeley, 31 July–8 August 1985. Information about the publication of its Proceedings may be obtained from the Office for History of Science and Technology, 470 Stephens Hall, University of California, Berkeley, CA 94720.

In 1972, course LVII of the International School of Physics "Enrico Fermi" at Varenna, Italy, was devoted to lectures on several aspects of the history of modern physics; its Proceedings were published as

1. **History of Twentieth Century Physics**, edited by C. Weiner (Academic, New York, 1977). Includes lectures by historians of physics and recollections of physicists. (I).

Other conferences have been devoted to particular topics or individual physicists, and are listed below in the appropriate sections.

IV. TEXTBOOKS AND EXPOSITIONS

A. History of science

For the history of physics since the 18th century, the best brief survey is the article by Pyenson (Ref. 17). To get an overview of the early developments one should read a book that presents physics as part of physical science or science as a whole. There are, of course, a few books devoted solely and somewhat comprehensively to the history of physics, but one of the first things to be learned from the history of science is that the area of science we now call physics did not develop in isolation from other sciences and indeed did not even exist as a coherent subject until the middle of the 19th century. Mechanics, the oldest part of physics, was strongly influenced by the Copernican Revolution in astronomy and by developments in branches of mathematics, such as analytic geometry and calculus in the 17th century. Optics, acoustics, and pneumatics were firmly established as quantitative sciences in the 17th century; heat, electricity, and magnetism had to wait until the 18th century. (Kuhn's article, Ref. 6, explains the historical relations between these different subjects.) It was the discovery of the conservation and transformation of different forms of energy, along with changes in the educational systems in France, Germany, and Britain, that allowed these separate branches of "natural philosophy" to be combined into a single science. Yet atomic physics and solid state, despite their roots in antiquity, did not become respectable members of the club until the 20th century, partly with the help of discoveries in chemistry. Astronomy again played a role in stimulating the growth of the newest subfield of physics, relativity/cosmology.

2. **Greek Science in Antiquity,** Marshall Clagett (Collier, New York, 1963), Chaps. 1, 2, 3, and 5 cover the origins of science and Greek mathematics; Chap. 6 is on Greek physics and Chap. 13 is on later Greek science. (E)

3. **The Exact Sciences in Antiquity,** Otto Neugebauer (Brown U. P., Providence, 1957), 2nd ed. (A)

4. **Physical Science in the Middle Ages,** Edward Grant (Wiley, New York, 1971), Chaps. I, II, III, IV, VI. (E)

5. **Birth of a New Physics,** I. Bernard Cohen (Norton, New York, 1985), revised and updated edition. Read especially Chaps. 1, 2, 5, and 7 (other chapters deal primarily with astronomy) and supplements 3–10, 14–16 which incorporate recent scholarship on Galileo and Newton. (E)

6. **Mathematical vs. Experimental Traditions in the Development of Physical Science,** T. S. Kuhn, J. Interdiscip. History 7, 1–31 (1976). Reprinted in Ref. 30. (I)

7. **The Revolution in Science 1500–1750,** A. Rupert Hall (Longman, New York, 1983). New edition of an authoritative text on the Scientific Revolution, revised to take account of recent research. (A)

8. **The Edge of Objectivity,** Charles Coulston Gillispie (Princeton U. P., Princeton, 1960). A sophisticated narrative account of major developments in physics and biology from the 17th through the 19th century, best appreciated by readers already familiar with the outlines of the standard history of science in this period. (I)

9. **The Major Achievements of Science,** A. E. E. McKenzie (Cambridge U. P., Cambridge, 1960). My own preference for a textbook for a course in the history of modern science. Provides in 350 pages a narrative account of astronomy, physics, biology, and chemistry from the 16th to the 20th centuries, with brief sections on geology and medicine and on the social-philosophical aspects of science. This is followed by about 200 pages of extracts from sources. (E)

10. **The Evolution of Physical Science,** Cecil J. Schneer (Grove, New York, 1960), reprinted by University Press of America, Lanham, MD, 1984. A readable, largely qualitative account of developments in the 17th through 19th centuries with brief chapters on relativity and atomic physics. (E)

11. **The Fabric of the Heavens: The Development of Astronomy** and **Dynamics** and **The Architecture of Matter,** Stephen Toulmin and June Goodfield (Harper, New York, 1961 and 1962), reprinted by University of Chicago, Chicago, 1977. An excellent survey of topics in the history of astronomy, physics, chemistry, and biology up to the 19th century, with brief excursions into modern physics. (E)

12. **The Ascent of Man,** J. Bronowski (Little, Brown & Co., Boston, 1973). Grand survey of aspects of the history of science and technology, based on the well-known television series. (E).

13. **A History of the Sciences,** Stephen F. Mason (Abelard-Schuman, New York, 1956, reprinted by Collier/Macmillan, New York, 1962). Widely used comprehensive textbook. (E)

14. **Science and Belief: From Darwin to Einstein** (Open University, Milton Keynes, England, 1981), American distributor: Taylor & Francis, Philadelphia. One of several courses prepared by the staff of the British "Open University" for correspondence courses. There are sections on evolution, thermodynamics and randomness, modern physics and the problem of knowledge, and selected topics in the social aspects of

science. See also the anthologies prepared to accompany this course: **Darwin to Einstein: Historical Studies on Science and Belief,** edited by Colin Chant and John Fauvel (Longman, New York, 1980), reprinting several articles by historians of science, and **Darwin to Einstein: Primary Sources on Science and Belief,** edited by Noel G. Coley and Vance M. D. Hall (Longman, New York, 1980). This is all first-rate material whose usefulness has been severely limited by its exorbitant price and inaccessibility in the United States. (I)

15. **A History of European Thought in the Nineteenth Century,** 4 Vols., John Theodore Merz (Blackwood, Edinburgh, 1904–1914). The first two volumes present a survey of scientific theories and discoveries, written in an elegant style that vividly conveys the 19th century perspective. (I)

16. **The World of the Atom,** 2 Vols., edited by Henry A. Boorse and Lloyd Motz (Basic, New York, 1966). The most comprehensive of the many anthologies of original sources; it concentrates on 19th and 20th century atomic physics. (I)

B. History of physics

17. **"History of Physics,"** Lewis Pyenson, in *Encyclopedia of Physics*, edited by Rita G. Lerner and George L. Trigg (Addison-Wesley, Reading, MA, 1981), pp. 404–414. Survey emphazing physics since 1800. (E)

18. **From Falling Bodies to Radio Waves: Classical Physicists and Their Discoveries** and **From X-Rays to Quarks: Modern Physicists and Their Discoveries,** Emilio Segrè (Freeman, New York, 1984 & 1980). These two volumes provide a lively, personalized but reasonably accurate account of the growth of modern physics; since the first volume is somewhat weak on 17th century mechanics, it should be supplemented by Cohen's book (Ref. 5). (I)

IV. CURRENT RESEARCH TOPICS

A. Subfields of physics

1. Fundamental concepts and philosophy of physics

19. **The Aim and Structure of Physical Theory,** Pierre Duhem (reprint of translation, Atheneum, New York, 1962). Originally written in 1905, the book is still worth reading as a provocative thesis on how science works, or should work. It includes a number of frequently quoted insights on topics such as "crucial experiments" and the difference between French and British physics. (I)

20. **The Nature of the Physical World,** A. S. Eddington (Cambridge U. P., Cambridge, 1930). One of the first and most readable popular presentations of the worldview resulting from the 20th century revolution in physics. (E).

21. **Concepts of Space, Concepts of Force, Concepts of Mass,** Max Jammer (Harvard U. P., Cambridge, 1954, 1957, 1961). Three monographs, each organized chronologically. (I).

22. **Forces and Fields,** Mary Hesse (Littlefield, Adams, Totowa, NJ, 1961). A historical and philosophical perspective over a broad chronological period. (I)

23. **The Question of the Atom,** edited by Mary Jo Nye (Tomash, Los Angeles, 1984). Anthology of original sources, including late 19th century debates on chemical atomic theory and the views of Maxwell, Boltzmann, Mach, Einstein, etc., culminating with Jean Perrin on Brownian movement. (I)

24. **"Irreversibility and Indeterminism: Fourier to Heisenberg."** Stephen G. Brush, J. Hist. Ideas 37, 603–630 (1976); reprinted in Ref. 93. (I)

25. **Metaphysics and Natural Philosophy: The Problem of Substance in Classical Physics,** P. M. Harman (Barnes & Noble, New York, 1982). (I)

26. **Physical Reality,** edited by Stephen Toulmin (Harper & Row, New York, 1970). Anthology of original sources, mostly from the early 20th century (Planck, Mach, Einstein, Bohr, etc.) (A)

27. **"Helmholtz' Kraft": An illustration of concepts in flux,"** Yehuda Elkana, Hist. Stud. Phys. Sci. 2, 263–298 (1970). An interpretation of a crucial event in the history of physics: the establishment of the law of energy conservation. (I) For a more extended treatment see Elkana's

book **The Discovery of the Conservation of Energy** (Harvard U. P., Cambridge, 1974).

28. **Energy: Historical Development of the Concept,** edited by R. Bruce Lindsay (Dowden, Hutchinson & Ross, Stroudsburg, PA, 1975). Anthology of original sources. (I)

29. **The Structure of Scientific Revolutions,** Thomas S. Kuhn (University of Chicago, Chicago, 1962), revised edition 1970. This is undoubtedly the best-known book on the history of science, though many historians and philosophers of science are reluctant to accept its conclusions. Most of the examples are from the physical sciences. (E)

30. **The Essential Tension,** Thomas S. Kuhn (University of Chicago, Chicago, 1977). A collection of historiographic and "metahistorical" studies, including Ref. 6 and an influential 1959 article "Energy conservation as an example of simultaneous discovery." (E)

31. **Method and Appraisal in the Physical Sciences,** edited by Colin Howson (Cambridge U. P., Cambridge, 1976). This book provides a good introduction to the influential "methodology of scientific research programmes" of the philosopher Imre Lakatos, including its application to the history of late 19th century atomism, Thomas Young's optics, Einstein's relativity theory, etc., with a critique by Paul Feyerabend. (I)

32. **"Mach, Einstein, and the Search for Reality,"** Gerald Holton, Daedalus **97**, 636–673 (1968). Concludes that Einstein eventually accepted the existence of an objective real world independent of the observer—a view that many thought he had eliminated from physics with his special relativity theory. Reprinted in Ref. 33. (E)

33. **Thematic Origins of Scientific Thought,** Gerald Holton (Harvard U. P., Cambridge, 1973). Holton's "thematic analysis of science" is an alternative to the views of Kuhn, Lakatos, and other philosophers; he applies this approach to the development of complementarity, relativity, and other topics in modern physics. Includes Ref. 32. (E)

34. **The Anthropic Cosmological Principle,** J. D. Barrow and F. J. Tipler (Oxford, U. P., New York, 1986). History and modern revival of teleology in physical science. (I)

2. Mechanics, including acoustics

35. **The Science of Mechanics in the Middle Ages,** Marshall Clagett (University of Wisconsin, Madison, 1959). Includes translated excerpts from original sources. (A)

36. **"The Science of Motion,"** J. E. Murdoch and E. D. Sylla, in *Science in the Middle Ages*, edited by D. C. Lindberg (University of Chicago, Chicago, 1978), pp. 206–264. (I)

37. **"The Principle of Inertia in the Middle Ages,"** Alan Franklin, Am. J. Phys. **44**, 529–545 (1976). A useful synthesis of research by historians of science. An expanded version has been published as a monograph under the same title (Colorado Associated University, Boulder, 1976). (E)

*38. **"Mechanics from Bradwardine to Galileo,"** William A. Wallace, J. Hist. Ideas **32**, 15–28 (1971). For further details on the background of Galileo's ideas see Wallace's books *Galileo's Early Notebooks: The Physical Question* (University of Notre Dame, Notre Dame, IN, 1977); *Prelude to Galileo* (Reidel, Boston, 1981); *Galileo and his Sources: The Heritage of the Collegio Romano in Galileo's Science* (Princeton University, Princeton, NJ, 1984). (I)

39. **Galileo Studies,** Stillman Drake (University of Michigan, Ann Arbor, 1970), Chaps. 5, 11, 12, and 13 discuss Galileo's contributions to mechanics. (I)

*40. **"Galileo's discovery of the law of free fall,"** Stillman Drake, Sci. Am. **228**(5), 84–92 (May 1973). Discovery of a manuscript that contradicts the usual view that Galileo erroneously assumed that the velocities of a falling body were proportional to distances. (E) For a more technical account see **"Mathematics and discovery in Galileo's physics,"** S. Drake, Historia Mathematica **1**, 129–150 (1974). (A)

41. **Force in Newton's Physics,** Richard S. Westfall (American Elsevier, New York, 1971). A treatise on the development of dynamics in the 17th century. (A)

*42. **"Newton's Second Law and the Concept of Force in the Principia,"** I. Bernard Cohen, Tex. Q. **10**(3), 127–157 (1967). Reprinted in *The Annus Mirabilis of Sir Issac Newton, 1666–1966*, edited by R. Palter

(MIT Press, Cambridge, 1970). (I)

43. **"History of Classical Mechanics,"** C. Truesdell, Naturwissenschaften **63**, 53–62, 119–130 (1976). A concise nonmathematical summary which includes thermodynamics, elasticity, and hydrodynamics up to the mid-20th century. (E)

44. **Essays in the History of Mechanics,** C. Truesdell (Springer-Verlag, New York, 1968). Deals primarily with the 18th century. (I, A)

45. **Historical Roots of the Principle of Conservation of Energy,** Erwin N. Hiebert (State Historical Society of Wisconsin, Madison, 1962), reprint, University Microfilms, Ann Arbor, MI. On 18th century mechanics (I)

46. **"The Mechanical Foundations of Elasticity and Fluid Dynamics,"** C. Truesdell, J. Ration. Mech. Anal. **1**, 125–300 (1952); **2**, 593–616 (1953). Critical-historical review of all major contributions to the subject in the past two centuries. (A)

47. **"Mathematics and Rational Mechanics,"** H. J. M. Bos, in *The Ferment of Knowledge*, edited by G. S. Rousseau and R. Porter (Cambridge U. P., New York, 1980), pp. 327–355. Comments on Truesdell's interpretation of 18th century mechanics. (I)

48. **The Evolution of Dynamics: Vibration Theory from 1687 to 1742,** J. T. Cannon and S. Dostrovsky (Springer-Verlag, New York, 1981). (A)

49. **"The Experimental Foundations of Solid Mechanics,"** James F. Bell, *Handbuch der Physik*, Volume VIa/1 (Springer-Verlag, New York, 1973). A detailed survey devoted primarily to 19th century work. (A)

50. **"The Main Concepts and Ideas of Fluid Dynamics in their Historical Development,"** P. F. Nemenyi, Arch. Hist. Exact Sci. **2**, 52–86 (1962). (E)

51. **"Resource Letter PhM-1: On Philosophical Foundations of Classical Mechanics,"** Mary Hesse, Am. J. Phys. **32**, 905–911 (1964). (E)

52. **"The Story of Acoustics,"** R. Bruce Lindsay, J. Acoust. Soc. Am. **39**, 629–644 (1966). (E)

53. **Acoustics: Historical and Philosophical Development,** edited by R. Bruce Lindsay (Dowden, Hutchinson & Ross, Stroudsburg, PA, 1973). Anthology of original sources, Aristotle to Lord Rayleigh. (A)

54. **Physical Acoustics,** edited by R. Bruce Lindsay (Dowden, Hutchinson & Ross, Stroudsburg, PA, 1974). Anthology of original sources, mid-19th to mid-20th centuries. (A)

55. **"A New Wave of Acoustics."** Robert Beyer, Phys. Today **34**(11), 145–157 (November 1981). Survey of some developments since about 1950. (I)

3. Gravity and celestial mechanics

56. **"Newton's Discovery of Gravity,"** I. Bernard Cohen, Sci. Am. **244**(3), 167–179 (1981). (E)

57. **History of Physical Astronomy,** Robert Grant (Johnson Reprint, New York, 1966). Though first published in 1852, this book remains a useful (and almost the only) detailed account of the development of theories of planetary motion in the 17th and 18th centuries. (I)

58. **"The Planetary theory of Laplace,"** A. Pannekoek, Pop. Astron. **56**, 300–312 (1948). (A)

59. **"The Great Inequality of Jupiter and Saturn: From Kepler to Laplace,"** Curtis Wilson, Arch. Hist. Exact Sci. **33**, 15–290 (1985). One of the most severe tests of Newton's gravitational theory was to account for discrepancies from Kepler's laws such as those in the observed motions of Jupiter and Saturn. Laplace's success in doing this was widely regarded as a brilliant confirmation of the "clockwork universe" idea. Wilson provides for the first time a detailed technical account of how it was done. (A)

60. **Mercury's Perihelion from Le Verrier to Einstein,** N. T. Roseveare (Oxford U. P., New York, 1982). For a brief summary see the same author's article "LeVerrier to Einstein: A Review of the Mercury Problem," Vistas Astron. **23**, 165–171 (1979). (A)

61. **"The Development of Theoretical Astronomy in the U.S.S.R.,"** Victor K. Abalakin, Vistas Astron. **19**, 163–177 (1975). (I)

62. **"Gravitation, Relativity and Precise Experimentation,"** C. W. F. Everitt, in *Proceedings of the First Marcel Grossman Meeting on General Relativity*, edited by R. Ruffini (North-Holland, New York, 1977), pp. 545–615. On the experiments of C. V. Boys (1889), Michelson & Morley (1881, 1887), R. V. Eötvös (1890–1922), etc., and the

current gyro relativity experiment; attempts to detect gravity waves. (A)

See Ref. 160 for further references to modern work.

4. Optics, electricity, and magnetism

63. **A History of Theories of Aether and Electricity,** E. T. Whittaker (Humanities, New York, 1973), revised edition. Volume I, The Classical Theories, is still the most comprehensive account of the subject up to 1900. Volume II, The Modern Theories, is partly obsolete and should be used with caution. (A)

64. **Conceptions of Ether,** edited by G. N. Cantor and M. J. S. Hodge (Cambridge U.P., New York, 1981). Includes a long introduction entitled "Major Themes in the Development of Ether Theories from the Ancients to 1900," and ten articles by historians and philosophers of science on topics in the history of ether theories from 1740 to 1900. (I)

65. **"The Science of Optics,"** David C. Lindberg, in *Science in the Middle Ages* (University of Chicago, Chicago, 1978), pp. 338–368. (I)

66. **Theories of Light from Descartes to Newton,** A. I. Sabra (Cambridge U. P., New York, 1981) 2nd ed. (I)

67. **"Kinematic Optics: A Study of the Wave Theory of Light in the Seventeenth Century,"** A. E. Shapiro, Arch. Hist. Exact Sci. **11**, 134–266 (1973). (A)

68. **The Effluvial Theory of Electricity,** R. W. Home (reprinted, Arno Press, New York, 1981). Discusses 18th century electrical theories. (I)

69. **Electricity in the 17th and 18th Centuries,** J. L. Heilbron (University of California, Berkeley, 1979). (A) For a shorter version see Ref. 161.

70. **Optics after Newton: Theories of Light in Britain and Ireland, 1704–1840,** G. N. Cantor (Manchester U. P., Manchester, 1984). (A)

71. **"Corpuscular optics and the wave theory of light: The science and politics of a revolution in physics,"** Eugene Frankel, Soc. Stud. Sci. **6**, 141–184 (1976). (I)

*72. **"The Physical Interpretation of the Wave Theory of Light,"** Frank A. J. L. James, Br. J. Hist. Sci. **17**, 47–60 (1984). (I)

*73. **"Speculation and Experiment in the Background of Oersted's Discovery of Electromagnetism,"** Robert C. Stauffer, Isis **48**, 33–50 (1957). Shows the influence of Naturphilosophie on the 1820 discovery of the action of a current on a magnet. (I)

74. **Early Electrodynamics,** R. A. R. Tricker (Pergamon, New York, 1965). Includes translations of papers by Biot, Savart, and Ampère. (A)

75. **The Origins of Field Theory,** L. Pearce Williams (reprinted, University Press of America, Lanham, MD, 1980). (E)

76. **Nineteenth-Century Spectroscopy,** William McGucken (Johns Hopkins, Baltimore, 1969). (A)

77. **Nineteenth-Century Aether Theories,** Kenneth F. Schaffner (Pergamon, New York, 1972). Includes papers by Fresnel, Green, Michelson and Morley, Larmor, etc. (A)

78. **"The Mutual Embrace of Electricity and Magnetism,"** M. Norton Wise, Science **203**, 1310–1318 (1979). On Maxwell's development of electromagnetic field theory. (I)

79. **"The Widening World of Magnetism,"** J. H. Van Vleck, Phys. Bull. **19**, 167–175 (1968). Survey of developments in the past 50 years. (I)

80. **"Optics: An ebullient evolution,"** Peter Franken, Phys. Today **34**(11), 160–171 (November 1981). Survey of recent developments, especially the laser. (A)

5. Heat, kinetic theory, properties of matter

81. **"The Discovery of Boyle's Law, and the Concept of the Elasticity of Air in the Seventeenth Century,"** C. Webster, Arch. Hist. Exact Sci. **2**, 441–502 (1965). In Boyle's own day, understanding the qualitative nature of gas pressure was more important than the quantitative law that now bears his name. (I)

82. **The Caloric Theory of Gases from Lavoisier to Regnault,** Robert Fox (Clarendon, Oxford, 1971). Treats the rise and fall of the caloric theory of heat. (A)

83. **From Watt to Clausius: The Rise of Thermodynamics in the Early Industrial Age,** D. S. L. Cardwell (Cornell U. P., Ithaca, 1971). Excellent work. (I)

84. **"A New Chart for British Natural Philosophy: The Development of Energy Physics in the Nineteenth Century,"** Crosbie Smith, Hist. Sci. **16**, 231–279 (1978). (I)

85. **"The Kind of Motion We Call Heat: A History of the Kinetic Theory of Gases in the 19th Century,** Stephen G. Brush (North-Holland, Amsterdam, 1976), paperback reprint 1986. (A)

86. **"The Kinetic Theory of Matter, 1845-1855,"** Eric Mendoza, Arch. Int. Hist. Sci. **32**, 184–220 (1982). Describes the range of phenomena that had to be explained—not just the ideal gas laws—in order to provide an adequate kinetic substitute for the caloric theory. (I)

*87. **"Kinetic atom,"** David B. Wilson, Am. J. Phys. **49**, 217–222 (1981). A survey of 19th century conceptions of the atom. (E)

88. **The Tragicomical History of Thermodynamics, 1822–1854,** C. Truesdell (Springer-Verlag, New York, 1980). Mathematical critique of the work of Fourier, Carnot, Clausius, etc. (A)

89. **Resource Letter EEC-1: On the Evolution of Energy Concepts from Galileo to Helmholtz,"** T. M. Brown, Am. J. Phys. **33**, 759–765 (1965).

90. **"Natural philosophy and thermodynamics: William Thomson and 'the dynamical theory of heat,'"** Crosbie Smith, Bri. J. Hist. Sci. **9**, 293–319 (1976). Discusses the interaction of scientific and religious factors in shaping the views of Thomson (later known as Lord Kelvin). (E)

91. **Kinetic Theory,** 3 Vols., Stephen G. Brush (Pergamon, New York, 1965–1971). Includes reprints and translations of original sources, Boyle to Maxwell, Boltzmann, Chapman, and Enskog. (E. I. A. Resp.)

92. **"The historical origins of the van der Waals equation,"** Martin J. Klein, Physica **73**, 28–47 (1974).

93. **Statistical Physics and the Atomic Theory of Matter from Boyle and Newton to Landau and Onsager,** Stephen G. Brush (Princeton U. P., Princeton, 1983). A survey of the history of kinetic theory of 1900, followed by detailed accounts of selected topics in modern physics (irreversibility, quantum statistics, superfluid helium, phase transitions). (I)

94. **Maxwell on Molecules and Gases,** edited by Elizabeth Garber, Stephen G. Brush and C. W. F. Everitt (MIT, Cambridge, 1986). Collection of published papers, manuscripts, and letters. Includes a survey of 19th century research on kinetic theory and gas transport properties. A volume on thermodynamics and statistical mechanics is in preparation. (A)

95. **Statistical Mechanics: A Survey of its One Hundred Years,"** Ryogo Kubo, in *Scientific Culture in the Contemporary World*, edited by V. Mathieu and P. Rossi (Scientia, Milan, 1979), pp. 131–57. Broad semihistorical survey from Boltzmann to Onsager. (A)

96. **The Quest for Absolute Zero,** Kurt Mendelssohn (Wiley/Halsted, New York, 1977) 2nd ed. A history of low-temperature physics. (I)

97. **"Resource Letter SH-1: Superfluid Helium,"** Robert B. Hallock, Am. J. Phys. **50**, 202–212 (1982). See also *Superfluid Helium: Selected Reprints*, edited by R. B. Hallock (American Association of Physics Teachers, Stony Brook, NY, 1983). (A)

98. **Helium 4,** Zygmunt M. Galasiewicz (Pergamon, New York, 1971). Includes reprints and translations of several research papers. (A)

99. **"The Beginnings of Solid State Physics. A Symposium held 30 April–2 May 1979,"** N. F. Mott *et al.*, Proc. R. Soc. London **A371**, 1–177 (1980). Includes "The Development of the Quantum Mechanical Electron Theory of Metals: 1900–28" by Lillian Hoddeson and G. Baym. (A)

100. **"The Golden Age of Solid-State Physics,"** Theodore H. Geballe, Phys. Today, **34**(11), 132–143 (November 1981). (I)

101. **"Resource Letter OEPM-1: on the Ordinary Electronic Properties of Metals,"** D. N. Langenberg, Am. J. Phys. **36**, 777–788 (1968). (A)
See also Refs. 24, 27, 28, 30 on irreversibility and energy conservation.

6. Quantum theory

102. **Men who made a New Physics,** Barbara Lovett Cline (New American Library, New York, 1969) first published in 1965 as *The Questioners*. Popular work, with chapters on Rutherford, Planck, Einstein, Bohr, etc., describing their theories. (E)

103. **The Story of Quantum Mechanics,** Victor Guillemin (Scribner, New York, 1968). Readable account which includes theories of elementary

particles up to about 1964, and philosophical implications. (I)

104. **The Conceptual Development of Quantum Mechanics,** Max Jammer (McGraw-Hill, New York, 1966). A comprehensive history. (A)

105. **"Max Planck and the Beginnings of the Quantum Theory,"** Martin J. Klein, Arch. Hist. Exact Sci. **1,** 459–479 (1962). A comprehensive technical account. (A) See also Klein's lectures on this subject published in Ref. 1.

106. **Black-Body Theory and the Quantum Discontinuity, 1894–1912,** Thomas S. Kuhn (Oxford U. P., New York, 1978). A controversial new interpretation; argues that Max Planck did not introduce energy quanta in 1900. See the next two items for further discussion. (A)

107. **"Kuhn and the Quantum Controversy,"** Peter Galison, Br. J. Philos. Sci. **32,** 71–84 (1981). Essay review, comparing Kuhn's interpretation (previous item) with that of M. J. Klein (Ref. 105) and suggesting a third alternative. (I)

*108. **"Revisiting Planck,"** Thomas S. Kuhn, Hist. Stud. Phys. Sci. **14,** 231–252 (1984). Summarizes and defends the new interpretation presented in Ref. 106. (I)

109. **"The genesis of the Bohr atom,"** J. L. Heilbron and T. S. Kuhn, Hist. Stud. Phys. Sci. **1,** 211–290 (1969). A new interpretation placing Bohr's 1913 paper in the context of his earlier work. (I) See also Heilbron's lectures published in Ref. 1, and a shorter version, "Bohr's first theories of the atom," J. L. Heilbron, Phys. Today **38**(10), 28–36 (October 1985). (I)

110. **"Einsten and the Wave-Particle Duality,"** M. J. Klein, Nat. Philos. **3,** 1–49 (1964). (I)

111. **The Tiger and the Shark: Empirical Roots of Wave-particle Dualism,** Bruce R. Wheaton (Cambridge U. P., New York, 1983). On the study of x and gamma rays; should be supplemented by Ref. 112. (A) See also his article "Louis de Broglie and the Origins of Wave Mechanics," Phys. Teach. **22,** 297–301 (1984). (E)

112. **The Compton Effect: Turning Point in Physics.** Roger H. Stuewer (Science History, New York, 1975). Developments in radiation physics that established the particle nature of light. (A)

113. **The Creation of Quantum Mechanics and the Bohr–Pauli Dialogue,** John Hendry (Reidel, Boston, 1984). (A)

*114. **"Schrödinger's Route to Wave Mechanics,"** Linda Wessels, Stud. Hist. Philos. Sci. **10,** 311–340 (1979). (I)

115. **"Erwin Schroedinger and the Wave Equation: The Crucial Phase,"** Helge Kragh, Centaurus **26,** 154–197 (1982–83). (A)

116. **"The Genesis of Dirac's relativistic theory of electrons,"** Helge Kragh, Arch. Hist. Exact. Sci. **24,** 31–67 (1981). (A)

117. **"Origins of Dirac's electron, 1925–1928," "Evaluations of Dirac's electron, 1928–1932,"** and **"Vindications of Dirac's electron, 1932–1934,"** Donald Franklin Moyer, Am. J. Phys. **49,** 944–949, 1055–1062, 1120–1125 (1981).

118. **"Weimar Culture, Causality, and Quantum Theory, 1918–1927: Adaptation by German Physicists and Mathematicians to a Hostile Intellectual Environment."** Paul Forman, Hist. Stud. Phys. Sci. **3,** 1–115 (1971). Claims that the adoption of indeterminism was influenced by cultural factors; this article is frequently cited by those who discuss the "social construction" of scientific ideas. See next two items for critical comments. (I)

119. **"Weimar culture and quantum causality,"** John Hendry, Hist. Sci. **18,** 155–180 (1980). (I)

120. **"Adaptation of scientific knowledge to an intellectual environment. Paul Forman's "Weimar culture, causality, and quantum theory, 1918–1927": Analysis and criticism,"** P. Kraft and P. Kroes, Centaurus **27,** 76–99 (1984). (I)

121. **Scientific Explanation and Atomic Physics,** Edward M. MacKinnon (University of Chicago, Chicago, 1982). Philosophical history of quantum mechanics. (I)

122. **The Philosophy of Quantum Mechanics,** Max Jammer (Wiley, New York, 1974). Detailed historical account of the interpretation of quantum theory. (A)

123. **"Resource Letter IQM-1: On the Interpretation of Quantum Mechanics,"** B. S. DeWitt and R. N. Graham, Am. J. Phys. **39,** 724–723 (1971). (E)

124. **"Quantum Historiography and the Archive for History of Quantum Physics,"** J. L. Heilbron, Hist. Sci. 7, 90–111 (1968). Describes the project to interview quantum physicists and preserve their letters and manuscripts; the project has stimulated much of the historical research on this subject during the past two decades. (A)

125. **"The Search for Unity: Notes for a History of Quantum Field Theory,"** Steven Weinberg, Daedalus **106**(4), 17–35 (Fall 1977). (I)

126. **"Plausibility and the Evaluation of Knowledge: A Case-Study of Experimental Quantum Mechanics,"** Bill Harvey, Soc. Stud. Sci. **11,** 95–130 (1981). On the "hidden variables" experiments and how they were interpreted. (I)

127. **Wave Mechanics,** Gunther Ludwig (Pergamon, New York, 1968). Includes translations of papers by de Broglie, Schrödinger, Heisenberg, Born, and Jordan. (A)

128. **Sources of Quantum Mechanics,** edited by B. L. Van der Waerden (North-Holland, Amsterdam, 1967). Includes translations of papers by Einstein, Heisenberg, Born, Pauli, etc. (A)

129. **Quantum Theory and Measurement,** edited by J. A. Wheeler and W. H. Zurek (Princeton U. P., Princeton, 1983). Anthology of sources. (A)

7. Nuclear and particle physics

130. **Discoveries in Physics,** David L. Anderson (Holt, Rinehart and Winston, New York, 1973). Includes chapters on the electron, nuclear fission, and the neutrino. (E)

131. **" 'Chance favors the prepared mind': Henri Becquerel and the Discovery of Radioactivity,"** L. Badash, Arch. Int. Hist. Sci. **18,** 55–66 (1965).

132. **The Self-Splitting Atom: The History of the Rutherford–Soddy Collaboration,** Thaddeus J. Trenn (Taylor & Francis, London, 1977). (I)

*133. **"The Discovery of Atomic Transmutation: Scientific Styles and Philosophies in France and Britain,"** Marjorie Malley, Isis **70,** 213–223 (1979). Discusses reasons why the Curies did not discover transmutation. (E)

134. **The Discovery of Radioactivity and Transmutation,** edited by Alfred Romer (Dover, New York, 1964). Anthology of orignal sources. (A)

135. **Nuclear Physics in Retrospect: A symposium on the 1930s,** edited by Roger H. Stuewer (University of Minnesota, Minneapolis, 1979). A collection of articles on nuclear physics in the 1930s by H. A. Bethe and other physicists. (I)

136. **Nuclear Forces,** D. M. Brink (Pergamon, New York, 1965). Includes reprints and translations of original sources. (A)

137. **"The Birth of Elementary-Particle Physics,"** Laurie M. Brown and Lillian Hoddeson, Phys. Today **35**(4), 36–43 (April 1982). Reprinted in Ref. 177. (E)

138. **"Resource Letter Neu-1: History of the Neutrino,"** Leon Lederman, Am. J. Phys. **38,** 129–136 (1970). (A)

139. **Particle accelerators: A brief history,** M. Stanley Livingston (Harvard U. P., Cambridge, 1969). (I)

140. **"The Disvovery of Fission,"** O. R. Frisch and J. A. Wheeler, Phys. Today **20**(11), 43–52 (November 1967). Reprinted in Ref. 177. (E)

141. **The Discovery of Fission,** Joan Warnow *et al.* (American Institute of Physics, New York, 1984). Audiocassette, script, and reprinted articles. (E)

142. **Otto Hahn and the Rise of Nuclear Physics,** edited by W. R. Shea (Reidel, Boston, 1983). Articles by S. R. Weart, R. H. Stuewer, F. Kraft, and others. (I).

143. **The Discovery of Nuclear Fission,** H. G. Graetzer and D. L. Anderson (Van Nostrand Reinhold, New York, 1971). Consists mostly of extracts from the original papers. (I)

144. **The Birth of Particle Physics,** edited by Laurie M. Brown and Lillian Hoddeson (Cambridge U. P., New York, 1983). Collection of papers by scientists and historians. (I)

145. **International Colloquium on the History of Particle Physics: Some Discoveries, Concepts, Institutions from the Thirties to Fifties** (Les Editions de Physique, Les Ulis Cedex, France, 1982). Collection of papers, mostly by physicists. (A)

146. **Constructing Quarks: A Sociological History of Particle Physics,** Andrew Pickering (University of Chicago, Chicago, 1984). (I)

147. **"Resource Letter NP-1: New particles,"** Jonathan L. Rosner, Am. J. Phys. **48,** 90–103 (1980). (A)

8. Relativity

148. **Genesis of Relativity,** Loyd S. Swenson, Jr. (Franklin, New York, 1979). Includes relevant 19th century background. (E)

*149. **"Einstein and the 'crucial' experiment,"** Gerald Holton, Am. J. Phys. **37**, 968–982 (1969). On the alleged influence of the Michelson–Morley experiment. For further discussion of Einstein's work on relativity theory see Ref. 33. (I)

150. **"The Lorentz theory of electrons and Einstein's theory of relativity,"** Stanley Goldberg, Am. J. Phys. **37**, 982–994 (1969).

*151. **On Einstein's invention of special relativity,"** Arthur I. Miller, in *PSA 1982.* Vol. 2, edited by P. D. Asquith and T. Nickles (Philosophy of Science Association, East Lansing, MI, 1983), pp. 377–402. See also the discussion of historiography of special relativity theory by other contributors to this symposium. (A)

152. **Albert Einstein's Special Theory of Relativity: Emergence (1905) and Early Interpretation (1905–1911),** Arthur I. Miller (Addison-Wesley, Reading, MA, 1981). (A)

153. **Special Theory of Relativity,** C. W. Kilmister (Pergamon, New York, 1970). Includes reprints and translations of original papers. (A)

154. **General Theory of Relativity,** C. W. Kilmister (Pergamon, New York, 1973). Includes reprints and translations of original papers. (A)

155. **"Physicists receive Relativity: Revolution and Reaction,"** and **"Einstein, Relativity and the Press: the Myth of Incomprehensibility,"** Jeffrey Crelinsten, Phys. Teach. **18**, 187–193, 115–122 (1980). (E)

156. **"How Einstein found his Field Equations, 1912–1915,"** John Norton, Hist. Stud. Phys. Sci. **14**, 253–316 (1984). (A)

157. **"Einstein and Hilbert: Two months in the History of General Relativity,"** John Earman and Clark Glymour, Arch. Hist. Exact Sci. **19**, 291–308 (1978). (A)

158. **The Measure of the Universe: A History of Modern Cosmology,** J. D. North (Clarendon, Oxford, 1965). Includes applications of general relativity theory. (A)

159. **"Experimental Tests of General Relativity: Past, Present and Future,"** C. W. F. Everitt, in *Physics and Contemporary Needs,* Vol. 4, edited by Riazuddin (Plenum, New York, 1980), pp. 529–555. (A)

160. **"Resource Letter GI-1: Gravity and inertia,"** P. W. Worden, Jr., and C. W. F. Everitt, Am. J. Phys. **50**, 494–500 (1982). See also *Gravity and Inertia: Selected Reprints* (American Association of Physics Teachers, Stony Brook, NY, 1983). (A)

B. Physics in particular times and places

1. Origins of physical science

161. **Early Physics and Astronomy,** Olaf Pedersen and Mogens Pihl (American Elsevier, New York, 1974). (I)

162. **Science and Civilization in China,** Vol. 4, **Physics and Physical Technology,** Joseph Needham (Cambridge U. P., Cambridge, 1962). See especially Chap. 26. (E)

163. **Science and Civilization in Islam,** S. H. Nasr (Harvard U. P., Cambridge, 1968), Chap. 4. (E)

164. **Medieval and Early Modern Science,** A. C. Crombie (reprint, Doubleday, Garden City, NY, 1959), Vol. 1, pp. 1–120; Vol. 2, pp. 1–220. (I)

2. Seventeenth and eighteenth centuries

165. **Elements of Early Modern Physics,** J. L. Heilbron (University of California, Berkeley, 1982). Physics in 17th and 18th century Europe, with a chapter summarizing his more comprehensive study of electricity (item 69). (I)

166. **"Natural Philosophy,"** Simon Schaffer, in *The Ferment of Knowledge,* edited by G. S. Rousseau and R. Porter (Cambridge U. P., New York, 1980), pp. 55–91. Survey of historical studies on 18th century physics. (E)

3. Nineteenth and early twentieth centuries

167. **Intellectual Mastery of Nature,** 2 Vols., Christa Jungnickel and Rus-

sell McCormmach (University of Chicago, Chicago 1986). A combined institutional-technical history of German physics from 1800 to 1925. (I)

168. **The History of Modern Physics 1800–1950,** edited by Gerald Holton and Katherine Sopka (Tomash, Los Angeles, and American Institute of Physics, New York, 1983). A series of reprints of classic books, anthologies of original sources, and historical monographs. (I,A)

169. **"The Invention of Physics,"** Susan Faye Cannon, in her book *Science in Culture: The Early Victorian Period* (Science History, New York, 1978), pp. 111–136. Argues that physics in the modern sense was invented by the French during the years 1810–30, then transferred to Britain. (E)

170. **"The Transmission of Physics from France to Britain: 1800–1840,"** Maurice Crosland and Crosbie Smith, Hist. Stud. Phys. Sci. **9**, 1–61 (1978). The later stage in this process is described in Smith's article " 'Mechanical Philosophy' and the emergence of physics in Britain: 1800–1850." Ann. Sci. **33**, 3–29 (1976). (E)

171. **Energy, Force, and Matter: The Conceptual Development of Nineteenth-Century Physics,** P. M. Harman (Cambridge U. P., New York, 1982). (I)

172. **Wranglers and Physicists: Studies on Cambridge Mathematical Physics in the Nineteenth Century,** edited by P. M. Harman (Longwood, Dover, NH, 1985). (A)

173. **"Editor's Foreword,"** Russell McCormmach, Hist. Stud. Phys. Sci. **3**, ix–xxiv (1971). On the study of scientific disciplines, exemplified by German physics in the 19th century. (E)

174. **"Physics circa 1900: Personnel, Funding, and Productivity of the Academic Establishments,"** Paul Forman, John L. Heilbron, and Spencer Weart, Hist. Stud. Phys. Sci. **5**, 1–185 (1975) (entire volume). (I)

175. **Rutherford and Physics at the Turn of the Century,** edited by Mario Bunge and William R. Shea (Science History, New York, 1979). (E)

176. **Inward Bound: Of Matter and Forces in the Physical World,** Abraham Pais (Oxford U. P., New York, 1986). (I)

177. **History of Physics,** edited by Spencer R. Weart and Melba Phillips (American Institute of Physics, New York, 1985). Reprints 47 articles from Physics Today, mostly on 20th century physics. (I)

178. **The Physicists,** C. P. Snow (Little, Brown, Boston, 1981). A personal account of 20th century atomic physics, written "largely from memory." This book was Snow's last heroic attempt to bridge the gap between his famous "two cultures" by explaining to the lay reader what it is like to be a scientist. (E)

179. **Crucial Experiments in Modern Physics,** George L. Trigg (reprint, Crane, Russak, New York, 1975). Detailed accounts of selected experiments in 20th century atomic and nuclear physics. (A)

180. **Landmark Experiments in Twentieth Century Physics,** George L. Trigg (Crane, Russak, New York, 1975). Detailed accounts of selected experiments in atomic physics, superconductivity, liquid helium, and elementary particles. (A)

181. **Adventures in Experimental Physics,** 5 vols., edited by Bogdan Maglich (World Science Communications, Princeton, NJ, 1972–76). First-person discovery accounts and reprints of original papers on several topics in 20th century physics and astronomy. (A)

182. **The Physicist's Conception of Nature,** edited by Jagdish Mehra (Reidel, Boston, 1973). Proceedings of a symposium honoring Paul Dirac, including several valuable historical papers on 20th-century physics. (A)

183. **The Solvay Conferences on Physics,** by Jagdish Mehra (Reidel, Boston, 1975). Summaries of papers presented at the conferences held from 1911 to 1973. (A)

184. **Haphazard Reality,** H. B. G. Casimir (Harper & Row, New York, 1983). A personal account of physics in the first half of the 20th century. (E)

185. **The New Physics,** Armin Hermann (Inter Nationes, Bonn-Bad Godesberg, 1979). Mostly on physics in Germany, 1905–55, richly illustrated with photos and reproductions of documents. (E)

186. **Physics Citation Index 1920–1929,** 2 Vols., Henry Small (Institute for Scientific Information, Philadelphia, 1981). An analysis of 16 major physics journals published during the decade in which modern quantum mechanics was established, enabling the user to find citations of any given paper (including those published in earlier decades). (A)

4. Twentieth century America

187. **American Physics in Transition: A History of Conceptual Change in the Late Nineteenth Century,** Albert E. Moyer (Tomash, Los Angeles, 1983). (I)
188. **"History of Physics,"** Albert E. Moyer, Osiris (new series) **2**, 163–182 (1985). On the historiography of 20th century American physics. (E)
189. **"A New Site for the Seminar: The Refugees and American Physics in the Thirties,"** Charles Weiner, Perspect. Am. Hist. **2**, 190–234 (1968). On the transfer of physics from Europe to the United States. (E)
190. **The Physicists: The History of a Scientific Community in Modern America,** Daniel J. Kevles (Knopf, New York, 1978). (E)
191. **"50 Years of Physics in America,"** Norman F. Ramsey *et al.*, Phys. Today **34**(11) (November 1981). Special issue to commemorate the 50th anniversary of the American Institute of Physics. Includes "The Last Fifty Years—A Revolution?" by Spencer R. Weart (reprinted in Ref. 177). (E)
192. **Physics and Society in Twentieth Century America,** Roger H. Stuewer (Department of Independent Study, Continuing Education and Extension, University of Minnesota, Minneapolis, 1982). Study guide, reprint booklet, and two audio cassettes (lectures by Feynman and Oppenheimer). (E)

5. Nonwestern countries

193. **"The emergence of Japan's first physicists: 1868–1900,"** Kenkichiro Koizume, Hist. Stud. Phys. Sci. **6**, 3–108 (1975). (I)
194. **"Physics on the periphery: A world survey, 1920–1929,"** L. Pyenson and M. Singh, Scientometrics **6**, 279–306 (1984).
195. **Cultural Imperialism and Exact Sciences: German expansion overseas 1900–1930,** Lewis Pyenson (Lang, New York, 1986). How German physicists and astronomers came to staff major research and teaching institutions in Argentina, the South Pacific, and China. (E)
196. **"Pioneer scientists in pre-independence India,"** William A. Blanpied, Phys. Today **39**(5), 36–44 (May 1986). The problems of conducting basic research in a less developed country are illustrated by the careers of six leading Indian physicists. (E)

C. Biographies and autobiographies

Because of the large number of published biographies of physicists and the relative ease of locating them (compared to works on topics in the history of physics), it does not seem useful to give a comprehensive list here. Instead, the reader seeking a biography of a particular physicist should simply consult the *Dictionary of Scientific Biography* (below) and use the bibliographies appended to each article for further information. (See Ref. 223 for brief biographical articles.) Only a few works published after the *Dictionary* are listed here.

197. **Dictionary of Scientific Biography,** 16 Vols., edited by Charles Coulston Gillispie (Scribner, New York, 1970–1980). A collection of authoritative articles on all major and many minor scientists of the past, which is at the same time (with the help of the detailed index in volume 16) a comprehensive encyclopedia of the history of science. A one-volume condensation is available under the title *Concise Dictionary of Scientific Biography* (Scribner, New York, 1981). Additional volumes, covering scientists who died in the last two decades, are in preparation. (I)
198. **Galileo,** Stillman Drake (Hill & Wang, New York, 1980). (E)
199. **Novità celesti e crisi del sapere: Atti del Convegno Internazionale di Studi Galileiani,** edited by Paolo Galluzzi (Barbèra, Firenze, Italy, 1984). Collection of papers on Galileo and 17th century science, some in English. See the Isis Critical Bibliography 1985, #1467, for listing of contents. (A)
200. **Never at Rest: A Biography of Isaac Newton,** R. S. Westfall (Cambridge U. P., New York, 1981). A very authoritative, comprehensive biography. (I) For a brief sketch see Westfall's article "The Career of Isaac Newton: A Scientific Life in the Seventeenth Century," Am. Scholar **50**, 341–353 (1981). (E)

201. **Lord Kelvin, The Dynamic Victorian,** Harold Issadore Sharlin and Tiby Sharlin (Pennsylvania State University, University Park, 1979). (E)
202. **The Demon in the Aether: The Life of James Clerk Maxwell,** Martin Goldman (Heyden, Philadelphia, 1983). (E)
203. **Ludwig Boltzmann, International Tagung anlässlich des 75. Jahrestages seines Todes, 5.-8. September 1981,** edited by Roman Sexl and John Blackmore (Akademische Druck-u. Verlagsanstalt, Graz, Austria, 1982). Collection of papers by historians and scientists, mostly in English. (I)
204. **Rutherford, Simple Genius,** David Wilson (MIT, Cambridge, 1983). (E)
205. **Marie Curie,** Robert Reid (Saturday Review/Dutton, New York, 1974). (E)
206. **Night Thoughts of a Classical Physicist,** Russell McCormmach (Harvard U. P., Cambridge, 1982). A fictionalized account of the response of a physicist trained in the 19th century to the revolution at the beginning of the 20th century. (I)
207. **'Subtle is the Lord': The Science and the Life of Albert Einstein,** A. Pais (Oxford U. P., New York, 1982). (I)
208. **Albert Einstein: Historical and Cultural Perspectives,** edited by Gerald Holton and Yehuda Elkana (Princeton U. P., Princeton, 1982). Essays on his scientific contributions and his impact on the 20th century. (E)
209. **Niels Bohr: A Centenary volume,** edited by A. P. French and P. J. Kennedy (Harvard U. P., Cambridge, 1985). Biographical memoirs and short articles on his scientific work and philosophical ideas. (I)
210. **Springs of Scientific Creativity,** edited by Rutherford Aris, H. Ted Davis, and Roger H. Stuewer (University of Minnesota, Minneapolis, 1983). Essays on Newton, Maxwell, Einstein, Schrödinger, and others. (E)
211. **"Women in Physical Science: From drudges to discoverers,"** Stephen G. Brush, Phys. Teach. **23**, 11–19 (1985). Includes references to information about several 20th century women physicists. (E)

D. Reference works

212. **"History of Physics,"** W. D. Hackmann, in *Information Sources in Physics*, edited by Dennis F. Shaw (Butterworths, Boston, 1985), pp. 209–232. (E)
213. **"History of Physical Science,"** Simon Schaffer, in *Information Sources in the History of Science and Medicine*, edited by Pietro Corsi and Paul Weindling (Butterworths, Boston, 1983), pp. 285–314. (E)
214. **The History of Classical Physics: A Selected, Annotated Bibliography,** R. W. Home (Garland, New York, 1984). (I)
215. **Literature on the History of Physics in the 20th Century,** John L. Heilbron and Bruce R. Wheaton (Office for the History of Science and Technology, University of California, Berkeley, 1981). Comprehensive bibliography. (A)
216. **The History of Modern Physics: An International Bibliography,** Stephen G. Brush and Lanfranco Belloni (Garland, New York, 1983). Intended to complement Ref. 215 by including publications dealing with the post-1950 period. (A)
217. **An Inventory of Published Letters to and from Physicists, 1900–1950,** Bruce R. Wheaton and J. L. Heilbron (Office for History of Science and Technology, University of California, Berkeley, 1982). (A)
218. **National Catalog of Sources for History of Physics. Report No. 1. A Selection of Manuscript Collections at American Repositories,** Joan Nelson Warnow (American Institute of Physics, New York, 1969). Information about further reports in this series may be obtained from the Center for History of Physics at AIP. (A)
219. **Isis Cumulative Bibliography: A Bibliography of the History of Science formed from Isis Critical Bibliographies 1–90, 1913–1965,** edited by Magda Whitrow (Mansell, London, 1971–84). The first two volumes list publications on the life and work of individual scientists, and on institutions; Vol. 3 is a subject index (Physics, pp. 134–177); Vols. 4 and 5 are organized by civilizations and chronological periods from prehistory through the 19th century. Vol. 6 is an author index. (A)

220. **Isis Cumulative Bibliography 1966–1975,** Vol. 1. **Personalities and Institutions,** edited by John Neu (Mansell/Merrimack Book Service, Salem, NH, 1980). (A)

221. **Resources for the History of Physics,** edited by Stephen G. Brush (University Press of New England, Hanover, NH, 1972). The first part of this book is largely superseded by the present Resource Letter, except for a list of films; the second part is a "Guide to Original Works of Historical Importance and their Translations into Other Languages." (A)

222. **The History of Modern Science: A Guide to the Second Scientific Revolution, 1800–1950,** Stephen G. Brush (Iowa State University, Ames, 1987). Includes suggested readings, synopses and bibliographies for several topics in the history of modern physics. (I)

223. **The Biographical Dictionary of Scientists: Physicists,** edited by David Abbot (Blond Educational, London, 1984). (E)

224. **"The historical investigation of science in North America,"** Frederick Gregory, Z. Allg. Wiss. **16**, 151–166 (1985). Includes discussion of recent trends, books, list of research centers, and degree programs. (A)

Acknowledgment to the *Journal of the History of Ideas*, **32**, William A. Wallace, "Mechanics from Bradwardine to Galileo," pp. 15–28, ©1971 The University of Rochester.

MECHANICS FROM BRADWARDINE TO GALILEO

By William A. Wallace

It is difficult to understand any movement, as the Schoolmen would say, without knowing its *terminus ad quem*.[1] The movement that gave rise to modern science is no exception. Assuming that science in its classical form arrived in Western Europe during the seventeenth century, to investigate how medieval and scholastic thought may have accelerated its arrival one must have clearly in mind what is to be understood by science. It will not do, for instance, to think of an interplay between pure theory and experiment, as though the seventeenth-century scientist was intent on elaborating a postulational system from which he could deduce empirically verifiable consequents. Such a view of science had to wait at least two more centuries; it belongs to a different thought-context than that of the seventeenth century.

No, the *terminus ad quem* of the movement that gave rise to modern science is simpler than that. At the risk of oversimplifying it, in this paper the concept of science will be narrowed to that of physics, and only the part of physics known as mechanics will be considered. More particularly still, discussion will be centered only on the part of mechanics that deals with what the Schoolmen called *motus localis,* or local motion, comprising the present-day disciplines of kinematics and dynamics.[2] From such a restricted viewpoint, seventeenth-century mechanics may be characterized by its attempts at a precise mathematical formulation of laws that regulate such natural phenomena as falling bodies, and by parallel attempts to measure and determine experimentally how well such a formulation corresponds to reality. In sum, the developing science of mechanics, the *terminus ad quem* of late medieval thought, may be seen as made

[1] The *terminus ad quem* is the goal or end that terminates a movement or change. Scholastics generally held that every motion or change is "specified," i.e., given its species, from its *terminus ad quem*. In this they were merely following Aristotle, *Physics*, Bk. 5, ch. 1, 225b 6–10; see the *Expositio* of St. Thomas Aquinas on this *locus*, lect. 1, n. 6, ed. P. M. Maggiolo (Turin, 1954), 319.

[2] Thus the study of statics is omitted. This portion of the science of mechanics has been well treated by M. Clagett and E. A. Moody in *The Medieval Science of Weights* (Madison, 1952), where they show that the contributions of the medieval period to statics resulted from the interpenetration of two Greek traditions, the Aristotelian and the Euclidean-Archimedean. The influences studied here lie predominantly within the Aristotelian tradition.

16   WILLIAM A. WALLACE

up of two elements: (1) the mathematical analysis of motion, and (2) its experimental verification. The problem that this conception of the science of mechanics suggests to the historian of science is this: To what extent did late medieval and scholastic physics contribute to either or both of these elements, and thus influence the development of mechanics in its seventeenth-century understanding?

As to the first element, that of mathematical analysis, the late medieval contribution is quite well established. With the appearance in 1328 of Thomas Bradwardine's *Tractatus de proportionibus* (Treatise on Ratios) a new and distinctive mathematical approach to the study of motion was inaugurated.[3] This was developed and refined by Bradwardine's successors for the next two centuries. Out of the development came the concept of instantaneous velocity, the use of fairly complex mathematical functions to correlate factors affecting motions, procedures for calculating distances traversed in uniformly accelerated motions, and the rudimentary notions of analytical geometry and the calculus. Historians of medieval science such as Maier,[4] Clagett,[5] Crosby,[6] Wilson,[7] and Grant[8] have sufficiently documented the extent of this contribution. One may argue on points of detail, but there is a general consensus that the late Middle Ages contributed substantially to the mathematical foundation on which seventeenth-century mechanics was built. What is more, most of this contribution was already implicit in Bradwardine's treatise or was contained in the extensions of this treatise by Heytesbury,[9]

[3] H. Lamar Crosby, Jr., ed. and trans., *Thomas of Bradwardine. His "Tractatus de Proportionibus": Its Significance for the Development of Mathematical Physics* (Madison, 1955).

[4] Anneliese Maier's *Studien zur Naturphilosophie der Spätscholastik*, 5 vols., are indispensable for a study of this contribution; for details see J. A. Weisheipl, *The Development of Physical Theory in the Middle Ages* (London and New York, 1959), 91. A good summary is given in Maier's first vol., *Die Vorläufer Galileis im 14. Jahrhundert* (Rome, 1949).

[5] Marshall Clagett, *The Science of Mechanics in the Middle Ages* (Madison, 1959), a valuable source book that contains editions and translations of the most important contributions.

[6] See fn. 3, above.

[7] Curtis Wilson, *William Heytesbury: Medieval Logic and the Rise of Mathematical Physics* (Madison, 1960).

[8] Edward Grant, *Nicole Oresme: 'De proportionibus proportionum' and 'Ad pauca respicientes'* (Madison, 1966).

[9] William Heytesbury composed his *Regule solvendi sophismata* (Rules for Solving Sophisms) at Merton College, Oxford, *c.* 1335. Ch. 6 of this work is entitled *De tribus predicamentis* (On the Three Categories); here he discusses the three Aristotelian categories in which motion can be found, treating at length of the velocity of local motion in the section entitled *De motu locali*. Apart from Wilson's work on Heytesbury (fn. 7), consult the extracts from the *Regule* in Clagett (fn. 5).

Dumbleton,[10] and Swineshead.[11] Thus the fourteenth-century Mertonians were definite contributors to the mathematical component that made classical mechanics possible.

With regard to the second component, experimentation and measurement, the historical origins are not so clear. Certainly the role of late medieval Aristotelians in its development has not been emphasized, and there has been a tendency to look elsewhere for its historical antecedents. The remainder of this paper will address itself to this problem of experimental origins, and will attempt to trace some factors contributing to its solution that seem to derive from Bradwardine and his successors. The thesis to be defended is that such factors were present, and that at least they set the stage, or established the climate of opinion, wherein experimentation would be sought as a natural complement to the mathematical formulation of laws of motion. Unlike the mathematical component, this experimental component (if one may call it such) was not clearly present in the work of the Mertonians, but it did evolve gradually, over two centuries, as their ideas came to be diffused on the Continent.

Why it was that a mathematical basis for seventeenth-century mechanics was apparent to the Mertonians whereas an experimental basis was not, poses an interesting question. The answer seems to lie in a certain ambivalence that was latent within the Mertonian analysis of motion. The tension that should have resulted from this ambivalence was not sensed immediately; had it been, perhaps the experimental component would have gotten off to as good a start as the mathematical. Yet there is evidence for maintaining that this latent tension did come to be recognized as the "calculatory" analyses developed at Merton College were propagated in France, Italy, and Spain during the fourteenth, fifteenth, and sixteenth centuries. The tension was gradually resolved, and in its resolution the way was prepared for an experimental investigation of nature that would complement a mathematical formulation of its laws.

To argue the thesis, one must be more explicit about the ambivalence present in the treatises of Bradwardine, Heytesbury, and

[10]John Dumbleton wrote his *Summa logicae et philosophiae naturalis* (Sum of Logic and Natural Philosophy) *c.* 1349, likewise at Merton College, Oxford. Part 3 is concerned with *De motu*, and gives rules for calculating the velocity of local motion. The work exists only in manuscript, and apart from the brief selection in Clagett (fn. 5), has never been edited or translated into English. Weisheipl analyzes portions of it in his unpublished dissertation, "Early Fourteenth-Century Physics of the Merton 'School' with Special Reference to Dumbleton and Heytesbury" (Oxford, 1956).

[11]Richard Swineshead, not to be confused with John or Roger Swineshead. Richard was likewise a fellow of Merton College; his *Liber calculationum* (Book of Calculations) was composed *c.* 1350, and contains many rules for calculating the velocities of motions. There are early printed editions of the work but no critical edition or translation, apart from excerpts in Clagett (fn. 5).

18   WILLIAM A. WALLACE

Swineshead. At first approximation, this may be identified as the
merging there of both Aristotelian and non-Aristotelian elements,
as a heterogeneous blending of what was called the *via antiqua* with
the *via moderna*.[12] Possibly this ambivalence can be concretized
by examining the concept of motion contained in the Mertonian
treatises, focusing in particular on the reality of motion and on the
causality involved in its production.

Bradwardine would undoubtedly have identified himself as an
Aristotelian, for the problem to which he set himself was to save
the rules given by Aristotle for comparing motions and deciding
on their commensurability. Yet in defining motion he and the other
fellows of Merton College implicitly abandoned Aristotle's analysis
in favor of that furnished by William of Ockham. A close study of
their writings shows that, rather than conceive motion as the *actus
entis in potentia inquantum huiusmodi*, as Aristotle had done, the
Mertonians regarded motion essentially as a ratio.[13] For them, the
formal cause of motion was velocity, or the ratio of space traversed
to the time elapsed.[14] Following Ockham, they even denied the

[12]The *via antiqua*, or "old way," was that of Aristotle and his commentators of
the Middle Ages such as Averroës, Albertus Magnus, Thomas Aquinas, and John Duns
Scotus. The *via moderna*, or "modern way," was that of William of Ockham; for a
discussion of Ockham's views as these relate to motion, see H. Shapiro, *Motion,
Time and Place According to William Ockham* (St. Bonaventure, N. Y., 1957).

[13]How this transition came about is not easy to explain, involving as it did a
rejection of such basic principles as the Euclidean condition for any ratio, viz, that it
must be between entities of a single kind. Some of the factors that perhaps account
for the transition have been documented in ch. 5 of Weisheipl's unpublished Oxford
dissertation (fn. 10); they relate to the various attempts to locate successive or
continuous motion in one or other of the Aristotelian categories, e.g., *passio* or *ubi*.
Associated with these attempts was the question whether motion should be viewed more
properly as a *forma fluens* or as a *fluxus formae*; on this, see Anneliese Maier, "Die
scholastische Wesensbestimmung der Bewegung als forma fluens oder fluxus formae
und ihre Beziehung zu Albertus Magnus," *Angelicum*, 21 (1944), a study enlarged in vol.
1 of Maier's *Studien* (fn.4), 9–25. Such problems were being discussed by Avicenna, Averroës, and Albertus Magnus appreciably before the fourteenth-century development, and
by John of Jandun, William of Alnwick, and John Canonicus early in the fourteenth
century. Attention was thereby focused on the relative (as opposed to the absolute)
character of motion, and the way prepared for viewing motion itself merely as an *ens
rationis* in the sense of a relation or a negation. Possibly the association of motion with
a ratio emerged from a thought-context in which motion was being implicitly subsumed
under the category of relation, which is precisely the category in which ratio would have
to be situated.

[14]John Dumbleton, for example, in his *Summa logicae et philosophiae naturalis*,
Part 3, chs. 22–25, identifies the matter of local motion as the distance traversed
(*spatio acquisita*) and implies that the formality of motion (*ut est forma realis vel
imaginata*) is the velocity with which that distance is traversed. In Dumbleton's view,
increase of local motion is nothing more than increase of velocity, which itself means
greater distances being traversed in equal times. Thus in Part 3, ch. 7, Dumbleton

physical reality of motion; in their analysis motion became nothing more than the object moved.[15] And, although they cited extensively the *loci* in Aristotle where he spoke of ratios or velocities of motion, in general they were insensitive to what for Aristotle was an important distinction, namely, that between motions that are natural, proceeding from a source within the body, and those that are violent, resulting

equates an increase or decrease of motion with an increase or decrease of ratio ("latitudo motus et proportionis inter se equaliter acquirunter et deperduntur"—MS Vat. lat. 6750, fol. 40vb). More explicit is the statement in an anonymous fourteenth-century *Tractatus de motu locali difformi*, contained in Cambridge, Caius College MS 499/268, fol. 212ra-213rb, whose author (possibly Richard Swineshead) identifies the *causa materialis seu materia* of motion as the *ipsum acquisitum per motum* (i.e., in the case of local motion, the space acquired or traversed), the *causa formalis* as a *transmutatio quedam coniuncta cum tempore* (i.e., the time rate at which the space is acquired or traversed), and the efficient cause as the *proportio maioris inequalitatis potencie motive super potenciam resistivam* (i.e., the ratio by which the motive force exceeds the resistance). Here not only the formal cause but also the efficient cause seem to be identified with ratios. Weisheipl, who has generously allowed me to use his reading of this text, translates the entire passage as follows: "The material cause of motion is whatever is acquired through motion; the formal cause is a certain transmutation conjoined with time; the efficient cause is a proportion [= ratio] of greater inequality of the moving power over resistance; and the final cause is the goal intended."— *Development of Physical Theory* (fn. 4), 76.

[15]Ockham, as is well known, thought that the succession involved in local motion could be adequately accounted for by the negation of all parts of the motion not yet acquired. Since such a negation is not a *res*, but only an *ens rationis*, there is no reality to motion over and above the existing *res permanentes* (for him, quantified substance and qualities). Ockham conceded that those who use the term "motion" as an abstract noun imagine that it signifies a distinct reality, but he regarded this as an error, the *fictio nominum abstractorum*; for related texts, see Shapiro (fn. 12), 36–53. Wilson finds essentially the same teaching in William Heytesbury: "For Heytesbury, the real physical world consists only of objects; point, line, surface, instant, time, and motion are *conceptus mentis*. These affirmations (or perhaps they are better termed 'negations') are in accord with the nominalist or terminist position, developed at length in William of Ockham's work on the *Physica*."—*William Heytesbury* (fn. 7), 24. Since *entia rationis* (or *conceptus mentis*) such as relations and negations were constantly invoked in these analyses of motion, it is not surprising that the reality of motion itself was denied and an ontological claim made only for the object moved. Yet this did not prevent highly imaginative mathematical analyses of various motions, particularly in terms of the ratios they involved. As Wilson observes, "It is of some interest, then, that the reductive tendency in nominalism—its tendency to deny real existence to what is not observable—does not operate as a prescription against speculation concerning the *imaginabilia*. Quite the reverse: in the discussion of hypothetical physical problems, Heytesbury and his contemporaries frequently multiply *formalitates* in the Scotian manner. The result is a kind of mathematical physics which at times runs strangely parallel to modern physics, but which neither seeks nor claims to have application to the physical world."—*ibid.*, 25. The impact of these nominalist analyses of Galileo's immediate precursors is discussed in W. A. Wallace, "The Concept of Motion in the Sixteenth Century," *Proceedings of the American Catholic Philosophical Association*, 41 (1967), 184–95.

20   WILLIAM A. WALLACE

from some type of externally applied force. The rules they formulated applied indifferently to both.

Despite the conceptual changes that these emphases implied, however, the Mertonians continued to speak of motion as having causes and effects. In fact, some writers have been intent on showing that the distinction between dynamics, which ostensibly studies motion from the point of view of the causes or factors producing it, and kinematics, which studies motion in terms of its effects or its spatiotemporal characteristics, was already known to the Oxford school.[16] It is here that the ambivalence of the Mertonian position lies. If motion is not something real, as Ockham himself was quick to point out, then there is actually no point in seeking out its causes or its effects.[17] Causal terminology becomes meaningless in such a context; a *flatum vocis* or an *ens rationis* is really no basis for differentiating dynamics from kinematics. One may speak of a ratio being the "cause" of another ratio, but what one really means by this is that the ratios are functionally related. And, as the writings of the Mertonians so abundantly show, their interest was ultimately in kinematics. They discussed all types of imaginary motions generally without reference to nature or even to artifacts; they spoke of abstractly conceived and mathematicized motive powers and resistances and examined every type of functionality to be found between them.[18] This explains how they could lay the mathematical foundation on which modern mechanics was to be based. But it also explains why they failed to lay any foundation for the experimental component of this science. The experimenter, according to the classical analysis at least, attempts to cause motions, and to study the effects of what he

[16]Crosby (fn. 3), 52–54, gives the evidence in support of this thesis, which he offers as a mild corrective to Maier's analysis.

[17]Thus Ockham rejected the motor causality principle, *Omne quod movetur ab alio movetur*, as applying to local motion, precisely on the grounds that "local motion is not a new effect"—Shapiro (fn. 12), 53.

[18]Even a cursory examination of the *Regule* and the *Liber calculationum* will show this. Wilson explains Heytesbury's frequent use of the phrase *secundum imaginationem* and his abstract, logical treatment in the work cited (fn. 7), 25. Again, as M. A. Hoskin and A. G. Molland point out in "Swineshead on Falling Bodies: An Example of Fourteenth-Century Physics," *The British Journal for the History of Science*, 3 (1966), 150–82, the author of the *Liber calculationum* uses impressive mathematical techniques to reach a null result that, for him, justifies an Aristotelian principle to which he already subscribes. They conclude: "The tractate therefore ends with the frustrating spectacle of an author using sophisticated techniques of applied mathematics in order to show that in the problem at issue mathematics is inapplicable" (154). Yet, paradoxically, it was the very development of these "inapplicable techniques" that provided the mathematical apparatus earlier identified in this paper as a major contribution of late medieval writers to the developing science of mechanics. Cf. Wilson as cited toward the end of fn. 15 above.

MECHANICS FROM BRADWARDINE TO GALILEO   21

himself causes, in an attempt to duplicate nature's operation in a laboratory situation.[19] The Mertonians may have paid lip service to the causes and effects of local motion, but for all practical purposes they did not believe in them, and so they lacked an important requisite for the experimentalist's mentality.

Where, then, were the later medieval and scholastic influences that could have generated an experimentalist attitude? A possible answer is that these resulted from gradual changes of mentality that came about during the fourteenth, fifteenth, and sixteenth centuries as the works of the Mertonians were studied and reevaluated in centers as diverse as Paris, Padua, and Salamanca.

The Paris development was the first chronologically, taking place within a few decades of the Mertonian contribution. Here the work of John Buridan,[20] Albert Saxony,[21] and Marsilius of Inghen[22] quickly led to an incorporation of Mertonian ideas within a more realistic framework. This development has been sufficiently studied by Duhem[23] and Maier,[24] and need not be detailed here. The Paris terminists, for example, developed the theory of *impetus* precisely because of their concern over the reality of motion, which made more meaningful for them the question of its causes and effects. The basic problems of dynamics were certainly broached by Thomas Bradwardine, but they were not taken seriously before John Buridan. The study of motion *quoad causam* and *quoad effectum* was probably mentioned by Richard Swineshead,[25] but it was Albert of Saxony and

[19]Such a mentality lay behind Newton's "Rules of Philosophizing," and also the elaborations of scientific methodology by Francis Bacon, John F. W. Herschel, and William Whewell. See R. M. Blake, *et al.*, *Theories of Scientific Method: The Renaissance Through the Nineteenth Century* (Seattle, 1960) and J. J. Kockelmans, ed., *Philosophy of Science: The Historical Background* (New York, 1968).

[20]The more important of Buridan's works, for purposes of this study, are his *Subtilissime questiones super octo physicorum libros Aristotelis* (Paris, 1509; reprinted Frankfurt a. M., 1964) and his *Questiones super libros quattuor de caelo et mundo*, ed. E. A. Moody (Cambridge, Mass., 1942). Significant excerpts are given in Clagett (fn. 5).

[21]For Albert's teachings consult his *Tractatus proportionum* (Paris, c. 1510), his *Acutissime questiones super libros de physica auscultatione* (Venice, 1516) and his *Questiones subtilissime. . . in libros de caelo et mundo* (Venice, 1520). Excerpts are again to be found in Clagett (fn. 5), while a summary of Albert's position on the reality of motion is given in my paper cited at the end of fn. 15, 189.

[22]Marsilius's teachings are contained in his *Questiones. . . super octo libros physicorum secundum nominalium viam* (Lyons, 1518; reprinted Frankfurt a. M., 1964) and his *Abbreviationes super octo libros physicorum* (Venice, 1521).

[23]Pierre Duhem, *Études sur Léonard de Vinci* (3 Vols; Paris, 1906–13), esp. vol. 3.

[24]Anneliese Maier, *Die Vörlaufer Galileis im 14. Jahrhundert*, 81–154, and *Zwischen Philosophie und Mechanik* (Rome, 1958), 59–144.

[25]The distinction is contained in *De motu* commonly ascribed to Richard Swineshead; the text is in Clagett (fn. 5), 245.

22   WILLIAM A. WALLACE

Marsilius of Inghen who took these terms seriously and effectively divided the study of motion into two areas, one *penes causam* and corresponding to dynamics, the other *penes effectum* and corresponding to kinematics.[26] This is not to say that either Albert or Marsilius abandoned nominalism in their attempts to treat motion realistically. No, they tried to be nominalist and realist at the same time, and they were not completely successful in resolving the latent contradictions that this implied.[27] But they were more consistently Aristotelian than were the Mertonians. They thought, for example, of applying "calculatory" techniques to the cases of falling bodies and the movement of the heavens. Such cases were used by them to illustrate the very types of motion that had been treated by the Mertonians only in kinematic fashion.[28] Thus they took the first step toward investigating the real world, the world of nature, with the new mathematical techniques.

The next step, as it appears, took place at the University of Padua in the mid-fifteenth century. Paul of Venice studied at Oxford in the latter part of the fourteenth century, and, on his return to Italy, propagated Mertonian ideas among his students.[29] Most important of these students for purposes of this study was Gaetano da Thiene, who wrote an extensive commentary on Heytesbury's *Regule*.[30] The difference between Gaetano's mentality and that of Heytesbury becomes clear on reading the following small part of Heytesbury's text, and then contrasting this with the corresponding portion of Gaetano's commentary. Heytesbury points out, at the beginning of his treatment on local motion, the distinction between uniform and nonuniform motion. He then explains how one goes about measuring a uniform velocity:

In uniform motion . . . the velocity of a magnitude as a whole is in all cases measured by the linear path traversed by the point which is in most rapid

[26]For Albert of Saxony, see his fourth question on the sixth book of the *Physics*, (fn. 21), fol. 66va. Marsilius of Inghen gives a similar distinction in his fifth question on the sixth book of the *Physics*, (fn. 5), fol. 68rb.

[27]For Albert of Saxony's difficulties, see my article cited at the end of fn. 15, 189. Maier gives a similar analysis of Marsilius of Inghen's apparently contradictory position in her *Zwischen Philosophie und Mechanik* (fn. 24), 139–40, esp. fn. 100.

[28]For a complete analysis and documentation of the various examples used by fourteenth- to sixteenth-century writers to illustrate the kinds of local motion discussed by the Mertonians, see W. A. Wallace, "The Enigma of Domingo de Soto: *Uniformiter Difformis* and Falling Bodies in Late Medieval Physics," *Isis*, 59 (1968), 384–401.

[29]A summary of Paul's teaching is contained in his *Summa philosophiae naturalis* (Venice, 1503), which was widely used as a textbook.

[30]Gaetano's commentary is to be found in Heytesbury's *Tractatus de sensu composito et diviso, Regulae cum sophismatibus, Declaratio Gaetani supra easdem*, etc. (Venice, 1494).

motion, if there is such a point. And according as the position of this point is changed uniformly or not uniformly, the complete motion of the whole body is said to be uniform or difform. Thus, given a magnitude whose most rapidly moving point is moved uniformly, then, however much the remaining points may be moving non-uniformly, that magnitude as a whole is said to be in uniform movement. . . .[31]

The language, as is easily recognized, is that of kinematics. Heytesbury is talking of moving bodies and moving points, but these he conceives very abstractly, and one is hard put to see how they apply in any way to the order of nature.

Commenting on this section, however, Gaetano's imagination takes a realistic and practical turn. To exemplify Heytesbury's reasoning he proposes the case of a rotating wheel that expands and contracts during its rotation.[32] He talks also of a cutting edge placed against a wheel that continually strips off its outermost surface.[33] Another of his examples is a wheel whose inner parts are expanding while its outer surface is being cut off.[34] Gaetano speaks too of a disk made of ice rotating in a hot oven; here the outermost surface continually disappears and the velocity at the circumference becomes slower and slower, whereas the inner parts expand under the influence of heat and their linear velocity increases.[35] Yet another of his examples is a wheel that rotates and has material gradually added to its circumference, as clay is added by a potter to the piece he is working. Here the velocity of rotation would be uniform but the linear velocity of a point on the circumference would increase, unless the entire wheel could be made to contract in the process, in which case the linear velocity of the outermost point might remain constant.[36]

These examples, it should be noted, are Gaetano's and not Heytesbury's.[37] Heytesbury's kinematic doctrine is, of course, important,

[31]The translation is from Clagett (fn. 5), 235–36.

[32]"Notandum quod illa conclusio habet veritatem primo propter corruptionem punctorum extremorum, ut dicit magister [Hentisberus]. Secundo propter condensationem forme circularis ab intra et rarefactionem ab intra. Tertio per condensationem ab intra et additionem ab extra. . ."—ed. cit. (fn. 30), fol. 38rb.

[33]"Que conclusio declaratur sic. Ponatur gladius supra rotam ut prius et dolet continue partem extremam rote. . ."—ibid.

[34]"Deinde volo quod quelibet pars citra ultimam remotam et dolatam rarefiat, ita tamen quod non transeant magnitudinem dolatam. . ." – ibid.

[35]"Adhuc posset considerari alius casus de aqua congelata et ponatur in furno calidissimo et continue volveretur. . ."—ibid.

[36]"Que conclusio probatur sic. Ponatur quod una rota moveatur et continue in superiori parte addantur alie partes sicut sit in rota figuli cui addiur glis circumquamque simul. . ."—ibid., fol. 38va.

[37]One would not know this, unfortunately, from Wilson's explanation of Heytesbury's teaching. He presents fifteenth-century examples as though these were in the original fourteenth-century text; see the work cited (fn. 7), 117–28.

24   WILLIAM A. WALLACE

for without it, Gaetano would have had no reason to seek its exempli-
fication. But the examples furnished by Gaetano are important
too, for they show that Gaetano was convinced that Heytesbury's
doctrine could be applied to the real world, and in fact was thinking
of cases that were realizable in materials close at hand after the
fashion of the experimenter. Gaetano did not perform experiments
or measurements (at least as far as is known), but he took another
step closer to their realization. And he, like Paul of Venice, was
a realist, perhaps more in the Averroist than in the Scotist sense, but
nonetheless unwilling to accept fully the nominalist philosophy of
nature.[38]

A third step in the evolution of an experimental component for
modern mechanics may be said to have taken place at the University
of Paris in the early sixteenth century. Here the school of John
Major, as exemplified particularly in the writings of John Dullaert of
Ghent, Alvaro Thomaz, and Juan de Celaya, focused attention once
more on the controversy between the realists and the nominalists.[39]
Seemingly eclectic in their philosophical views, these thinkers
actually sought a *via media* that would be acceptable to partisans of
the old controversy.[40] And, in so doing, they attempted to incorporate
the entire Mertonian tradition, as reworked particularly by Alvaro
Thomaz, into the *Physics* of Aristotle.[41] Dullaert[42] and Celaya[43]
thought that the proper place to do this was in their questions on
the third book of the *Physics,* where Aristotle, significantly, treats
the definition of motion. The tracts *De motu* they produced in this
context are recognizably closer to modern mechanics than those
of any of their predecessors. Certainly they included both dynamical
and kinematical questions, and exemplified these with cases drawn
both from nature and from artifacts along lines suggested by
Gaetano da Thiene, whose work they knew and, in the case of Celaya,
even cited.[44]

[38]Commenting on Heytesbury's nominalism as implied in the statement: ". . . in
rerum natura non est aliquid quod est instans ut instans, nec tempus ut tempus, aut
motus ut motus. . . ." Gaetano writes: "Hoc dixit quia credidit quod motus non
distingueretur realiter a mobili. . . ." *ed. cit.* (fn. 30), fol. 26ra, 28vb.

[39]For details, see Hubert Élie, "Quelques maîtres de l'université de Paris vers l'an
1500," *Archives d'histoire doctrinale et littéraire du moyen âge*, 18 (1950-51), 193-243.

[40]This point is developed in my paper cited at the end of fn. 15.

[41]Alvaro Thomaz, a Portuguese, was the "calculator" *par excellence* of the sixteenth-
century Paris group, as can be seen from even a rapid perusal of his *Liber de triplici
motu. . .* (Paris, 1509). Grant gives a brief appraisal of this treatise in his work on
Oresme (fn. 8), 56n-58n, 70-72, 319-20.

[42]*Questiones super octo libros physicorum Aristotelis necnon super libros de caelo
et mundo* (Lyons, 1512); an earlier edition appeared at Paris in 1506.

[43]*Expositio. . .in octo libros physicorum Aristotelis, cum questionibus. . .secundum
triplicem viam beati Thomae, realium, et nominalium* (Paris, 1517).

[44]Celaya refers to Gaetano in the edition cited (fn. 43), fol. 95ra. That Dullaert

MECHANICS FROM BRADWARDINE TO GALILEO   25

The writings of these Paris masters exerted a great influence in Spain within a few decades, and this led to what may be regarded as the final stage in the preparation for an experimental mentality.[45] The figure who best characterizes this development is Domingo de Soto, who himself had studied at Paris under Celaya, and further developed the doctrines while teaching at Alcalá and Salamanca.[46] Soto was not himself a physicist in the modern sense. He was primarily a theologian and a political theorist, but by avocation he happened to be also a teacher of physics. The circumstances under which he composed his physics course, it seems, required him to be both a simplifier and an exemplifier. Again, by philosophical heritage he was a moderate realist; he wished to steer a middle course between the *nominales* and the *realissimi,* as he called them, acknowledging elements of truth in these extremes.[47] Both his practical and his ideological bent, under such influences, put him yet another step closer to the mentality of the seventeenth-century scientist.

An illustration may serve to make the point. The "questionaries" on the *Physics* used at Paris while Soto was a student there all employed the Mertonian terminology with regard to uniform and difform motions. For some curious reason, when exemplifying these motions most writers used a system of classification that may be traced back to Albert of Saxony.[48] This included, among others, motions that are uniform with respect to time and difform with respect to the parts of the moving object, and motions that are difform with respect to time and uniform with respect to the parts of the moving object. The first was by now commonly exemplified by a wheel or by a heavenly sphere, which rotate uniformly with respect to time but whose parts move with greater velocity as they are located farther from the center of the pole and toward the outermost periphery. The second was similarly exemplified by the falling body, whose velocity of fall increases with time, but all of whose parts move with the same velocity at any instant. With very few exceptions, the authors

knew of Gaetano's work is evident from his furnishing the same examples and the same method of dividing the types of local motion; details are given in my paper cited in fn. 28.

[45]Here are omitted many details relating to the transmission of the Paris teaching to Spain; these are contained in W. A. Wallace, "The 'Calculatores' in Early Sixteenth-Century Physics," *The British Journal for the History of Science,* 4 (1969), 221-32.

[46]For biographical details, see V. Beltrán de Heredia, *Domingo de Soto: Estudio biográfico documentado* (Salamanca, 1960). Soto wrote many logical, philosophical, and theological works; here interest is focused on his *Super octo libros physicorum Aristotelis questiones* (Salamanca, 1545?; first complete ed. 1551).

[47]Soto spoke frequently of the *nominales* and the *reales,* meaning by the latter term Scotists. In question 2 on book 2, however, he refers to the *realissimi,* and likewise in question 1 on book 4. Soto's middle position is outlined in my paper cited at the end of fn. 15, 193-94.

[48]Details are given in my paper referenced in fn. 28.

26   WILLIAM A. WALLACE

before Soto who attempted to illustrate uniform or difform motions did so with examples that employed this two-variable schema.[49] They always spoke of variations that take place both with respect to time and with respect to the parts of the object moved, and spoke of either being uniform or difform, in all the possible combinations.

Soto's advance here, it would seem, was one of simplification. He thought of discussing motion that is uniform merely with respect to time or uniform merely with respect to the parts of the object moved, and gave simple illustrations of these. He exemplified also motions that are difform with respect to time alone, and then went further to seek examples from nature illustrating how some motions are uniformly difform with respect to time, whereas others are difformly difform in the same respect. In other words, Soto substituted a one-variable schema for a two-variable schema, and restricted himself to one variable at a time when furnishing realistic (as opposed to imaginary) examples. This simple device, apparently, was what enabled him to adumbrate Galileo's work in the association of uniform acceleration with actual falling bodies. Some eighty years before Galileo—and, as far as is known, this is a "first"—Soto explicitly indicated that falling bodies accelerate uniformly over the time interval of their fall, and that the Mertonian "mean-speed theorem" can be used to calculate the distances they traverse in so doing.[50]

Did Soto ever measure the distances covered by a falling body to see if his exemplifications were correct? Certainly he did not. There are indications in his writings that he performed what later thinkers would call "thought experiments," particularly relating to the vacuum, but he seems not to have done any measuring or experimenting himself.[51] What is significant about his contribution is that he laid the groundwork, that he prepared the ideas, that he simplified the examples, so that someone else might see that here was a case that is experimentally tractable, and finally put a mathematical law of motion to empirical test.

The final chapter in this development, of course, was written at the turn of the century in northern Italy—at Padua, fittingly perhaps, in view of the work done there earlier by Gaetano da Thiene. And other influences were undoubtedly present, apart from those deriving from the Mertonians and the Schoolmen of the

[49]The more significant authors, all of whom are discussed in the same paper (fn. 28), include Gaetano da Thiene, Angelo da Fossombrone, John Dullaert of Ghent, Diego Diest, and Diego de Astudillo.

[50]Seemingly Duhem was the first to point this out in the third volume of his work on Leonardo da Vinci (fn. 23); some details are given in my paper cited in fn. 28.

[51]Discussion of this point will be found in an article by Charles B. Schmitt, "Experimental Evidence for and against a Void: The Sixteenth-Century Arguments," *Isis*, 58 (1967), 352–66.

fifteenth and sixteenth centuries. These have been adequately discussed elsewhere.[52] They include other traditions in mechanics, such as the study of dynamical problems deriving from the *Questions of Mechanics* of the Aristotelian school, the Archimedean study of statics and hydrostatics in the light of precise geometrical principles, the Alexandrian concentration on theoretical mechanics typified in the works of Hero and Pappus, the medieval science of weights associated with the name of Jordanus Nemorarius, and an on-going technological tradition with its roots in antiquity but developing rapidly in the sixteenth century through the efforts of civil and military engineers. To the last-named tradition may be assimilated the work of craftsmen and mathematicians done largely outside the universities—Niccolò Tartaglia and Giovanni Battista Benedetti come to mind in the latter category—who provided the proximate materials for the development of an experimentalist mentality.

All of these traditions somehow merged in the complex personality, activity, and literary productivity of Galileo. Analyzing this productivity in one of his celebrated theses, Alexandre Koyré discerned three stages in Galileo's intellectual development: the first stage, that of his youthful work at Pisa, was concerned mainly with the philosophy of motion of the later Schoolmen under the influence of his Pisan professor Francesco Buonamici; the second stage, exemplified by his preliminary attempts at Pisa to write a definitive treatise *De motu,* was concerned with the anti-Aristotelian impetus mechanics of Benedetti; and the third stage, associated with his move to Padua, was one in which he became more markedly Archimedean than Benedetti and laid the foundations for the *Two New Sciences* of his declining years.[53] In this analysis, the least that can be claimed for the movement we have been tracing throughout this paper is that it provided the point of origin, the springboard, for Galileo's distinctive, but later, contributions. In his *Juvenilia,* of course, Galileo does cite the *doctores Parisienses,*[54] Heytesbury,[55] Swineshead,[56] Gaetano da Thiene,[57] and Soto.[58] Through Buonamici's *De motu libri X* it seems certain that Galileo knew also of Albert of Saxony's *Tractatus propor-*

[52]John H. Randall, Jr., *The School of Padua and the Emergence of Modern Science* (Padua, 1961); Stillman Drake and I. E. Drabkin, ed. and trans., *Mechanics in Sixteenth-Century Italy* (Madison, 1969).

[53]A. Koyré, *Etudes galiléennes,* I. A l'aube de la science classique (Paris, 1939), 10.

[54]*Opere di Galileo Galilei,* ed. A. Favaro, Vol. I (Florence, 1890), 35, 138.

[55]*Ibid.,* 172.   [56]As "Calculator," *ibid.,* 172.

[57]*Ibid.,* 72, 172; the remaining entries in the index of Vol. I (422) ascribed to "Caietanus Thienensis" are erroneous, since they refer to the Dominican Thomas de Vio Caietanus.

[58]*Ibid.,* 144, 146—not, however, where Soto is discussing falling motion.

28   WILLIAM A. WALLACE

tionum and, through Achillini, whom Buonamici cites,[59] of the kinematical contributions of Bradwardine, Heytesbury, and Swineshead.[60] Whether he relinquished these authors and their ideas totally after leaving Pisa, as Koyré implied, or whether he returned to them from time to time as he worked out the teaching of the *Two New Sciences,* is still under debate and the subject of active research by historians of science.[61]

Obviously there is no easy answer to the question of late medieval and scholastic influences on the thought of Galileo. The line of argument pursued in this paper, however, suggests a modest conclusion. Bradwardine's *Tractatus de proportionibus* and its successors laid the mathematical foundations that made the seventeenth-century accomplishment in northern Italy a possibility. Less noticeably, perhaps, they introduced the problematic of how motions can be conceived and analyzed mathematically, and at the same time studied in nature or in artificially contrived situations. Scholasticism may have been in its death throes by the time the full solution to this problematic could be worked out, but withal the Schoolmen were not completely sterile in the influences they brought to bear on its statement and eventual resolution.

The Catholic University of America, Washington, D.C.

[59]F. Buonamici, *De motu libri X* (Florence, 1591), 528–29.

[60]Alessandro Achillini, *De proportione motuum,* in *Opera omnia in unum collecta* (Venice, 1545), 190v–195r.

[61]Koyré's thesis has recently been challenged by Raymond Fredette, *Les DE MOTU "plus anciens" de Galileo Galilei: prolégomènes* (Montreal, 1969), a doctoral dissertation in the Institut d'études médiévales, Faculté de philosophie, Université de Montréal.

UNPUBLISHED MANUSCRIPT records Galileo's first discovery of the law of free fall, before October, 1604. It is reproduced by courtesy of the National Central Library in Florence, where it is preserved as *f. 152r* in Volume 72 of the Galilean manuscripts. The crucial diagram is the large one at the top; the lower diagram and the bottom paragraph refer to calculations for horizontal motion.

Galileo's Discovery of the Law of Free Fall

It has been thought that he erroneously assumed that the velocities of a falling body were proportional to distances. A new manuscript shows that he treated them correctly as being proportional to time

by Stillman Drake

The modern era in physics opened with the publication of Galileo's *Discourses on Two New Sciences* in 1638. It reported basic discoveries he had made 30 years earlier. In describing his book to a friend in January, 1639, the old and blind author dictated these words (here, of course, translated from the Italian): "I assume nothing but the definition of that motion of which I wish to treat and whose properties I then demonstrate.... I declare that I wish to examine the essentials of the motion of a body that leaves from rest and goes with speed always increasing...uniformly with the growth of time.... I prove the spaces passed by such a body to be in the squared ratio of the times.... I argue from supposition about motion defined in that manner, and hence even though the consequences might not correspond to the events of natural motion of falling heavy bodies, it would little matter to me.... But in this, I may say, I have been lucky; for the motion of heavy bodies, and the properties thereof, correspond point by point to the properties demonstrated by me."

Not even Galileo's severest critics attribute his discovery of the law of free fall to sheer luck; hence it may seem odd that I, one of his most fervent admirers, should do so now. I believe Galileo meant his remark quite literally. Evidence for this belief exists in an early manuscript of his that has never before been published in full. That document unfolds a fascinating story of scientific discovery through a combination of error, good luck, persistence and mathematical ingenuity.

Historians of science have searched the writings of earlier men for the possible origins of Galileo's analysis of accelerated motion because no documentary evidence in his own hand had been found. Nothing surviving from classical antiquity offers a plausible source. In the 14th century, however, there were some very interesting developments in the application of mathematics to physical questions. In particular William Heytesbury and Richard Swineshead of Merton College, Oxford, and Nicole Oresme of Paris analyzed accelerated motion. The roots of medieval investigations lay in a theological problem—the increase of charity in a man—and in its philosophical implications, which spread into the general problem of rate of change. The results that were achieved are impressive, and it seems strange that their authors never thought to apply them to the problem of the changing speed of a freely falling object.

Medieval English writers adopted an arithmetical approach, from which they developed the mean-degree or mean-speed theorem, often called the Merton rule. Under this rule the speed at the middle instant was taken as being representative. Uniform motion at this mean speed over a fixed time was declared to be equivalent to uniformly accelerated motion from rest over that same time. It followed that in any uniformly accelerated motion from rest, one-fourth of the total distance was traversed in the first half of the time. This fact yielded the ratio 3 : 1 for the distances traversed in the later and earlier halves. Oresme proved the rule geometrically, and in another manuscript he extended the ratio to the progression 1, 3, 5, 7 and so on for equal times. Not even Oresme, however, connected uniform acceleration with free fall, nor did any medieval writer announce that the distances covered are proportional to the squares of the times, a fact that is deducible from the above progression. The prevailing view among historians of science was recently summarized by Edward Grant in his book *Physical Science in the Middle Ages:*

"Oresme's geometric proof and numerous arithmetic proofs of the mean-speed theorem were widely diffused in Europe during the fourteenth and fifteenth centuries, and were [then] especially popular in Italy. Through printed editions of the late fifteenth and early sixteenth centuries, it is quite likely that Galileo became reasonably familiar with them. He made the mean-speed theorem the first proposition of the Third Day of his *Discourses on Two New Sciences,* where it served as the foundation of the new science of motion."

Puzzles nevertheless remained. I again cite Grant's words: "The Mertonians arrived at a precise definition of uniform acceleration as a motion in which an equal increment of velocity is acquired in any whatever equal intervals of time, however large or small." Yet "Galileo, as late as 1604, mistakenly assumed that velocity is directly proportional to distance rather than to time, as he later came to realize." If we were to assume that the medieval writings were Galileo's source, it would be hard to explain why he accepted and extended the earlier results while rejecting the only definition on which they were based. Similarly, if he came on the medieval writings later in life, why did he still make no use of the Merton rule in his proof of the proposition mentioned, either in his notes or in his book?

I once suggested two possible alternative sources for Galileo's times-squared law. The first was that he might have discovered that spaces traversed in equal times follow the odd-number rule by roughly measuring the distance traveled by an object rolling down a gently inclined plane, using the first distance as his unit. I thought he might have made

Note: This was written when only a single page of Galileo's working papers could be safely identified with discovery of the law of fall. Other pages subsequently removed the element of luck and show how pendulums were involved in the discovery.

GALILEO'S CALCULATIONS	EXPLANATION
(1) 4 miles with 10 degrees of speed in 4 hours	Galileo first assumed that one more degree of speed is consumed in each mile of distance traversed: the speed is of one degree in the first mile, two in the second, three in the third and four in the fourth. That gave him 4 miles with 10 degrees (1 + 2 + 3 + 4) of speed consumed. The time elapsed was arbitrarily put at one hour for each different degree of speed.
(2) 9 miles with 15 degrees of speed in 5$^?$ hours	5$^?$ means that the time was to be examined. Although Statement 2 was intended to follow from Statement 1, and the next increment of speed to be added was indeed five degrees, the number of miles to be added was one and not five. Had his intention been to add a greater increment of speed for each hour, he would have written "9$^?$ miles" instead of "5$^?$ hours," and Statement 1 would still have contradicted Statement 2. (An unsuccessful attempt to get a ratio of overall speeds appears in the top right corner of f. 152r. In it Galileo tried multiplying the ratios of times and distances together.)
(1a) 4 miles with 10 degrees of speed in 4^6 hours (3) 4 miles with 15 degrees of speed in 4 hours (3a) 8 miles in 8 hours at 15 degrees of speed	To get a ratio of distances and times Galileo substituted 6 hours for the 4 hours in Statement 1. With this change, in order to cover 4 miles in 4 hours it is necessary to travel at 15 degrees of speed. Galileo noted the implication that if "15 degrees" is an overall rate, then in order to cover twice the distance (8 miles) it would take twice as long (8 hours), which would contradict Statement 2.
(4)	In an effort to resolve the contradiction Galileo drew a vertical line and first lettered it A, B and C. He divided the distance AB into four units; AC is nine units. D he added so that the distance AD would be the mean proportional between them at six units ($\sqrt{4} \times \sqrt{9} = 2 \times 3 = 6$). He let this represent the time taken in traversing AC and let AB represent the time taken in falling from A to B.

(5) through AB speed 10
    through AC as . . . 15

distance	speed acquired	time through
AB 4		AB – 4
	BE 20	
AD 6		
AC 9	CF 30	AC – 6

Using the mean proportional AD, Galileo tabulated new times for the two original distances. A new working hypothesis had emerged in which the original time through AB at 10 degrees of speed was again 4 hours, but the time through AC at 15 degrees of speed had become 6 hours, the mean proportional of the distances from the rest. It happens that if two objects fall from rest through distances that have the ratio 4 : 9, their respective average speeds do have the ratio 10 : 15. In other words, like the times (4 hours and 6 hours), the average speeds have the ratio 2 : 3. The entries for BE and CF, made later, represent the acquired speeds at the *end* of the falls AB and AC and are exact doubles of the speeds previously assigned *through* those distances.

(6)

Galileo next drew the line AE and placed point E so that BE would represent the speed acquired at B. He wrote out his conclusion, assuming that the ratio of the acquired speeds would be the same as the ratio of the overall speeds through falls from rest. He expected that other termini of horizontal lines representing acquired speeds would fall on line AE, just as the terminus of BE did. When he calculated the placement of point F by the ratio he had developed in his conclusion (BA is to AD as BE is to CF), he found that F lay not on line AE but on a parabola through points A and E. The values BE 20 and CF 30 later added to his tabulation in Statement 5 would increase the horizontal scale of his diagram by about five to one as compared with the vertical scale.

As BA to AD, let DA be to AC, and let BE be the degree of speed at B; and as BA to AD, let BE be to CF; CF will be the degree of speed at C. And since as CA is to AD, so CF is to BE, then as the square of AC to the square of AD, so will be the square of CF to the square of BE; and further, since as the square of CA to the square of AD, so CA is to AB, the square of CF will be to the square of BE as CA is to AB. Therefore the points E and F are in a parabola passing through A.

GALILEO'S CALCULATIONS (*left column*) for the law of free fall are explained in detail (*right column*). Portions of Galileo's calculations in color do not appear on f. 152r but are inserted for clarity. The term "degrees of speed" is an arbitrary one and is used in much the same way that doctors refer to degrees of burn injury. Galileo's conclusion is analyzed in the chart on page 88.

this discovery incidentally in testing an earlier (and mistaken) belief of his that acceleration is only a temporary event at the very beginning of motion.

Alternatively, it seemed to me that Galileo might have arrived at the odd-number rule by pure reasoning, as Christiaan Huygens did many years later. For example, suppose that acceleration adds an equal increment of distance in each equal time. Then in the sequence of numbers representing spaces the ratio of the first number to the second number must be the same as the ratio of the first two numbers to the second two numbers, which in turn must be the same as the ratio of the first three numbers to the second three numbers, and so on. Why must these equalities be preserved? Because we have arbitrarily chosen to use a certain unit of time, and we might have used its double or its triple instead.

Since the number representing each distance must be uniformly larger than the preceding one, the numbers will have to be in arithmetic progression. Does such a progression exist? It is certainly not the progression of consecutive integers. The ratio of the first two integers (1 and 2) is 1 : 2. The ratio of the first two integers added together (1 + 2) to the second two added together (3 + 4) is 3 : 7, which is clearly not the same as 1 : 2.

The progression of the odd integers alone, however, does follow the rule we have set up. First, each number is uniformly two greater than the preceding one (1 + 2 = 3, 3 + 2 = 5, 5 + 2 = 7, and so on). Second, the sum of the first two numbers (1 + 3) has the ratio of 1 : 3 to the sum of the next two numbers (5 + 7), and 1 + 3 + 5 has the same ratio (1 : 3) to 7 + 9 + 11, and so on. Moreover, the progression of odd integers is the only arithmetic progression that meets these conditions, as Galileo pointed out in 1615 to the same friend to whom he was writing in 1639. (There are plenty of other number sequences, such as 1, 7, 19, 37, 61, 91 . . . , that meet the ratio test above, but in no such sequence is each number uniformly greater than the preceding one.)

Alas! My two earlier suggestions must now be rejected along with the notion that Galileo got his idea for the law of free fall from medieval writings. The document I shall present shows no more trace of experimental evidence or of arithmetical reasoning than it does of the Merton mean-speed rule. Moreover, that document cannot be dated after October, 1604, when Galileo wrote to his friend Paolo Sarpi in Venice clearly

stating the times-squared law and also worked out a curious proof of it. In the letter he remarked that he had known the law for some time but had lacked any indubitable principle from which to prove it. Now, he believed, he had found a very plausible principle, which was borne out by his observations of pile drivers, namely that the speed of an object is proportional to the distance it has fallen. In the ordinary meaning of the words this statement is simply false.

The proof Galileo gave Sarpi has been a headache to historians of science for a long time. In his later *Discourses on Two New Sciences* Galileo correctly stated that the speed of a falling object is proportional to the time of fall. We could forgive him for having started with the wrong assumption and having found the right one before he had published his account. That does not, however, seem to be what happened. Apparently he had reached the right assumption earlier and had then written out for Sarpi a demonstration based on the wrong one.

Even that is not the worst of the puzzle. Galileo candidly admitted in his last book that he had long believed it made no difference which assumption was chosen. Now, if Galileo ever actually believed such seeming nonsense, there should be among his notes at least one instance where he made some obvious mistake traceable to the wrong choice. Yet no such mistake is found, even in notes that can be positively dated to the long period during which he adhered to the proof composed for Sarpi. (Even 10 years later he had a version of that proof copied for use in a book on motion that he intended to publish.) Galileo was either pulling the reader's leg in 1638 by confessing to a pretended error, or we are missing some essential point in what he said in 1638 and have been misinterpreting his earlier demonstration concerning the pile drivers, although both seem so clear as to be incapable of being misunderstood.

The answer to all these puzzles now shows that Galileo was no less candid in 1638 than he had been ingenious in 1604. What we have been missing is at last revealed by a document designated *f. 152r* in Volume 72 of the Galilean manuscripts preserved at the National Central Library in Florence [*see illustration on page 84*]. I do not think anyone could have guessed the real answer. Furthermore, if anyone had guessed it, his conjecture would have been laughed at in the absence of a document in its support.

How could such a document have es-

caped notice for so long? All Galileo's notes were edited and published around the turn of this century. The distinguished editor, Antonio Favaro, omitted only those sheets (and portions of sheets) that contain nothing but diagrams and calculations whose meaning was uncertain. There are many such sheets; Galileo was a paper saver, and until the time of his death he kept many calculations he had made 40 years earlier. So when Favaro published *f. 152r*, he retained only two coherent sentences at the right center and bottom center of the sheet, together with a much modified version of the diagrams. These extracts meant little unless they could be dated. Favaro despaired of putting the 160 sheets of Volume 72 in the order of their composition after they had been chaotically bound together long ago.

Through the generosity of the John Simon Guggenheim Memorial Foundation and the University of Toronto, I was enabled to spend the first three months of 1972 studying the manuscripts in Florence. My purpose was to attempt a chronological arrangement in order to annotate a new English translation of the *Discourses on Two New Sciences*. Like Galileo, I was lucky. It turned out that the watermarks in the paper he had used provided an essential clue. He had lived at Padua until the middle of 1610 and then had moved back to Florence. It was therefore reasonable to expect that there would have been a change in his source of paper. An inspection of his dated correspondence revealed that there was no duplication of watermarks between the letters written from the two cities. Watermarks on the undated sheets thus separate the earlier ones from the later ones.

It also happens that 40 of the 160 sheets in Volume 72 were copies in the handwriting of two of Galileo's pupil-assistants in Florence. Every one of the 40 sheets shows the same watermark, one that also appears on Galileo's letters between 1615 and 1618. It is evident that the copies were made at his house under his direction in the writing of a book on motion. Most of the originals survive, and their watermarks confirm the Paduan origin of the theorems copied, whereas the copies supply some early theorems of which the originals are lost.

A probable order of Galileo's notes thus began to emerge. The watermark evidence sorted out Galileo's early work. Additional clues were provided by his handwriting. Finding such clues presented a harder problem to the skilled editor of 50 years ago than to the relative amateur of today. He could have compared samples of handwriting only by leafing back and forth in a bound volume, whereas I could work with a Xerox copy made from microfilm. I could therefore not only place sheets side by side but also cut them up for the closer comparison of individual entries.

At that stage Galileo's pages of calculations began to fit in order, enabling me to recognize the origin of the undistinguished-looking *f. 152r*. It is pretty certain that the document is Galileo's first attempt at the mathematics of acceleration. His train of thought has now been traced out, and it is summarized in the chart on page 86.

The first two lines on the sheet, written neatly across the top and centered, certainly have nothing to do with any actual experiment. The units—miles, hours and "degrees of speed"—are quite arbitrary. Acceleration in free fall was

GALILEO'S CONCLUSION IN EUCLIDEAN RATIOS	GALILEO'S CONCLUSION REPHRASED IN MODERN TERMINOLOGY

	New definitions: s_1 denotes distance *AB*
	                    s_2 denotes distance *AC*
	            $\sqrt{s_1 s_2}$ denotes *AD*, the mean proportional of *AB*, *AC*
	                    t_1 denotes the time *AB* for an object to fall through *AB*
	                    t_2 denotes the time *AD* for an object to fall through *AC*
	                    v_1 denotes the velocity (speed) *BE* acquired by an object at the end of the fall s_1
	                    v_2 denotes the velocity (speed) *CF* acquired by an object at the end of the fall s_2

By construction of the diagram. *AB* : *AD* : : *AD* : *AC*.	$s_1/\sqrt{s_1 s_2} = \sqrt{s_1 s_2}/s_2$ by the definition of a mean proportional.

Let *BE* be the degree of speed at *B*. Then *CF* will be the degree of speed at *C* if we let
  (7a)   *AB* : *AD* : : *BE* : *CF*.
When each of two ratios is equal to a third, they are equal to each other, so that
    AD : *AC* : : *BE* : *CF*.
	Let v_1 be the velocity of an object at the end of s_1. Then v_2 is the velocity at the end of s_2 if we let
	  (7b) $s_1/\sqrt{s_1 s_2} = v_1/v_2$.
	From the above statements it can be said that
	    $\sqrt{s_1 s_2}/s_2 = v_1/v_2$.

It follows that if both sides are squared,
    $(AD)^2 : (AC)^2 : : (BE)^2 : (CF)^2$.
	It follows that if both sides are squared,
	    $(\sqrt{s_1 s_2})^2/s_2{}^2 = v_1{}^2/v_2{}^2$.

Furthermore, since *AD* is the mean proportional of *AB* and *AC* and
    $(AD)^2 : (AC)^2 : : AB : AC$,
and since ratios equal to the same ratio are equal, then
  (8a) $(BE)^2 : (CF)^2 : : AB : AC$.
	Furthermore, since $(\sqrt{s_1 s_2})^2/s_2{}^2 = s_1 s_2/s_2{}^2 = s_1/s_2$, then this statement with the previous one gives
	
	  (8b) $v_1{}^2/v_2{}^2 = s_1/s_2$.

Therefore the points *E* and *F* are on a parabola passing through *A*.
	Although Galileo did not expressly mention time in his conclusion, he had tabulated the times t_1 and t_2 (*AB* and *AD*) through the falls s_1 and s_2 (*AB* and *AC*) in such a way that they exhibited the ratio
	    $t_1/t_2 = s_1/\sqrt{s_1 s_2}$,
	and that is precisely how he habitually spoke of times and calculated with them. Thus it is not an exaggeration to say that a direct implication of Statement 7b is
	    $t_1/t_2 = v_1/v_2$.
	This, together with Statement 8b, at once implies
	  (9) $s_1/s_2 = t_1{}^2/t_2{}^2$.
	In other words, the ratio of the distances traversed by two falling objects is equal to the ratio of the squares of the times from rest. Therefore if the velocities, which are proportional to the times, are plotted against the distances on Galileo's original diagram, the result would be a parabola, as he said. (A parabola results whenever one variable is proportional to the square of the other.)

GALILEO'S CONCLUSION is expressed in his own terms of Euclidean ratios (*left column*) and is rephrased in modern notation (*right column*). Euclidean ratios, such as "*AB* : *AD* : : *AD* : *AC*," are read in the manner "*AB* is to *AD* as *AD* is to *AC*." The order of the quantities in some of his ratios has been changed for consistency when this does not interfere with his train of thought.

what basically interested Galileo, but his first step was to seek a general rule of proportionality for uniform growth of distances, times and speeds. In his first working hypothesis he assumed for the sake of argument that one degree of speed was gained for each unit of distance fallen. Thus he assumed that four miles were traversed with one degree of speed consumed in the first unit of distance, two in the second, three in the third and four in the fourth. This gave him four miles with 10 degrees (1 + 2 + 3 + 4) of speed consumed. He arbitrarily put one hour as the time elapsed for each different degree of speed. Consistent times for each speed (or distance) could not be, and did not need to be, determined in the initial working hypothesis.

Galileo next wrote: "9 miles with 15 of speed in 5? hours." The question mark is Galileo's, not mine; it indicates that this time was to be examined, as we would today write "x hours." I mention the point because in the original manuscript the number looks more like a 1 than a 5, as 5's written in those days often do. This second statement was doubtless meant to be a part of the same working hypothesis as the first, Galileo's purpose being to have two different examples in order to compare ratios. In actuality the two statements are not even consistent. If the increment of speed to be added next was five degrees, bringing the total to 15, then the number of miles to be added was one, not five. I believe the ambiguity of the rule "One more degree of speed for each additional mile" tricked Galileo into adding five miles for the new five degrees of speed. The slip was careless, but it was not fatal; far from it. As James Joyce remarked, a man of genius makes no mistakes; his errors are portals of discovery.

Now that Galileo had two distances, two speeds and two times to work with, he proceeded to apply the Euclidean theory of proportion, which was the only device he trusted for the application of mathematics to physics. His first step was to reduce both hypothetical motions to the same speed in order to compare ratios of distances and times. Accordingly he wrote 6 above the 4 in the phrase "4 hours" of his first statement. In this way he redefined "10 [degrees] of speed" so that four miles at that speed would take six hours rather than four. In order to cover the four miles in four hours, then, a body would have to travel at 15 degrees of speed. Galileo noted this fact to the left of his original statements and continued with its implication: If the meaning of "15

degrees of speed" is "going 4 miles in 4 hours," then in order to cover eight miles at 15 degrees of speed the same body would take eight hours. (This would be true of any motion, however irregular, if the meaning of "speed" is fixed by total time and total distance.) That, however, immediately contradicted his earlier hypothesis that nine miles were traversed at 15 degrees of speed in only five hours.

Here it will be useful to recall that all Galileo was trying to do at this stage was to find a mathematically consistent rule for the use of the phrase "degrees of speed." He was not yet concerned with attaching any physical meaning to the phrase, a task that would seem pretty pointless to him until its use in ratios was first made possible. In order for the phrase to be useful in ratios, any given overall speed must carry the moving body through proportional distances in proportional times.

Faced with an apparent contradiction from a working hypothesis that he (mistakenly) thought he had expressed consistently in his first two statements, Galileo did not go back to see whether or not he had made an error. If he had discovered his error and corrected it, he would have ended up with the consistent but useless formula "4 miles at 10 degrees of speed in 4 hours; 5 miles at 15 degrees of speed in 5 hours." Such a formula would have equalized the ratios of distances and of times under acceleration, contradicting good sense and the basic idea of acceleration itself, and it would have told him nothing at all about ratios of speeds. In any case he did not check back. Instead he looked for the source of the trouble by drawing a vertical line and lettering it A, B and C to represent distances from rest. He indicated four units from A and B. The distance from B to C is supposed to be five units, making AC equal to nine units, as is shown in his tabulation.

Quite by accident, in my opinion, the two distances in Galileo's working hypothesis were both square numbers: 4 and 9. If Galileo had had any previous inkling of a rule involving squares and square roots when he wrote his first two statements, he would surely have put in the numbers 2 and 3, either for the speeds or for the times. I have already noted three ways in which he might have suspected that a square-root law existed, and there may be still other ways. His choice of numbers here, however, seems to me to exclude all of them. I believe that he first discovered the squaring relation in precisely this search

for consistent ratios, and that he confirmed it afterward by experimental test.

It is the fact that 4 and 9 both happen to be square numbers that accounts for Galileo's addition of point D between B and C in his diagram. The distance AD is equal to six units. So far Galileo's problem had been one of conflicting ratios, and this was a difficulty that could be forever eliminated by introducing continued proportion. To the mind of any mathematician of the time, continued proportion was immediately suggested by the squares of two integers. Between any two such numbers there is always an integral mean proportional that is the product of the two square roots. The ratio of the lesser square to the mean proportional is then equal to the ratio of the mean proportional to the greater square.

In Galileo's case the mean proportional is equal to the product of the square root of 4 and the square root of 9, that is, the product of 2 and 3, which is 6. The ratio 4 : 6 is equal to the ratio 6 : 9, that is, 2 : 3. Galileo entered point D on his vertical line six units distant from A just because it created a continued proportion. It then occurred to him to use the mean proportional to solve the puzzle of the time ratios. If the distance AB represented the time in his first statement (4 hours), then AD (6) represented the time in the second statement (originally 5? hours.) From the mean proportional a new working hypothesis had emerged: The original time through the shorter distance AB at 10 degrees of speed was again four hours, but now the time through the longer distance AC at 15 degrees of speed had become six hours, the mean proportional of the distances from rest.

From that day forward Galileo had in his hands the principal analytical device that he applied in all his reasoning about free fall. There is no logic to the foregoing procedure except for the logic of discovery. Galileo perceived that this arrangement of numbers would preserve a consistency of ratios, and that was the necessary first move. Whether or not the rule would also agree with observable facts he left until later; in this, as he later remarked, he had been lucky.

I have said that he was just lucky in having two square numbers to start with. If that is so, can anything else be said of the ratio 10 : 15 chosen for the speeds assigned to those distances at the outset of the inquiry? When two objects fall from rest through distances that have the ratio 4 : 9, their respective average speeds do have the ratio 10 : 15. In other words, the speeds, like the times, are in the ratio 2 : 3. If, however, Galileo had started with any other two numbers, even two squares other than 4 and 9, then neither the ratio 10 : 15 nor the ratio of numbers obtained by adding up "degrees of speed consumed"

would have agreed with the ratio of the smaller number to the mean proportional between the two numbers selected.

Two such lucky coincidences may strain the reader's credulity. All I can say is that I have not succeeded in finding any better—or indeed any other—

reconstruction that will account reasonably well for all the entries on $f.$ $152r$. Moreover, coincidences among very small numbers are not as improbable as they may seem.

Another line from point A stretches off to the left at an angle from the original vertical line. Galileo probably intended it to represent the speeds acquired in acceleration. These speeds at the end of the falls AB and AC happen to be exact doubles of the speeds properly assigned *through* those distances. Galileo seems to have not yet been aware of that relation when he placed point E on his slanted line so that the distance BE would represent the speed acquired at B and wrote out his long conclusion. He did assume, however, that the ratio of acquired speeds would be the same as the ratio of his "overall" speeds through falls from rest. He apparently drew the slanted line in the expectation that other termini of horizontal lines representing acquired speeds would fall on it just as the terminus of BE did. Here a surprise was in store for him. When he calculated the length of CF (the speed acquired through AC) according to the ratio he had developed in his conclusion, he found that point F lay not on the slanted line AE but on a parabola through points A and E.

Galileo's conclusion on $f.$ $152r$ is in a way the starting point of the modern era in physics. In it he correctly related the acquired speeds to the mean proportional of distances from rest, and he obtained explicitly or implicitly all the essential rules governing acceleration in free fall. Galileo's conclusion is analyzed in his terms of Euclidean ratios of proportion and in 20th-century terminology in the chart on page 88.

Galileo's conclusion on $f.$ $152r$ did not mention time as such. Here I wish to point out once more that the only acceleration he ever applied was the acceleration of free fall. Thus whereas we generally think of acceleration as a function of two variables, distance and time, Galileo's acceleration was completely determined by either one. A rule for speeds in terms of distances, assuming one universal constant acceleration, left no freedom of choice of times. The relations Galileo expressed were thus complete and correct; $f.$ $152r$ implied all the significant relations of distance, time and speed that exist in free fall.

Although Galileo did not expressly mention time in his conclusion, he had explicitly tabulated the times through the two distances from rest using a mean-proportional relation between those distances, and that is precisely how Galileo

---

COMPARISON OF SARPI PROOF WITH PROOF IN $f.$ $152r$

Galileo arranged matters so that the points representing speeds on his diagram would fall on a straight line rather than on a parabola. All he had to do was replace his original ratio $v_1{}^2/v_2{}^2 = s_1/s_2$ with the ratio $V_1/V_2 = s_1/s_2$, where $V_1 = v_1{}^2$ and $V_2 = v_2{}^2$. In his original proof of the law of free fall for Paolo Sarpi, Galileo often applied the phrase "degrees of speed" to speed acquired at a point but not to speed through a distance. He considered speed through a distance in the Sarpi proof as being proportional to V^2 (instead of proportional to v, as he had done in $f.$ $152r$). Thus he gave the phrase two senses: In $f.$ $152r$ "speed through a distance" was an overall speed, a kind of average treated as being proportional to terminal speed. It is denoted w below. In the Sarpi proof "speed through a distance," which is denoted W, represents Galileo's provisional idea of a "total speed," for which we have no word or concept.

Using these convenient denotations (as well as the previous ones s, t and v), Galileo's two treatments can be displayed along with their modern counterpart:

$f.$ $152r$	Modern usage	Proof for Sarpi
$v_1/v_2 = w_1/w_2$	$v_1/v_2 = t_1/t_2$	$V_1/V_2 = s_1/s_2$
$w_1/w_2 = t_1/t_2$	$s_1/s_2 = v_1 t_1/v_2 t_2$	$W_1/W_2 = V_1{}^2/V_2{}^2$
(8b) $s_1/s_2 = v_1{}^2/v_2{}^2$	$s_1/s_2 = v_1{}^2/v_2{}^2$	W "in contrary proportionality to t," that is $\sqrt{W_1}/\sqrt{W_2} = t_1{}^2/t_2{}^2$
(9) $s_1/s_2 = t_1{}^2/t_2{}^2$	$s_1/s_2 = t_1{}^2/t_2{}^2$	$s_1/s_2 = t_1{}^2/t_2{}^2$

Galileo's choice of $V_1/V_2 = s_1/s_2$ in the Sarpi proof does not mean that he had to reject $v_1{}^2/v_2{}^2 = s_1/s_2$ (Statement 8b). It did mean, however, that *velocità*, defined for Sarpi as being proportional to distances traversed, could not be the same physical entity as the "speed" that had been implicitly defined as being proportional to time (the times-squared law) in $f.$ $152r$. The basic relations that can be derived from Galileo's two treatments are:

For $f.$ $152r$: $v_1/v_2 = t_1/t_2$ $v_1{}^2/v_2{}^2 = s_1/s_2$ $s_1/s_2 = t_1{}^2/t_2{}^2$

For Sarpi: $V_1/V_2 = t_1{}^2/t_2{}^2$ $V_1/V_2 = s_1/s_2$ $s_1/s_2 = t_1{}^2/t_2{}^2$

Galileo sought a proof for only the third and last relation, which remained unchanged by his new definition of "velocity." He believed that he could derive this law as well from his new definition $V_1/V_2 = s_1/s_2$ as from the old one $v_1/v_2 = t_1/t_2$, and he did. Later he found an experimental reason for adopting only the latter (older) definition of "velocities" in free fall.

"Contrary proportionality" in the proof for Sarpi relates the root of one variable to the square of the other. Since Galileo concluded from $W_1/W_2 = s_1{}^2/s_2{}^2$ that $s_1/s_2 = t_1{}^2/t_2{}^2$ by invoking "contrary proportionality," then his phrase meant that $\sqrt{W_1}/\sqrt{W_2} = t_1{}^2/t_2{}^2$. This relation is not derivable from $f.$ $152r$ alone, since there $w_1/w_2 = v_1/v_2 = t_1/t_2 = \sqrt{s_1}/\sqrt{s_2}$. Yet the relation does arise from $f.$ $152r$ and the proof for Sarpi *taken together*, that is, from the simultaneous assumption that $W_1/W_2 = V_1{}^2/V_2{}^2$ and $w_1/w_2 = v_1/v_2$. Hence it was *only* if Galileo had already perceived the fact that speeds in free fall are proportional to times ($v_1/v_2 = t_1/t_2$), and retained that knowledge when he was writing the proof for Sarpi, that he could validly invoke "contrary proportionality." Only on this same basis can we understand the fact that in all his notes and books Galileo never made use of the assumption that our ordinary velocity is proportional to distance ($v_1/v_2 = s_1/s_2$), as he has often been accused of doing.

PROOF OF THE LAW OF FREE FALL that Galileo wrote in a letter to his friend Paolo Sarpi can be analyzed in modern terms. The quantities s_1, s_2, v_1, v_2, t_1 and t_2 retain the same meaning that they have in the chart on page 88. To Galileo the relation $v_1/v_2 = \sqrt{s_1}/\sqrt{s_2}$ seemed anything but certain. He was dissatisfied that the velocities of falling bodies, when plotted against the distances they had fallen, followed a parabola instead of a straight line. The Sarpi proof redefined "velocity" by taking its square in order to make the velocities fall along a straight line and to justify Galileo's times-squared law of free fall ($s_1/s_2 = t_1{}^2/t_2{}^2$). The proof does not, as has been supposed up until now, begin by assuming that the velocities of falling bodies are proportional to the distances through which they have fallen.

habitually spoke of times and calculated with them in the rest of his notes and published books. Hence I do not think it is exaggerating to say that a direct implication of *f. 152r* is that the ratio of the speeds at two points in fall from rest is also the ratio of the elapsed times of fall. In modern terminology we write this ratio as $v_1/v_2 = t_1/t_2$, where v_1 and v_2 are velocities at two points and t_1 and t_2 are the respective times that have elapsed from rest. This ratio at once implies that the ratio of the distances is equal to the ratio of the times squared.

That Galileo himself saw the implication is indicated by the fact that the times-squared law was the very proposition he offered to prove for Sarpi, saying he had known it for some time but had lacked any indubitable principle from which to demonstrate it. The obvious conclusion to be drawn is that in 1604 Galileo did not consider a simple statement equating ratios of times and speeds under acceleration to be of such a nature as to be acceptable as an "indubitable principle."

On the contrary, in writing for Sarpi he took as his principle the seemingly contradictory assumption that speeds acquired are directly proportional to distances fallen from rest. Now, if we take the word "speed" here in the modern sense, which it had in *f. 152r* and was to have later in *Discourses on Two New Sciences,* then the principle assumed for Sarpi was not only false but also inconsistent with Galileo's own conclusion in *f. 152r,* where it is not the speeds but the squares of the speeds that are proportional to the distances. Thus it must seem at this point that we have no choice but to say that Galileo first knew the right answer and then turned his back on it in favor of the wrong one, and that only years later did he return to the position that had been implied in his very first try at the mathematics of acceleration. That is even worse than what all historians of science have been saying up to the present.

I said "It must seem at this point" because things are not going to turn out that way at all. We have already seen what a very chancy business Galileo's investigations on *f. 152r* had been. His conclusion entirely lacked objective evidence, nor is there any reason that we should think Galileo remained unaware of the flimsy character of his first stab at a perplexing problem. Even if we now know, in a kind of absolute way, that his first result was the right one, there was nothing sacred about it to Galileo, who had arrived at it merely as

an exercise in proportionality. He would have had every right to turn his back on it for something better, if indeed he had done so, which he did not. All that the result on *f. 152r* represented to Galileo was an internally consistent meaning for "speed," and for all he knew it might just be one among many.

The result that the ratio of the distances was proportional to the ratio of the velocities squared through those distances $(s_1/s_2 = v_1{}^2/v_2{}^2)$ was anything but certain. There was as yet no way to measure a speed directly, and the square of a speed was hard to even make physical sense of. Even if the result was rewritten so that the ratio of the speeds was proportional to the ratio of the square roots of the distances $(v_1/v_2 = \sqrt{s_1}/\sqrt{s_2})$, it did not even look probable. How would a speed go about adjusting itself to a quantity represented by a distance through which it had already passed?

Galileo's other relation, that the distances fallen were proportional to the squares of the times through those distances $(s_1/s_2 = t_1{}^2/t_2{}^2)$, may also have been puzzling, but it differed from the others in one respect: it could be physically verified. Galileo proceeded to verify it, probably in the way I formerly thought he might have discovered it, and in his letter to Sarpi he mentioned the odd-number rule that holds for successive spaces in equal times. Hence when he wrote to Sarpi, Galileo was far more certain of the truth of the times-squared law than he was of the validity of the derivation that had led him to it.

The most surprising implication of the documents, however, is this: Galileo never did adopt distance-proportionality for speeds in acceleration except as a temporary working hypothesis, and it led him at once to time-proportionality. We have been wrong in supposing that he rejected time-proportionality in favor of distance-proportionality in his proof for Sarpi late in 1604, in spite of its wording. Moreover, what he told the public in 1638 was the literal historical truth: For a long time he had thought it was a matter of indifference whether speeds in free fall were defined as proportional to the elapsed times or were related to the distances traversed. He merely neglected to add the statement, perhaps supposing that it was as obvious to others as it was to him: "Provided that the rest of your mathematics remains unaltered by your choice of assumption." It now remains for us to see how Galileo himself managed this, and to consider what it tells us about what he

thought was the physical meaning of the word "speed."

We can start by noting what seems at first to be the most sophisticated thing in *f. 152r*: Galileo's reference to the fact that under the normal definition, points representing speeds would fall on a parabola when they are plotted against distances of fall. If we had no idea of the date of this entry and only knew about Galileo's later achievements respecting the parabolic trajectory, we might imagine that this remark greatly pleased Galileo. Although his work on the trajectory was still four or five years in the future when he wrote his conclusion in *f. 152r,* he already knew a great deal about parabolas. His first paper in 1587 dealt with the centers of gravity of paraboloids, and he applied for a chair of mathematics on the strength of it. The parabola as such did not dismay him. In Galileo's lifelong view, however, nature always acts in the simplest way. Since the simplest rule would put the point *F* on the straight line *AE,* I fancy that Galileo wrote the last words of *f. 152r* not with the joy of discovery but with something like disgust, and that he regarded them as casting serious doubt on the reliability of this chancy speculation about acceleration in free fall.

It was not hard, however, to rearrange matters so that the points representing speeds acquired by a falling body would fall on a straight line. All Galileo had to do was replace his mean-proportional treatment of speeds in his conclusion with a linear treatment, requiring only a change in his definition of "speed." The substitution would simply make $s_1/s_2 = V_1/V_2$, where the new "velocity" V_1 represents $v_1{}^2$ and V_2 represents $v_2{}^2$. The relation between velocity and distance would then become linear instead of parabolic. Careful examination shows that Galileo did exactly this when he composed the demonstration for Sarpi. He also found a physical justification for the new definition of velocity.

One thing that historians of science have all overlooked is that no definition of "speeds" in acceleration had ever been clearly made in terms of ratios of distances and of times. Archimedes had done it only for the case of uniform motion. Galileo took it from there, seeking a ratio definition for the ever changing speeds under acceleration. The Merton rule that represented overall velocities by their mean speeds threw no light on the problem of these ratios but ingeniously circumvented it. When Galileo tackled it, no physical measure for each and every speed in acceleration had ever been assigned. He therefore felt free to define the measure of speed in any way he pleased as long as experience bore him out, or at least did not contradict him. I might add that it was a long time, probably more than 20 years, before Galileo realized what the medieval writers had always assumed, namely that there does exist a uniform motion equivalent to any uniformly accelerated motion from rest. No trace of this realization is to be found in his manuscript notes. It first appeared as Theorem I in Book II of the Third Day in his *Discourses on Two New Sciences*, a theorem that, with all due respect to Professor Grant, does not employ any mean speed to represent accelerated motion in free fall.

In his letter to Sarpi, Galileo observed that pile drivers strike twice as hard when the weight falls twice as far. That remark shows that what Galileo meant by his word *velocità* in the Sarpi proof could not be the same as what he had meant by *gradus velocitatis* in *f. 152r*. In *f. 152r* the speeds would be in the ratio of $\sqrt{2} : 1$ at the end of falls in the ratio of $2 : 1$. In the conclusion of *f. 152r* Galileo essentially stated that the velocities were proportional to the square roots of the distances from rest ($v_1/v_2 = \sqrt{s_1}/\sqrt{s_2}$), so that speeds in the ratio of $\sqrt{2} : 1$ would be found at the end of falls whose lengths were in the ratio of $2 : 1$. In the letter to Sarpi, however, Galileo appealed to observations of pile drivers. The effect of a pile driver, involving kinetic energy, is governed not by speed but by its square. Hence the effect does not support a distance-proportionality assumption in the sense that we have always imputed to Galileo's words. What it does support is the mean-proportional relation between distances and speeds derived in *f. 152r* and altered to linear form by Galileo in the Sarpi proof by a simple redefinition of "velocity." In modern terms the Sarpi proof is equivalent to the proof of *f. 152r* with the substitution of V_1 for v_1^2 and V_2 for v_2^2. Galileo did this, I believe, in order to place the acquired *velocità* along a straight line rather than on a parabola. With the transformation Galileo's earlier ratio $s_1/s_2 = v_1^2/v_2^2$ becomes $s_1/s_2 = V_1/V_2$ [*see illustration on page 90*].

Galileo's choice of the ratio $s_1/s_2 = V_1/V_2$ when he wrote to Sarpi did not mean, as historians have all previously supposed, that he had to reject his original relation $s_1/s_2 = v_1^2/v_2^2$. It did mean, however, that the *velocità* defined for Sarpi as being proportional to distance could no longer be the same physical entity as the "speed" implicitly defined as proportional to time in *f. 152r*. The times-squared law, however, remained unchanged by Galileo's new definition of *velocità*. He believed that he could derive the times-squared law as well from $s_1/s_2 = V_1/V_2$ as he could from $v_1/v_2 = t_1/t_2$. In fact he could, and he did. Later he found an experimental reason for adopting only the second (older) relation and defining "velocity" as we do.

It is easy enough to say at this point: "But there must have been some criterion of choice at every stage, since in fact the speeds in free fall do not increase according to distances fallen but do increase according to elapsed times." This statement, however, assumes a physical definition of "speed." What we know as "speeds" do indeed increase in that way. In Galileo's proof for Sarpi, however, he chose to use something else, based on his observations of pile drivers, which is the square of our notion of speed. Let us call Galileo's choice *velocità*. We could not very well argue that *velocità* fails to increase according to the distance fallen, basing our argument on the fact that *velocità* in Italian means, after all, the same thing as "speed" in English. It would be necessary to change the usual methods of measuring time and distance to make *velocità* behave like our speed. Galileo made no alteration in his relations of time and distance, however, so that his *velocità* behaved differently. The worst we can say of his definition is that we prefer another one: the same one that he himself later adopted. Galileo reasoned that the effect of a pile driver could change only when the falling weight acquired a different speed. In effect he decided, for a time, to think of "*velocità*" as "whatever it is that changes the striking power of a body falling from different heights." This quantity he could measure, and it does behave like our v^2. Later he found a way of directly observing speed in our sense of the word, but he had no such way in 1604.

In his proof for Sarpi, Galileo invoked a "contrary proportionality" between speeds and times, a curious relation that equates the ratio of the roots of one variable to the ratio of the squares of the other. With this tool he concluded that the square root of what he calls the "total speed" was proportional to the time squared [*see illustration on page 90*]. Did Galileo have any ground for asserting such a relation? He certainly presented none in the proof for Sarpi, since time was not mentioned there until he made his appeal to contrary proportion-

ality. Nor can such a relation be derived from *f. 152r* alone.

Yet the relation does arise from the two documents together, and it could only arise because Galileo assumed that what we call "speed" remained proportional to time even when his new *velocità* was made proportional to distance. It was only because he assumed this when he wrote the proof for Sarpi that he could validly invoke "contrary proportionality" as he did. And that he did assume it is borne out by the solid fact that throughout the 160 sheets of notes on motion, written over a period of 30 years, there is not one single instance in which Galileo ever made use of the assumption that speeds in the ordinary sense are proportional to distances in free fall.

This, then, is the new picture: Galileo obtained the result that speeds in free fall are proportional to times elapsed from rest in the course of his very first try at the mathematics of accelerated motion, probably in the middle of 1604. He never abandoned that conception, although for a time he altered his definition of *velocità* in the interests of elegance and the simplicity of natural phenomena, supported by reasoning about an observed phenomenon of kinetic energy. Ultimately a classical experiment, still unpublished (*f. 116*), induced him to reject the alternative definition.

We do not have, and we do not need, any special name for the physical quantity *velocità* of the Sarpi proof. If we did give it a name, that word (say "punch") would occur frequently in our discussions of energy, and it would seem to us that this newly named physical entity should enter such discussions directly, instead of as the square of something else. We might think of a falling body's velocity as "how fast it will go horizontally if deflected," and its "punch" as "how hard it will hit vertically." Whether there really are two physical entities (velocity and "punch") or whether there is only one entity (velocity) that enters into some calculations as itself (v) and into others as its square (v^2 or V) would be hard to decide.

This question was bound to crop up in some form after Galileo, convinced that nature had forced his hand, suppressed his alternative definition. In one form the question arose late in the 17th century, when Leibniz christened the neglect of v^2 ("punch") as "the memorable error of Descartes." After decades of heated argument the entire problem was recognized to be a merely semantic controversy.

Reprinted with permission from *The Texas Quarterly*, Vol. **X,** (3), I. Bernard Cohen, "Newton's Second Law and the Concept of Force in the Principia," pp. 127–157, ©Autumn 1967 University of Texas Press. Reprinted in *The Annus Mirabilis of Sir Isaac Newton, 1666-1966*, edited by R. Palter, MIT Press, Cambridge, ©1970.

I. BERNARD COHEN

Newton's Second Law and the Concept of Force in the *Principia*

INTRODUCTION[1]

Newton's *Principia* appeared in 1687 in three "books," preceded by two preliminary sections: "Definitions" and "Axioms or Laws of Motion." My assignment here is to discuss the Second Law of Motion in relation to Newton's concept of force. In this presentation, I shall begin with the *Principia* as it stood in 1687, and then turn for further illumination to certain revisions either projected by Newton or actually incorporated in later editions.

Let me remind you that in the *Principia* there are three Axioms or Laws of Motion. The third, which shall not concern us here, posits the equality of "action" and "reaction." The first is the Law of Inertia.

The Second Law specifies or quantifies how "motion" is produced, or altered, either in magnitude or in direction or in both simultaneously. Since the main burden of the *Principia* is to analyze the motions of bodies, and the changing motions of bodies, in terms of the forces causing such actions, the Second Law of Motion in its manifold applications lies at the very heart of the Newtonian system of physical thought.

In the physics of gross bodies, Newton had to deal with three groups of "forces" which may change "motion": percussion, pressure, and the varieties of centripetal force. The first two of these are contact "forces," exerted when one body actually touches another; but the third includes both contact "forces" and those that act at a distance. (In the *Principia,* Newton's examples include (1) the whirling of a stone in a sling [contact] and (2) gravity, the planetary force, and magnetism [action-at-a-distance].) Another way of dividing these "forces" is in terms of the duration of their action. Percussion alters the motion of a body in an instant, as when one billiard ball strikes another or a racquet strikes a tennis ball. But pressure and centripetal force act continuously during a finite time-interval. A third way of grouping these "forces" is to note that percussion and pressure produce a change in "motion" in which an act associated with the "force" may be independently observed; whereas in the case of centripetal force acting at a distance, we are able to see only a change in "motion" and from this observation alone deduce that a force has been acting. Finally, pressure and certain types of centripetal force may be quantified independently of the alteration of motion; e.g., the downward "force" accelerating a stone in free fall may be determined by weighing the stone. But the quantification of percussion comes from the observed change of "motion" and cannot ordinarily be gained independently.

All these differences belie the commonality of a single concept of "force." We may well understand that Newton's predecessors and contemporaries did not have a simple and unified system of physics to encompass all these varieties of the alteration of the "motion" of bodies. Thus, we shall see below, it is a clear sign of Newton's genius that he was able to produce a physics of central forces upon the more generally acceptable base of the contact "forces" of impact or percussion.

In the presentation that follows, it will be seen how Newton made the transition from the physics of impulses or instantaneous "forces" to the physics of continuous forces. Some attention will be given to two forms of the Second Law of Motion,

$$(1.1) \quad \vec{F} = m\vec{A} \ or \ \vec{F} = m\frac{d\vec{V}}{dt},$$

which is the way we currently write the law, and yet another,

$$(1.2) \quad \text{"force"} = k \cdot \Delta(m\vec{V}),$$

which is closer to the way in which Newton enunciated the Second Law in the *Principia*, although of course Newton did not write the Law as an equation. We shall see that although equation (1.2) represents the Second Law as formally stated by Newton, he also knew and used the law in the more familiar version represented by equation (1.1).

THE SECOND LAW OF MOTION

The Second Law reads (in Whiston's literal translation, 1716) as follows: "The Mutation of Motion is proportional to the moving Force impress'd; and is according to the right Line in which that Force is impress'd." Newton is thus dealing with "mutatio motus," or "change in motion," or, to refer back to Definition II, a change in the "quantity of motion," a quantity measured by the product of mass and velocity: hence with the alteration of what we denote today by momentum. There is no ambiguity. Newton does not say here, "*rate of* change in quantity of motion" or "change in quantity of motion *per second*," despite the many scientists and historians who have alleged that this is what he meant. The "force" in this statement of the Second Law is clearly what we have come to know as an "impulse."

In the presentation which follows, I shall mean by impulse a "force" acting *for a very brief time* indeed, as when a steel ball at the end of a pendulum strikes another, or when a tennis racquet strikes a ball. I shall refer to such "forces" or impulses as *instantaneous*, in order to contrast them with continuous "forces" such as gravity and other "centripetal forces." (But of course they are not absolutely instantaneous in the sense of requiring no time whatsoever for their action!) We may now restate Newton's Second Law: every change of momentum of a body is proportional to the impulse which produces it, and occurs in the direction of the line of action of the impulse.

A fundamental question that we must explore, therefore, is whether this impulse form of the law—however much it may appeal to us because it is both dimensionally sound and physically true—is a proper rendition of what Newton intended. Is it in this form that he applied the law? and if the answer to this question is "Yes," we must then ask whether Newton ever got to the law as we know it:

$$\vec{F} = m\vec{A} \text{ or } \vec{F} = \Delta(m\vec{V})/\Delta t,$$

where $\vec{F}$ is force in our modern sense and is consequently a measure of (or measured by) the rate of change of momentum.

The answer to the first question is simple. We have only to examine the first instance in which the law is applied: in Corol. 1 to the Laws of Motion, immediately following the statement of the three Laws. Here it is Newton's intention to show how effects of two "forces" may be combined. The corollary reads:

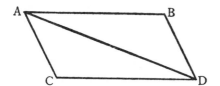

A body, acted on by two forces simultaneously, will describe the diagonal of a parallelogram in the same time as it would describe the sides by those forces separately.

The proof begins by considering the motion of a "body in a given time, by a force M impressed apart in the place A": it will, in this time, "with an uniform motion be carried from A to B." There is no doubt that in referring to the "force impressed," Newton has in mind an impulse; it cannot be a force acting continually while the body moves from A to B, since Newton says that after the force has acted the body will have a "uniform motion." If another blow N were "impressed apart in the same place," then Newton supposes that in the above-mentioned time, the body "should be carried from A to C." Now, "by both forces acting together," the body will in this given time move along the diagonal AD of the parallelogram ABDC.

In the paragraph following the enunciation of Law III, Newton gives the following example: "If a body impinge upon another, and by its force change the motion of the other, that body also (because of the equality of the mutual pressure) will undergo an equal change, in its own motion, towards the contrary part [i.e., in the opposite direction]." Here is a case of impact, where a body, "by its force," produces a change in the quantity of motion or momentum of another body. Yet another example of Newton's direct use of impact to illustrate force in the Second Law occurs in Prop. I, Book I, of the *Principia*, and is discussed below.

TWO LAWS OF MOTION RATHER THAN ONE[2]

In the *Principia*, the First Law reads: "Every body perseveres in its state of resting or of moving uniformly in a right line, unless it is compelled to change that state by forces impress'd thereon."[3] Many students of Newton have speculated on the reason why there is such a Law I in addition to Law II, since it seems to us that the former is simply a special case of the latter when the impressed force is zero.

There are at least three historical bases for Newton's having stated Laws I and II separately. First, in chronological importance, is the influence of Descartes. Descartes's *Principia philosophiae*, as I have shown elsewhere, was the undoubted source of Definition III and Law I (even to the use of the phrase "quantum in se est"), which appear there as the substance of Descartes's "prima lex naturae" and "altera lex naturae." Descartes's presentation of what was to become Newton's First Law is divided into two parts (two "leges naturae"), the first of which deals with the body's tendency to maintain a *state* of motion or of rest, while the second specifies that inertial motion must be rectilinear.[4] The effect of forces, as in collisions, or in the circular motion of a body whirling in a sling, is dealt with in the succeeding laws. Certainly this procedure must have made an impression, conscious or unconscious, on Newton, who in his own *Principia philosophiae* also embodied the principle of inertia in a separate law from that specifying how forces may alter an inertial motion, just as Descartes had done in his *Principia*.

Second, and perhaps even more important, is the influence of Huygens, whom Newton admired so much that he named the new concept of "centripetal force" in his honor.[5] Part Two of Huygens' *Horologium oscillatorium* (1673) is entitled, "On the falling of bodies and their motion in a cycloid" (*De descensu gravium & motu eorum in cycloide*). Huygens begins with a set of Hypotheses, just as Newton did a decade or so later, in the tract *De motu* (written on the eve of the *Principia*) and in the beginning of Book Three of the *Principia* in 1687, "De mundi systemate." The first two Hypotheses in Huygens' *De descensu gravium* are so close in intent to Newton's first two Axioms or Laws of Motion as to permit no doubt whatsoever that they are the source of Newton's first two laws; and thus they help us to see why Newton too had a separate Lex I and Lex II.

The third historical basis for separating Law I and Law II is the fact that Law I embodies a radical departure from traditional physics to the extent that it declares motion (if uniform and rectilinear) to be a "state" and not a "process"—in the Aristotelian sense. This Law I, equating dynamically the "state" of resting and the "state" of moving uniformly in a right line, is thus a clear announcement at the outset that the *Principia* is constructed on the primary axiom of the new inertial physics of Galileo and Descartes, of Kepler,[6] and of Gassendi. In view of the newness of this inertial physics, a separate statement of Law I, without the complicating factors of the quantitative relation of force to change of motion or of the relative directions of force and change of momentum, was obviously of a heuristic value that by far transcended the narrow logical question of whether under certain circumstances Law I might to some degree be implied by Law II as a special case.

In the modern formulation of Newtonian dynamics, in which the concept of a continuous force $\vec{F}$ is primary and the concept of impulse ($\vec{F} \cdot dt$) is derived from it, Newton's First Law does indeed appear to be a special case of the Second Law. But if impulse be primary, and Law II be stated for impulses and changes in momentum (rather than *rate of* change in momentum), then Law I would be a special case of Law II only for impulses, but not for continuous forces. It is certainly possible that Newton had this distinction in mind since the explanatory paragraph following Law I refers only to continuous forces as examples, and not once to instantaneous forces. From Newton's illustrations of these two Laws of Motion, it would seem that these two laws—as given in the *Principia*—are not really as closely related to one another as is often supposed by those who do not take account of the fact that Newton's Law II, as stated, is restricted to impulsive forces, whereas Law I is not.

This possible distinction between the first two Laws of Motion is related to the fact that in the *Principia* Newton had to deal with two distinct types of dynamically significant forces: the quasi-instantaneous impulse (as the force of percussion: a blow, an impact), and the continuously acting force. In nature, whether in the world at large or in the specially contrived experiments in the laboratory, a single impulse or a set of discrete impulses produces an obvious change in the state of a body. The Second Law expresses quantitatively just how great this change of momentum must be for a given impulse, and it further specifies that the direction of the change in momentum must be the same as the direction of the impulsive or instantaneous force. As we shall see below, Newton was quite aware that continuous forces produce continuous accelerations, and he even applied in the *Principia* the rule that in these circumstances, there is an acceleration proportional to the force, if the force is constant. But since this rule was not stated explicitly in Law II, it must have appeared to Newton to be implied by the Definitions or to be a consequence (as, in the limit) of the Second Law for impulses, or possibly to be so obvious as hardly to be worth mentioning.

But it is far from obvious as to *when* continuous forces are acting. A continuous force is not *seen to act* in the sense that is usually true of instantaneous impulses. When an impulsive force acts on a body we tend to see both an action (cause) and a change of motion (effect). That is, when a blow is given to a ball by a tennis racquet, or an impact is given by one billiard ball to another, we see the physical act of one body striking another, which we interpret as the giving of an impulse, and we see a second phenomenon, the concomitant change in the state of motion or of rest in the struck body. How different this is from the common action of continuous forces, gravity, or electric and magnetic attraction or repulsion. We certainly don't *see* the sun act on Mars in the sense that we see a tennis racquet strike a ball. Indeed, the very existence of such forces within the Newtonian framework may only be inferred, and inferred specifically from the change in the state of motion (or of rest) in the affected body. In this sense the First Law of Motion contains a test for determining whether a continuous force is acting: this test consists of discovering whether or not the "state" of the body is changing. This is the second part of Law I: ". . . unless it is compelled to change that state [of resting or of moving uniformly in a right line] by forces impressed thereon." (I have reference here to action-at-a-distance only. In the case of a stone being whirled at the end of a spring we may both have the feeling of a force acting and see the effect of the spring being extended.)

Let us turn now to the evidence contained in the paragraph following each of the first two Laws of Motion. In the discussion of Law I, Newton does not refer to a body at rest remaining at rest unless acted on by a force, or even tending to "persevere" in its state of rest. Rather he gives a succession of examples in which a body is in motion and in which the body cannot "persevere" in that state because of the action of an external force. The first example is a projectile. There are two basic ways in which Newton says a projectile may be "compelled" to change its "state of uniform motion in a right line": having its uniform linear motion altered because the projectile is (1) "retarded by the resistance of the air," and (2) "impelled downward by the force of gravity." We may observe incidentally that Newton presents here the identical circumstances of Huygens' *De descensu gravium*: first a pure inertial motion (in Hypoth. I), which is then altered (in Hypoth. II) by air resistance and gravity.

The resistance of the air appears also as a cause for altering inertial motion in the remaining two examples, the rotation of a top and the motions of planets and comets which meet "with less resistance in more free spaces" and so "preserve their motions both progressive and circular for a much longer time." These examples of rotating tops (and the "progressive and circular" motions of planets and comets) to illustrate *linear* inertia is, to say the least, startling. But Newton is not adopting a Galilean position of circular inertia; quite the contrary! Circular motion, or any curvilinear motion, is inertial in Newtonian physics only to the degree that it is a combination of two components, of which one is a continuing tangential inertial motion and the other is a constantly accelerated motion caused by a force which makes the moving body leave its otherwise linear path to follow the curved trajectory.

Let us turn now to the Second Law. The discussion (in Motte's translation) reads:

> If any force generates a motion, a double force will generate double the motion, a triple force triple the motion, whether that force be impress'd altogether and at once, or gradually and successively [*successive et gradatim*]. And this motion (being always directed the same way with the generating force) if the body moved before, is added to or subducted from the former motion, according as they directly conspire with or are directly contrary to each other; or obliquely joyned, when they are oblique, so as to produce a new motion compounded from the determination of both.

The character of the "force" in question is determined by Newton's statement that the quantity of motion generated by any "force" is the same "whether that force be impress'd altogether and at once, or gradually and successively." In our interpretation we must be wary of the adverb *"gradatim,"* which means literally "by degrees" or "step by step." Hence Newton's two modes of action are (i) a single stroke or act and (ii) a finite set of distinct stages, degrees, or steps ("successive et gradatim"). There is no implication intended by Newton of any continuous action (in the sense that today we talk of a "gradual" slope: meaning an even, moderate inclination).

Thus the paragraphs following the two laws do seem to suggest that the "forces" in Law I are continuous, just as the "forces" in Law II appear to be instantaneous (i.e., very short-lived) forces. This would accord with the need for being able to specify whether or not a continuous (noncontact) force such as gravity, electrical attraction (or repulsion) or magnetic attraction (or repulsion) may be acting,[7] to the degree— as mentioned above—that we cannot see the "cause" as is possible when a blow is given to one object by another. We have, of course, no way of knowing whether Newton intended his readers to draw such a conclusion from his examples. Furthermore, the First Law is still valid even if we there interpret "force" as impulse, and thus give a consistent interpretation to Law I and Law II.[8] In this case the two laws are one: the first part a qualitative declaration of the new physics, the second part a quantitative statement of the magnitude (and direction) of the change of motion produced by an impact. This is true of impulses, but not of continuously acting forces.

IMPACT, IMPULSE, AND MOMENTUM: NEWTON'S DEBT TO DESCARTES

The paragraph in the *Principia* explaining Law II (printed above) concludes with a method of adding the new motion to a previously existing motion. The method, as stated, appears at first to be that of Corol. 1 to the Laws of Motion.[9] But there is one major difference. In Corol. 1 it is first considered that each of two forces may act separately to produce a motion in a given body starting at rest; then both forces act simultaneously *on the very same body at rest*. But in discussing Law II, Newton posits a force acting on a body already in motion. The difference may appear conceptually slight, but it proves to be most significant.

In the "Waste Book," and in a tract written much earlier than the *Principia* entitled "The lawes of motion," first published by A. R. and M. B. Hall and more recently analyzed by John Herivel, Newton discussed this proposition in a form more like the above-mentioned discussion of Law II than Corol. 1. That is, he first allows the body to be put into motion by one of the blows; and then *while the body is in a resulting steady state of motion*, Newton has it receive a second oblique blow. The parallelogram is formed and the proof proceeds from there on as in Corol. 1. This difference between the early presentation in this tract and the later one (in Corol. 1 to the Laws of Motion in the *Principia*) has been called to our attention by Dr. Herivel, who has quite properly emphasized its significance. But I cannot agree with his tentative suggestion that this is in any way directly related to Galileo's derivation of the parabolic path of a projectile.[10] In any event, as Dr. Herivel himself has warned us, there are at least two fundamental differences between Galileo's presentation of projectile motion and Newton's early version of Corol. 1. First, Galileo posits a "force" (gravity) only in the downward direction, there being none in the horizontal. Second, Galileo's downward force is continuous ("constant"), producing an acceleration, whereas the two "forces" of Newton's proposition are impulses, producing changes in momentum almost instantaneously. If I seem to be dwelling overly long on what may appear to be a minor point of scholarship, may I anticipate the sequel: I have found in this early form of Corol. 1 a neglected source of Newton's fundamental concepts of motion.

In seeking for the origins of this early version of Newton's Corol. 1, we must keep in mind that in Newton's day there were two outstanding examples of the change in "quantity of motion" (or in momentum) produced by a force acting instantaneously or quasi-instantaneously. One is the well-known type of impact or percussion that occurs when a freely moving gross body (say a ball) strikes *another* that may be either at rest or in motion. Modern commentators have generally tended to ignore the other instance, which I hope to show is of even greater importance for our understanding of Newton's Second Law of Motion, and which takes us to the source of Newton's Corol. 1. I have in mind Descartes's ingenious explanation of refraction in his *Dioptrique*. Descartes begins his explanation in terms of the motion of a tennis ball, using the analogy in order to show the validity of Snel's Law as a general principle of the mechanics of bodies; but not necessarily arguing therefrom that all optical phenomena arise from the properties and motions of "particles of light." Descartes's example thus takes us from the motion of gross bodies to the possible motions of sub-microscopic particles.

Descartes has just been considering reflection. As in the figure, let us—following Descartes—imagine a tennis player driving a tennis ball along the line AB. In the case of reflection, the path is altered at the point B so that the tennis ball will bounce up in the direction BF. Descartes then asks us to suppose that at B it encounters not the earth, but a cloth CBE, which the ball may rupture and pass through, losing a part of its velocity, say, one half. Descartes analyzes the motion into vertical and horizontal components; only the vertical component is diminished by the action of the cloth. The result is that the subsequent path will be along the line BI. Next, Descartes considers that CBE represents not a cloth but the surface of water, which acts as the cloth has done, so that the path is again AB + BI.

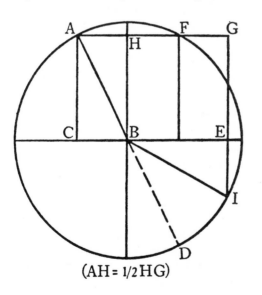

(AH = 1/2 HG)

But in order to have an analogue of the refraction of light, the path must be bent *toward the normal* as a ray goes from air into water, *not away from* the normal. Descartes, therefore, must have some action at the interface between the two media to augment the vertical component of the motion rather than to diminish it. Hence, he makes "yet another supposition"; he assumes that at the instant when the ball reaches the point B, it will be struck again by the racquet, "which increases the force of its motion. . . ." I shall not discuss here the significance of this demonstration in either the history of optics or the scientific thought of Descartes, because to do so would necessitate a discussion of the general problem of the corpuscular theory of light and the related question of the relative speed of light in more or less dense media (and of the finite speed of light). This would take us too far afield from our main topic.

The conditions of Descartes's proof are identical to the early form of Corol. 1 in the tract on "The lawes of motion" and in the explanation of Law II in the *Principia*. There can be no doubt that Descartes was Newton's source, since in Newton's optical lectures, the part presumably written in 1669,[11] Newton refers explicitly to Descartes's discussion of the Law of Refraction (or Law of Sines) and gives a succinct résumé of this proof. He even gives testimony of his acquaintance with the whole argument and not merely the results, in the statement: "The Truth whereof the Author [Descartes] had demonstrated not inelegantly, provided he had left no room to doubt of the physical Causes, which he assumed."

Those who first encounter statements that Descartes had a major formative influence on Newton's dynamics are apt to be skeptical, especially when they remember such anti-Cartesian statements as are found in the conclusion of Book Two of the *Principia,* and in the concluding General Scholium, in both of which Newton bitterly attacks the vortical "philosophical romance" of Descartes's *Principia*. But with regard to the Cartesian discussion of refraction, the situation is entirely different. Here we may actually see Newtonian physics in an earlier Cartesian dress. I would submit that it can hardly be a mere coincidence that the conditions of the early form of Newton's Corol. 1 fit so precisely the example given by Descartes. There is nothing like it in the writings of Galileo or of Kepler, nor does anything of this exact sort occur in Wallis' *Mechanica*. How fascinating to see this early version of Corol. 1 as a transformation of a statement of Descartes's, later transformed again in the tract *De motu* (on the eve of the *Principia*) to appear as we find it in the *Principia* itself! Since the Cartesian proof of the Law of Refraction makes use of "forces" that are unambiguously impulses, in the form of blows given to a ball by a tennis racquet, this source of Corol. 1 gives yet another confirmation of the view that the Second Law deals (exclusively) with so-called instantaneous "forces," or impulses.

One other short note on the sources of Newton's Second Law. Newton thought of the Second Law in relation to impact not merely through the influence of Descartes's *Dioptrique* but also because of the problem of impact, of determining just what motions will ensue when one object strikes another in both elastic and non-elastic types of collisions. This topic had occupied the attention of major scientists of the seventeenth century. Important early studies on percussion were made by Galileo, by Baliani, and by Torricelli; and the "Leges Naturae" in Descartes's *Principia* (after the first two dealing with inertia) discussed one by one several varieties of the impact of one body on another. Later in the century the topic was studied with especially rewarding consequences by Huygens and by Wallis and Wren. Wallis' paper in the *Philosophical Transactions* not only introduced Newton to the idea of conservation of momentum, but undoubtedly suggested to Newton—either consciously or unconsciously—the transformation of the Cartesian "Leges Naturae" into "Leges Motus," the Laws of Motion, as Wallis' paper was entitled. Yet another writer on percussion was Mariotte.

In presenting the problem of collision of bodies, in the Scholium following the Laws of Motion in the *Principia*, Newton referred to his predecessors (Huygens, Wallis, Wren, Mariotte) and discussed his own views concerning the imperfect elasticity of even the hardest bodies in nature. No reader could escape the conclusion that the physics of collisions and impacts was central to the dynamical thinking of Isaac Newton. Once again it was Descartes who was of major influence. For Descartes had stressed, in his *Principia*, that the force of a body in motion is measured by the product of its magnitude and its velocity, transformed by Wallis into weight and velocity, and finally transformed by Newton into mass and velocity. On impact, the whole force acts quasi-instantaneously,[12] and if we superimpose Wallis' concept of conservation on Descartes's concept of the force of a body in motion, then the description of the force of impact becomes almost the impulse-momentum form of Newton's Second Law as found in the *Principia*.[13]

CONTINUOUS FORCES AND THEIR ACCELERATIONS: THE CONCEPT OF "IMPRESSED FORCE"

I have said earlier that Newton was perfectly aware that continuous forces produce continuous accelerations, that if such forces are constant they produce constant accelerations, and that such accelerations are in a fixed ratio to the forces for any given mass, where the mass is the measure of the proportionality of forces to accelerations according to the law

$$\vec{F} = k \cdot m\vec{A}$$

We shall see some of the evidence to support this contention in the present section.

But first we must pay a little attention to "impressed force," a concept which occurs in both the first Law (". . . a viribus impressis . . .") and the Second Law (". . . vi motrici impressae . . ."). It is surprising to find Newton using, in his *Principia*, the expression "vis impressa," which we customarily associate with the late medieval writers on dynamics. But "vis impressa" also appears commonly in the seventeenth century, e.g., in the writings of Galileo and Wallis, in a somewhat new sense, and Newton's own use of this concept is highly original. In Definition IV, he says that an "impressed force" is an "action exerted upon a body, in order to change its state," which—as always—means a "state, either of resting, or of uniform motion in a right line." Then, in an explanatory paragraph, Newton makes it clear that this "force consists in the action only" and it "remains no longer in the body once the action is over." Next he says that every new state acquired by a body is subsequently "maintained by *vis inertiae* only."

Now if the "impressed force" *is* the "action," as Newton says so explicitly ("vis impressa est actio . . ."), whereby the state of a body is changed, then the magnitude of this change must be a measure of the "force" through its action. Newton is unambiguous on this point that the "force" or "vis impressa" consists of nothing but that "action"; for he says, in the explanatory sentence, "Consistit haec vis in actione sola. . . ." That an "impressed force" should be the "action" of altering the "quantity of motion," i.e., of altering the momentum, is yet another example of the way in which the Definitions of the *Principia* logically anticipate the Axioms or Laws of Motion.

In this Definition IV, however, Newton does not give any hint as to whether a "vis impressa" need be an instantaneous blow (or a sequence of such discrete blows), or a continuous force acting during a finite time-interval. Under both types of "force" there would be an "action" of changing the body's "state of resting or of moving uniformly in a right line." Actually, Newton had both types of force in mind, since, in the final sentence of the discussion of Def. IV, Newton says specifically that these "impressed forces" may be "of different origins, as from percussion, from pressure, from centripetal force."

By concentrating on the "action" rather than the force, on the change in state or change in momentum, rather than the cause of such change, Newton was able to deal with attractions and centripetal forces as if they had the same physical reality as percussion and pressure. He certainly believed personally in the reality of attractive "forces" acting over vast distances through the celestial spaces, and just as strongly as in the apparently observable "forces" of percussion, but he was surely aware that this would hardly be the case for the majority of his contemporaries. And so Newton couched his Definition in terms that would be unexceptionable, stating merely that by "impressed force" (as contrasted with "inherent force") he meant nothing more than the "action" of producing any observed change of state.

Now it must seem odd that, having defined "impressed force" for both impulsive and continuous forces, Newton then deals in the remaining Definitions V, VI, VII, and VIII only with centripetal, or continuous, forces and does not return to impulsive forces until Law II. By "centripetal force" he means "that by which bodies are drawn or impelled, or any way tend, towards a point as to a centre." As examples, he mentions the magnetic force, and "that force, whatever it is, by which the planets are continually drawn aside from the rectilinear motions, which otherwise they would pursue, and made to revolve in curvilinear orbits. . . ."

Of course, one reason why Newton gives so much space to centripetal force is that the very concept of such a force was a novelty. We have, in fact, seen that Newton not only held it to be his own invention but even explained how he had named it in honor of Christiaan Huygens. For almost all readers, this would have been their first encounter with even the expression "centripetal force," although they would have been familiar with the somewhat older "centrifugal force." Furthermore, centripetal force (or any form of continuous force) differs in one marked regard from impulses. An impulse may be measured directly by the momentum it produces, and so the equation

$$\vec{\Phi} = k \cdot \Delta(m\vec{V})$$

is true for all combinations of Φ and m. But in the case of a continuous force, the problem is not so simple. Let us assume a constant continuous "force" F, say the gravitational attraction of the sun. Then, the "action" $\Delta(m\vec{V})$ does not depend only on the "quantity of matter" or the mass m, but also on the distance of the mass m from the center of the sun. But even if we specify that distance, so that the "force" F is now determined and fixed, we still cannot say how much the change in momentum will be, since this will depend on the time during which the "action" occurs. Hence one cannot meaningfully discuss the "action" of a centripetal force: this is unspecified without bringing in other factors.

Newton therefore turns, once he has defined what he means by centripetal force, to three aspects of such forces, the "absolute quantity," (Def. VI), the "accelerative quantity" (Def. VII), and the "motive quantity" (Def. VIII). Each of these is a "measure" of centripetal force. Generally speaking, Newton means the "accelerative quantity" or "accelerative force" when he discusses the properties or effects of a "centripetal force." It is defined as follows: "The accelerative quantity of a centripetal force is the measure of it [i.e., is its measure], proportional to the velocity which it generates in a given time."

To illustrate what he means, Newton first deals with the force or "virtue" ("virtus") of any one and the same magnet (thus, as Whiston noted, keeping the "absolute quantity" constant); this is "greater at less distance and less at greater." Then Newton turns to the "gravitating force" ("vis gravitans")—"greater in valleys, less on tops of exceeding high mountains; and yet less (as shall hereafter be shown), at greater distances from the body of the Earth." Finally, Newton says: "At equal distances, it [the gravitating force] is the same everywhere, because (taking away, or allowing for, the resistance of the air) it equally accelerates all falling bodies whether heavy or light, great or small."

Just as this "accelerative quantity of a centripetal force" (or "accelerative force," as Newton also calls it) is "proportional to the velocity which it generates in a given time," so there is a "motive force" (or "motive quantity of a centripetal force") which is "proportional to the motion [i.e., quantity of motion or momentum] which it generates in a given time." It is to be observed that in each of these two "measures" of a centripetal force, the "accelerative quantity" and the "motive quantity," one must take account of the rate, i.e., use the words "in a given time," which—as we saw above—was not the case for impulses, as in the statement of the Second Law.

The definition of motive force sounds all but like the modern form of the Second Law,

$$\vec{F} = k \cdot m\vec{A} \;\; or \;\; \vec{F} = k \cdot \Delta(m\vec{V})/\Delta t,$$

that is, force is proportional to the rate of change of momentum. And, indeed, the very choice of words leads us mercilessly from the abstract logic of arbitrary definitions to the physical laws of central forces. According to Newton's physics, whenever there is a change of motion, there must be a force. If the direction of the change of momentum is directed to a point, as in circular motion or the motion of free fall, the motion may be called central, and we assume the existence of a causal agent or a force that is central or centripetal. Let a measure of such a force in a central motion be the change in momentum in a given time, says Newton, or the rate of change in momentum. Call this, he says, the "motive quantity of a centripetal force" or, for short, the "motive force." At once it is suggested to us—by the very name—that there *is* a "force" producing the motion, that this quantity has an independent physical existence, and that in particular, in the case of falling bodies or the motion of the moon encircling the earth, this "motive force" is gravity. And Newton himself, in discussing this Definition, does not talk explicitly about motions generated in given times—as we might have expected from the Definition itself—but discourses at once about weight.

He first says that "weight is greater in a greater body" and "less in a less body." To the uninitiated reader, it is not at all obvious how this property of weight may be related to motion generated in a given time; but it actually is, as is apparent to anyone who has already mastered the Newtonian physics. For Newton is basically only saying again that weight is proportional to quantity of matter, so that if weight is considered as the motive force in Def. VIII, it follows that the velocity generated in any given time is independent of the "body," i.e., its mass or weight.[14] This "centripetency toward the centre" or "weight" is known independently of the motion of descent, for—says Newton—"It is always known by the quantity of an equal and contrary force just sufficient to hinder the descent of the body."

I think that there can be no better index to Newton's revolutionary position in dynamics than his insight into the equivalence of gravitational and inertial mass (to use our contemporary expressions anachronistically) and his recognition of the significance of Galileo's experiments with falling bodies as a demonstration of this fact. That is, he understood the equivalence of mass as the resistance to being accelerated and of mass as the cause of weight or the source of the downward accelerating force of free fall. Indeed, it was Newton who first saw the meaning of Galileo's experiments precisely because it was Newton who first understood the relation of mass to weight. This required, among other things, a true sense for the Second Law of Motion for continuous forces. This may be seen in the way in which Newton discusses mass and weight in the third paragraph of explanation following Def. VIII (the "motive quantity of a centripetal force").

The "quantity of motion" (or momentum), Newton says, arises from the product of celerity and "quantity of matter" (or mass); hence the "motive force" (measured by the rate of change of momentum) arises from the accelerative force (measured by the rate of change of velocity, or by the acceleration) multiplied by the mass. But at or near the surface of the Earth, the "accelerative gravity or gravitating force" ("gravitas acceleratrix seu vis gravitans") is a constant, the same for all bodies; it follows from the Definition that "the motive gravity or the weight" ("gravitas motrix seu pondus") must be as the mass.[15] In modern numerical language, this is a recognition that the local constant small g is not only the experimentally determined 980 cm/sec^2 but is also the constant of proportionality 980 dynes/gram. Now the quantitative value of the acceleration of free fall varies. As Newton says, "if we should ascend to higher regions," we would find "the accelerative gravity [to be] less" and hence the "weight [is] equally diminished," because—as he says explicitly—the weight must always be "as the product of the body [multiplied] by the accelerative gravity." As examples, he mentions "regions, where the accelerative gravity is diminished into one-half," and hence "the weight of a body two or three times less, will be four or six times less." A thought experiment of magnificent proportions!

Newton's insight into mass and weight has thus led him not only to the position that the acceleration of free fall is the proportionality factor between these two quantities. It has also opened up the possibility of equating the "motive force" of free fall with the static (and thus independently observable and measurable) force of weighing. Newton has thus arrived at the unambiguous relation that

$$\vec{W} = m\vec{g}$$

where W is the weight at some one particular place and g is the acceleration at that place. This is not, therefore, a mere "law" for one fixed and specified "weight" of a body, but rather for the acceleration produced by a centripetal weight-force that is expressly said to vary. But no matter how it varies, the mass ("body" or "quantity of matter") remains constant, and under all circumstances the weight-force is proportional to the acceleration produced. Since Newton showed that the falling of bodies on Earth was essentially equivalent to the falling of the Moon, and so on out to the falling of the planets toward the Sun, this Second Law was certainly envisaged as applying to all gravitational forces and thus to all forces of attraction, and by extension to all continuous forces of nature. Surely, there can be no doubt that

Newton thus knew and stated explicitly a Second Law of Motion in a form equivalent to

$$\vec{F} = k \cdot (\Delta m\vec{V})/\Delta t = k \cdot m(\Delta\vec{V}/\Delta t),$$

although he never wrote such an equation.

In his advance towards this modern form of Law II, Newton used the experimental result that in free fall at any given place, the acceleration is constant for bodies of different weight.[16] He also assumed (though without explicit postulational statement) that the dynamic force pulling a body down to Earth in free fall is identically the same as the static weight, when standing still. He stated unambiguously that although the weight of a body is constant only at any one place, the mass is invariant under all conditions. Since the "vis inertiae" had been said to be proportional to the mass and to measure a body's resistance to acceleration, it followed at once that weight at any one place is proportional to mass, and that at some other place there is a new constant of proportionality, which is the acceleration. But such a result was based on the above-mentioned specific assumptions or postulates; and the identification of weight (considered as an independent quantity) as the postulated motive force could not be done universally, e.g., not for attractions generally. That is, in the case of the motion of the Moon, or of the planets, or of comets, there is no independent "force" as there is for falling bodies on or near the Earth; we cannot stop the motion of the heavenly bodies and, by some kind of balance, determine their "heaviness" or "weight" toward the body at their center of motion.

I would submit that Newton's procedure in thus producing a form of the Second Law out of Def. VIII prior to the formal statement of Law II among the Axioms not only shows again how he anticipated the Axioms or Laws of Motion in the prior Definitions. It equally demonstrates that any form of the Second Law is basically a kind of Definition.

IMPULSES AND CONTINUOUS FORCES

I have shown above that in the introductory section of the *Principia* devoted to "Axiomata sive Leges Motus," the Second Law is stated for impulses and changes in momentum, but that in the preliminary section of "Definitiones," the discussion following Definition VIII presents a form of the Second Law (restricted to gravity) as a proportion between a continuous force and an acceleration (the constant being the mass of the body concerned). It remains to show how Newton conceived the relation between impulses and continuous forces, between the types of "Second Law" applying to the two.[17]

Newton's method of making the transition from impulse to continuous force may be seen in a dramatic fashion in the opening Proposition I of the first Book of the *Principia,* stated in all three editions as follows: "The areas which revolving bodies describe by radii drawn to an immovable centre of force do lie in the same immovable planes, and are proportional to the times in which they are described."

The proof proceeds in three steps. First Newton considers a purely inertial motion: uniform and rectilinear. In any equal time-intervals it will carry a body through equal spaces. Hence, with regard to any point S not in the rectilinear path of motion, a line from S to the moving body sweeps out equal areas in equal times. At once, such is the clarity of the Newtonian revelation, an intimate—but hitherto unsuspected—logical connection is disclosed between Kepler's Law of Areas and Descartes's Law of Inertia.

Next, Newton applies the theorem from Descartes's *Dioptrique,* as follows. While a body is pursuing a uniform motion, it is given a blow, or, as Newton puts the matter, "a centripetal force acts at once with a great impulse." By an appeal to Corol. 1 to the Laws of Motion, Newton shows that area is still conserved. But in actual fact it is not Corol. 1 which is used, since the blow is given to a body in motion; this is the condition for that earlier form of this corollary which we saw Newton find in Descartes's *Dioptrique.* Under a regular succession of such blows toward one and the same point, each blow separated from the next by a constant time-interval, area with respect to that point (or center of force) is still conserved. Then Newton proceeds to the limit, whereupon the sequence of impulses becomes a continuous force and the series of line segments becomes a smooth curve. This proof may contain a hint that Newton conceived the Second Law for continuous forces as obtainable in the limit from the Second Law as stated in the *Principia,* just as in the limit a sequence of impulses of increasing frequency becomes a continuous force. But I am not aware of any direct evidence to support the view that Newton was aware that a continuous force F might be related to the impulse of Law II by being multiplied by the factor of time Δt. We are apt to be misled here because today we take force F to be primary, rather than impulse Φ and we derive the latter from the former by saying that

$$\vec{\Phi} = \vec{F} \cdot \Delta t$$

and that Newton's impulsive force is one that occurs for very small values of Δt. Hence for us, the equation

$$\vec{F} = k \cdot m \frac{\Delta \vec{V}}{\Delta t}$$

leads at once to

$$\vec{F} \cdot \Delta t = k \cdot m(\Delta \vec{V}) = k \cdot \Delta(m\vec{V}).$$

For us, in other words, Newton's "impulse" $(\vec{\Phi})$ is the limit of $(\vec{F} \cdot \Delta t)$ for small values of Δt, whereas for Newton, as $\Delta t \to 0$, $\Phi \to F$. That F and Φ have different physical dimensions was not a problem for Newton,[18] nor did he ever declare explicitly that these two types of "force" differed by an assumed (or "built-in") factor of time (Δt). Yet, although this topic is not the subject of explicit discussion in the *Principia,* it arises implicitly from the very fact that Newton saw the need of measuring the action of a *continuous force* (a centripetal force) by the rate of its action (Def. VII & VIII), whereas an *instantaneous force* may be measured by the total action without respect to time.

I would guess that Newton did not fully resolve the logical sequence from impulse to continuous force further than the intuitive conception that a continuous force may be conceived as the limit of a sequence of impulses of infinite frequency. Since Newton had been led by his Definitions to a Second Law for the continuous force of weight, he probably assumed it was *obvious* that the Second Law for impulses implied a Second Law for continuous forces. As stated in the *Principia*, Law II postulated (for impulses) the proportionality of a change in momentum and a "motive force impressed." I have stated earlier the importance of recognizing that by "force impressed" Newton meant the "action" only; so far as I know, no one has pointed out that the Second Law deals with such "action" for the good and proper reason (as Newton quite correctly understood) that in the case of impact or percussion, one cannot quantitatively determine the magnitude of the force or impulse as an independent entity. Hence we see why it was logically necessary for Newton to deal with the action, with the net change in momentum, as dictated by Law II. But for the continuous force, pure "action" does not suffice. For instance, in uniform circular motion through 360°, the net change in momentum is zero, exactly as if the body in question had not moved at all. And yet a force must have been acting since the motion is non-inertial, being non-linear. And in general, the amount of force would depend on the time, thus being more like energy expended than a force which in uniform circular motion ought to be of constant magnitude. And so we see again and again why for continuous forces Newton introduces "time" along with "action," so as to deal with the rate of change in momentum, rather than the change in momentum.

I suggest that in Newton's confusing use of "force" ("vis") for both the continuous force and the instantaneous force, he has given us the key to his signal achievement! For the fact of the matter is, as I read the history of dynamics in the seventeenth century, that it was Newton who for the first time conceived a system of physics embracing simultaneously the action of impulsive forces and centripetal forces. Newton, in a sense, thus combined the Keplerian and the Galilean modes of scientific thought: the action of celestial forces as postulated by Kepler and the physics of percussion as studied by Galileo. Hence the Second Law may serve as a particularly fascinating index to Newton's achievement in the *Principia* because it reveals to us how Newton was able to generalize his physics from the phenomenologically based dynamics of collisions and blows to the debatable realm of central forces, of gravitational attraction, and hence of continuous forces generally.

NEWTON'S REVISIONS OF THE SECOND LAW OF MOTION

Although Newton printed the three Laws of Motion almost without alteration in all the editions of the *Principia*,[19] he was not wholly satisfied with the Second Law and essayed a series of restatements of the law itself and the explanatory paragraph.

Newton's Second Law and the Concept of Force in the Principia   143

I have found among Newton's manuscripts no similar attempts to rework either the First Law or the Third Law, although there exists a manuscript fragment containing a discussion of the Third Law in relation to attraction. Newton evidently found less ambiguity or less possibility of misinterpretation in these laws than he did in the Second Law, or perhaps originally he had merely been more felicitous in expressing himself in Laws I and III.

In Newton's own copies of the first and second editions of the *Principia,* there are no manuscript changes in the statement of any of the Laws of Motion; and the only entry relating to these laws directly is a comment on the Second Law which Newton entered into his interleaved copy of the second edition (1713), a comment which appears in a variant form in the third edition (1726) in the Scholium following the Laws of Motion. The projected revisions of the Second Law occur in loose sheets of manuscript, preserved in the Portsmouth Collection in the University Library, Cambridge. Although the absolutely trivial change in verbiage in the First Law has been discussed again and again, the proposed alterations of the Second Law have not—to my knowledge—hitherto been printed or even mentioned in print.

Such manuscript texts are particularly valuable in showing us the care with which Newton was apt to revise the *Principia,* rewriting and reworking a part over and over again in an attempt to express himself as accurately and felicitously as possible. Furthermore, successive drafts often enable us to specify Newton's intentions, and thus they lend authority to the interpretations we may propose.

The attempts to reformulate the Second Law may all be dated within five years of the publication of the *Principia,* in the years 1692–93, a period of intense revision of the *Principia* by Newton. I shall present them more or less in chronological order. One sequence, all of which is later cancelled, begins as follows:

> Vis omnis [in corpus liberum *add. & del.*] impressa motum sibi proportionalem a loco quem corpus alias occuparet in plagam propriam generat.
>
> Every force impressed [on a free body *add. & del.*] generates a motion proportional to itself from the place the body would otherwise occupy into its own region [i.e., in its own direction].

On the next line, Newton starts afresh, writing: "Vis omnis impressa motum sibi proportionalem g (Every impressed force g[enerates] a motion proportional to itself"). At once we observe that Newton has introduced three conceptual innovations: (1) he writes of a motion that is generated, rather than a change of motion; (2) he no longer speaks of the new motion as merely being along the right line in which the force acts, but has added a point of origin; and (3) he has recast the sentence to make it read that every force generates a motion, rather than that every change in motion is proportional to (and hence produced by) a force.

But at this point Newton perhaps recognized that his sentence construction differed from that which he had adopted in the *Principia* where the original Lex II (like Lex I & Lex III) is given in the accusative-infinitive construction of "oratio obliqua," whereas the sentences just quoted use the nominative-indicative construction of "oratio recta." In the next version, Newton transfers the intended revision to the form appropriate to "oratio obliqua"; he uses two verbs, one for each part of the law, but both are in the infinitive mood rather than in the indicative, and with the subject consequently in the accusative:

> Motum genitum vi motrici impressae proportionalem esse & a loco quem corpus alias occuparet in plagam vis illius fieri.
>
> That a motion generated is proportional to the motive force impressed and occurs from the place which the body would otherwise occupy into the region of [i.e., in the direction of action of] that force.

And now, furthermore, he has reverted to the form of presentation in the *Principia*, at least to the extent of writing about a motion generated being proportional to the "force," rather than a "force" generating a motion; but it is still a motion *generated* and not a *change in motion*.

Newton's next attempt to express Law II is similar to the preceding one in using the indicative mood and in stating that an "impressed force" generates a motion rather than that a motion is generated by an "impressed force":

> Vis impressa motum sibi proportionalem a loco quem corpus alias occuparet in plagam propriam generat.
>
> An impressed force generates a motion proportional to itself from the place which a body would otherwise occupy into its own region [i.e., in its own direction].

Finally, the last statement in this sequence returns again to the style of "oratio obliqua," at first starting out "Mutationem motus . . . " as in the *Principia*:

> [Mutationem motus *del.*] Motum omnem novum quo status corporis mutatur proportionalem esse vi motrici impressae & fieri a loco quem corpus alias occuparet in plagam qua vis imprimitur.
>
> [That the change in motion *del.*] That every new motion by which the state of a body is changed is proportional to the motive force impressed & is made from the place which that body would otherwise have occupied toward the region [i.e., in the direction] in which the force is impressed.

In these versions, we may see that Newton has removed the ambiguity of the law as printed, by replacing the opening words of the latter, "Mutationem motus," by "Motum genitum." For while "change" ("Mutationem motus") may seem possibly to imply the concept of "rate of change," such expressions as "Motum genitum" or "Motum omnem novum" clearly denote a magnitude, without any possible suggestion of a rate. There is even less ambiguity on this point in the earlier revision, in which it is said: "Every force impressed generates a motion proportional to itself. . . ."[20]

This sequence of alterations may be of interest to the general reader primarily as a characteristic exercise of Isaac Newton in a creative mood, trying again and again to make language an exact expression of his changing thoughts. But the revision attempted on the other side of this same sheet is of a wholly different character and disturbs any simple view of Newton's Second Law as an unambiguous conception of its author, merely relating impulses to the momenta they may produce. Here Newton begins innocently enough with what looks at first glance to be only another version, as follows:

Lex II

Lex II. Motum [in spatio vel immobili vel mobili *del.*] genitum proportionalem esse vi motrici impressae & fieri secundum lineam rectam qua vis illa imprimitur. Law II. That a motion generated [in a space either immobile or mobile *del.*] is proportional to the motive force impressed and occurs along the right line in which that force is impressed.

With the deletion, this form of Lex II differs from that of the *Principia* only in the improvement in the opening pair of words: "Motum genitum" being used rather than "Mutationem motus."

But it is in fact the deletion that commands our attention. For that expression, "in a space either immobile or mobile," appears as a key to the problem with which Newton was wrestling. In the *Principia,* Law II deals with "mutatio motus" without specifying whether the body on which the force is impressed is at rest or in motion. In the discussion, Newton explains how to proceed if the body is moving initially: the old motion is altered by adding or subtracting the new.[21] But now Newton essays a revision in which he would reduce all conceivable applications of Law II to bodies at rest. A moving body could then be considered to be at rest within its space which moves. It would then be necessary only to assume (or to postulate explicitly) a principle of relativity: that the action of forces (in the sense of the Second Law) is identical in a moving space and in a space at rest. Thus, in the manuscript, the first version of the comment states (in translation) :

[And this motion if the body was resting before the impressed force must be computed in an immobile space according to the determination of the impressed force, but if the body was moving before must be computed in its own mobile space in which the body without the impressed force would be relatively at rest] And the same force will generate the same motion in a uniformly moving space as in an immobile space. Let A, B be two bodies[22]

Evidently Newton was thinking of some situation such as Galileo's famous example of the ship, in which motions occur within the ship at rest just as they do with respect to the ship if the latter is in uniform motion. For instance, apart from the effects of wind or air resistance, a body let fall will always fall straight down with respect to the ship, whether the ship be at rest or in uniform motion; even though if the ship is in uniform motion, the path will simultaneously appear to be a vertical line to an observer on the ship and a parabola to an observer on the shore. In the above extract Newton refers explicitly to a "uniformly moving space," but he did not do so in the sequel. For he replaced the above paragraph by another reading (in translation):

And this motion has the same determination as [*lit.,* with] the impressed force and happens from that place in which the body, before the force was impressed upon it, was at rest either truly or at least relatively: and therefore if the body was moving before the impressed force, it is either added to its conspiring motion or is subtracted from it if contrary or is added obliquely to it if oblique and is composed with it in accordance with the determination of both.

Newton then proceeds to deal exclusively with the problem of the composition of two oblique motions, that is, the situation in which a force produces a new motion "that is neither parallel to, nor perpendicular to, the original motion to which it is to be added." The text reads (in translation) as follows:

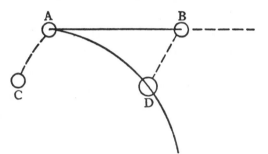

If body A was moving before the impressed force and with the motion which it had in A uniformly continued could have described the distance AB in a given time and meanwhile be urged by the impressed force into a given region: it will have to be thought that the place in which the body is relatively at rest moves together with the body from A to B and that the body through the impressed force is thrown out from this mobile place and departs from it in the direction of that impressed force with a motion which is proportional to the same force. And so if the force is determined, for example, toward the region of the right line AC and in that given time could have impelled the body deprived of all motion from the immobile place A to a new place C, draw BD parallel and equal to AC and the same force in the same time in accordance with this Law will impel the same body from its own mobile place B to a new place D. Therefore the body will move in some line AD with a motion which arises from the motion of its own relative place from A to B and the motion of the body from this place B to another place D, that is from the motion AB which the body shared before the impressed force and the motion BD which the impressed force generates by this Law. From these two motions joined according to their determinations will arise the motion of the body from the line AD. And

Thus Newton has more or less anticipated Corol 1 to the Laws, giving a proof of the method of composition of two motions. But again it is not exactly the example of Corol. 1 of the *Principia,* in which one considers either that one of two impulses may act separately on a body presumed at rest, or act simultaneously; here once more, the body is initially in uniform motion and then is given an oblique blow, as in the theorem on refraction in Descartes's *Dioptrique*. Observe that while the original motion is said explicitly to have been uniform (". . . uniformiter continuato . . .") the resulting trajectory is curved and not a straight line, indicating that the new motion is accelerated and not uniform. This trajectory AD, as drawn by Newton in three separate versions of the figure, is unmistakably a curve; and to a degree that by far exceeds the possibilities of a freehand straight line with a bit of a curve in it. No! In each case the trajectory is a deliberately drawn and obvious curve (or "curved line"), rather than the expected diagonal of a parallelogram. But if the resultant trajectory is such a curve, then the "force" must be continuous, producing a constant acceleration, as gravity does, rather than an impulse generating a motion which then continues uniformly.[23]

Nevertheless, and despite the figure, there can be no doubt from the text itself that Newton's "impressed force" is impulsive or instantaneous, and not continuous. For Newton says specifically: "The body through the impressed force is thrown out from this mobile place and departs from it in the direction of that impressed force with a motion which is proportional to the same force." Only for impulses can the "motion" (i.e., "quantity of motion," or momentum) be proportional to the "force." If the "impressed force" were not an impulse, the "motion" generated would not be simply proportional to the "force," but would rather be jointly proportional to the force and the time in which the force acts, as we have seen above. Therefore we must conclude either that Newton was careless in drawing the figure, or that he had—without explicit declaration—taken the step of assuming that the Second Law (stated explicitly for impulses and momenta) implies another Second Law for continuous forces and constant accelerations. We shall see, in a moment, that this is what Newton more or less does in the third and final edition of the *Principia,* stating that the parabolic path of projectiles may be derived from the Second Law in this manner.

Newton's discussion helps to clarify one feature of the several revised versions of the Second Law printed earlier in this section: that Newton specified the direction of the new motion as "from the place the body would otherwise occupy" along the direction of the force. Newton must have had in mind a "place" in a possibly mobile space. And indeed this phrase occurs again explicitly in two further versions of the Second Law which appear in the manuscript below the above commentary. They read (in translation) as follows:

> That motion from a place which a body would otherwise be occupying is proportional to the motive force impressed and is directed toward its region.
>
> That every new motion by which the state of a body is changed is proportional to the motive force impressed and is made from the place which the body would otherwise be occupying into the region [i.e., in the direction] which the impressed force aims for.

The discussion or commentary which comes next is much the same as the previous one and I shall not, therefore, quote it *in extenso*. But there are two possibly significant differences. The first is that in discussing the motion produced by the "force," Newton's first illustration has the "force" acting along the line of the original motion (initially in the same sense as the motion and then in a contrary sense). The body would have moved from A to a had there been no "force," and from A to B had the "force" been acting on the body at rest. Hence in these two cases, the resultant mo-

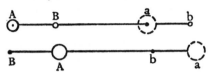

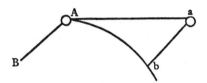

tions from A to b are shown by Newton in the diagrams reproduced here. Newton then deals with the oblique case, in part as follows:

When the translation of the body from a to b will be parallel and equal to the translation of the same body from A to B which the same force in the same time with the same direction would have been able to generate by acting on that body deprived of all motion.

The situation is equivalent to that of the previous version, i.e., ". . . the force . . . in that given time could have impelled the body deprived of all motion from the immobile place A to a new place C. . . ." But Newton now is saying that a "force" produces a displacement (or "translation") in a given direction *during a given time*. This is not a statement that the "force" has been acting during the given time, for in that case it would be generating a "motion," not a "translation," proportional to the time. In this version Newton has included the opening sentence of the original discussion of Lex II, as printed in the *Principia*, that if "any force generate a motion," a double "force" will generate a double motion, and a triple "force" a triple motion, whether the "force" be "impressed" "at once" or "by degrees and successively." We have seen that this would be true only for impulses and does not have a clear significance for continuous forces, which always generate a motion not proportional to their magnitudes (assuming a constant mass) but proportional to their magnitudes and the time in which they act. Nevertheless, the diagram illustrating the final trajectory is again a curve, as would be the case only for a continuous force and not for an impulse.

There is one further manuscript revision of the Second Law and comment. It occurs on a burnt sheet of paper. Unlike the previous versions, which are rewritten and worked over, this appears in a fair copy in Newton's hand (*ca.* 1693) with only

Newton's Second Law and the Concept of Force in the Principia   149

an occasional crossing-out and revision. This version of the law reads (in translation) as follows:

Lex. II.

Motum omnem novum quo status corporis mutatur vi motrici impressae proportionalem esse, & fieri a loco quem corpus alias occuparet in metam [*replacing* regionem *del., which in turn replaces* plagam] qua vis impressa petit.

That every new motion by which the state of a body is changed is proportional to the impressed motive force, and occurs [is made] from the place which the body would otherwise occupy into the region [i.e., in the direction] which the impressed force aims for.

There follows a comment, which apparently is like the others, including the ultimately curved trajectory.

We have no ancillary documents to tell us why, in the early 1690s, Newton was contemplating such an alteration of the statement of the Second Law of Motion,[24] but the nature of the diagram, in all the different versions, would seem to indicate that Newton was planning to arrive ultimately at a derivation of a parabolic path, or other curved path, as a consequence of two independent motions: one linear, uniform, and inertial; the other, the result of a continuously acting "force," presumably producing a constant acceleration. For projectiles, the constant acceleration needed to produce the curved path, when combined with an inertial motion, was known to him for the continuous force of gravity—as a consequence of intuition and the Definitions, as we have seen above. But it would be hard indeed for any critical reader to find a justification for this result in the Second Law of Motion as stated in the *Principia*—at least to do so without making further assumptions which are far from obvious to us.

Possibly, the reason why Newton rejected the proposed revisions of the Second Law and comment was that he could not thus fully justify the parabolic path of projectiles—not, in any event, on the basis of the stated Second Law of Motion. But he did not give up so easily. In his own interleaved copy of the second edition of the *Principia,* Newton wrote out an addition to the comment on Law II, including a hint as to how to get a Second Law for the continuous force of gravity from the stated Second Law for discrete impulses. But he did not print it at that place in the third edition, but moved it in a slightly altered form to become part of a longer insert into the Scholium concluding the section of "Axiomata sive Leges Motus." This particular addition was evidently intended to explain the statement made in the first two editions (without comment) that by using

the first two Laws and the first two Corollaries, *Galileo* discovered that the descent of bodies varied as the square of the time *(in duplicata ratione temporis)* and that the motion of projectiles was in the curve of a parabola; experience agreeing with both, unless so far as these motions are a little retarded by the resistance of the air.

In the first two editions Newton then passed directly onward to the statement, "On the same Laws and Corollaries depend those things which have been demonstrated concerning the times of the vibration of pendulums . . . ," and then to the application of "the same, together with Law III" whereby Wren, Wallis, and Huygens "did severally determine the rules of the impact and reflection of hard bodies. . . ."

Now, of course, Galileo did not at all use "the first two Laws" to prove that "the descent of bodies varied as the square of the time."[25] But whether he did or not, there can be no question that if gravity does produce a constant and uniform acceleration, so that the distances traversed downward in free fall (from rest) are proportional to the times; then the trajectory of a projectile, as Newton says, determined by "the motion arising from its projection . . . compounded with the motion arising from its gravity," will be in the shape of a parabola.[26] Neither Newton's misinterpretation of Galileo's procedure nor his correct deduction of the parabolic path are as interesting as his reasoning on how uniform gravity produces a velocity proportional to the time, according to the Second Law of Motion: the missing factor in the versions of the Second Law printed earlier in this section.

Newton's presentation is contained in a single sentence that was arrived at only after many trials. It is certainly a strange circumstance that this sentence, in which Newton has carefully stated the basis for the application of the Second Law to continuous forces, has been so badly translated as to hide Newton's meaning and his very intention from English readers and to cause otherwise well-informed scholars to use this sentence in attributing wrongly to its author a dual use of the word "force" in a single sentence.

The sentence reads, in the Latin and in an English translation, as follows:

Corpore cadente gravitas uniformis, singulis temporis particulis æqualibus æqualiter agendo imprimit vires æquales in corpus illud, & velocitates æquales generat: & tempore toto vim totam imprimit et velocitatem totam generat tempori proportionalem.	When a body is falling, uniform gravity, by acting equally in every equal particle of time, impresses equal forces upon that body and generates equal velocities: and in the whole time impresses a whole force and generates a whole velocity proportional to the time.[27]

The sense appears to be that because gravity is uniform, it acts "equally" in "every equal particle of time," in every instant, in the smallest "atomic" units into which time can be conceivably divided, in units— that is to say—so small that the action of gravity is "instantaneously" impulsive. Now without going into the question as to whether such "instants" or "particles" of time are infinitesimally small or finite though tiny, it follows from the Second Law as stated that in each such "particle of time," gravity—by so "acting equally"—"impresses equal forces upon that body" in the sense of Def. IV and so "generates equal velocities." During a "whole time," made up of the sum of such "particles of time," the "whole velocity" generated will be the sum of velocities generated within the separate "particles of time," and hence "proportional to the [whole] time." In short, if $\Delta \vec{v}$ be the velocity generated in one "particle of time," then in N such "particles of time" the total velocity generated will be $N \cdot \Delta \vec{v}$.

I have referred above to the existence of an earlier version of this idea in the interleaved (and annotated) copy of the second edition of the *Principia* that Newton had in his own library. It reads:

Actio gravitatis uniformis in corpus grave est ut tempus agendi, et vis impressa est ut actio illa et velocitatem generat tempori proportionalem. Et spatium cadendo descriptum est ut velocitas et tempus conjunctim seu in duplicata ratione velocitatis.	The action of uniform gravity on a heavy body is as the time of acting, and the impressed force is as that action and generates a velocity proportional to the time. And the space described by falling is as the velocity and time conjointly or in the duplicate ratio [i.e., as the square of] the velocity.

Yet another version, found among Newton's manuscripts, reads:

Cum vis a gravitate in corpus grave impressa sit ut tempus, corpus illud duplo tempore acquiret dupam [*sic*] vim et duplam velocitatem & describet quadruplam altitudinem cadendo, duplam scilicet ob duplam velocitatem & iterum duplam	Since the force impressed by gravity on a heavy body is as the time, that body in double the time will acquire double the force and double the velocity and will describe four times the height [distance] in falling, double that is because of the double velocity and again double [breaks off].

At about the same time that Newton wrote out the above versions in Latin, he made a similar statement in English, in a document apparently intended for use by Samuel Clarke in replying to Leibniz in the famous correspondence that took place between them. Newton's statement, written by him in English, reads:

Galileo argued that uniform gravity by acting equally in equal times upon a falling body would produce equal velocities of descent in those times, or that the whole force imprest, the whole time of descent & the whole velocity acquired in falling would be proportional to one another, but the whole descent or space described would arise from the time & velocity together & there be in a compound ratio of them both, or as the square of either of them.[28]

The projected revisions of the Second Law and the basis for using it to derive a curved trajectory for the motion of projectiles have enabled us to see how Newton justified the application of the Second Law in the *Principia* to continuously acting forces of gravity. We have seen Newton state more than once in the *Principia* that such a result was implied by the Second Law, although in point of fact it seemed to follow only from the particular nature of Def. VIII and was not ever proved to be a valid extension of Newton's Second Law save as an intuitive extension of the law from impulses to continuous forces, a step which he never justified by rigorous logic or by experiment. But Newton's intuition served him well, for it enabled him to apply correctly a law which originated in the seventeenth-century problem of impact to the universal problem of the gravitational forces governing the motions of the heavenly bodies. Newton's genius enabled him to transfer the concept of real forces, seen in impact, and in terrestrial weight, to imagined forces such as the forces on moons, tides, planets, and comets, and thereby to construct the first satisfactory dynamics of the world-system. This was the decisive step in constructing the physical system of the *Principia*.

NOTES

[1] The writer would like to express his gratitude to the National Science Foundation for a grant which has supported the research on which this article is based. A longer, and more extensively annotated, version of this paper is scheduled to appear as one of the chapters in the writer's *Newton's principles of philosophy: inquiries into Newton's scientific work and its general environment* (Cambridge: Harvard University Press, 1968). The article printed here in the *Texas Quarterly* is an abbreviated version of the final form of the paper prepared for the book in which the proceedings are to be published.

[2] A section is omitted from this version, but is to be included in the book form of this paper, on the concept of Force.

[3] Lex I: "Corpus omne perseverare in statu suo quiescendi vel movendi uniformiter in directum, nisi quatenus a viribus impressis cogitur statum illum mutare." This is the First Law as found in the first (1687) and the second (1713) editions; a minor change was made in the third (1726) edition.

The translations of extracts from Newton's *Principia* are for the most part given here and throughout this article in the versions of Andrew Motte (1729), unless otherwise specified; but I have occasionally altered these translations, since my purpose here generally is to present an accurate English and not to record the peculiarities of Motte's style.

[4] These read as follows:

"First law of nature: that each thing, in so far as it can of and by itself [i.e., by its own force], perseveres always in the same state; and that which is once moved always continues so to move."

"Second law of nature: that all motion is of itself in a straight line; and thus things which move in a circle always tend to recede from the center of the circle that they describe."

[5] In a MS containing "Remarks on Leibnitz' 1st Letter to the abbé Conti," Univ. Lib. Cambr. MS Add. 3968, §28, Newton wrote out the following brief statement on the rise of dynamics:

". . . Galileo began to consider the effect of Gravity upon Projectiles. Mr Newton in his Principia Philosophiae improved that consideration into a large science. Mr Leibnitz christened the child by [a] new name as if it had been his own calling it *Dynamica*. Mr Hygens gave the name of vis centrifuga to the force by wch revolving bodies recede from the center of their motion [.] Mr Newton in honour of that author retained the name & called the contrary force vis centripeta. . . ." For the context of Newton's remarks see A. Koyré & I. B. Cohen: Newton and the Leibniz-Clarke correspondence, *Archives Internationales d'Histoire des Sciences*, 1962, 15:62–126, Appendix Three, pp. 122–123.

[6] In an annotated copy of the second edition of the *Principia* in his own library (and now in the library of Trinity College, Cambridge), Newton wrote out a statement of the difference between his concept of inertia and Kepler's. This interesting contrast occurs in the margin, alongside the statement of Law I, and reads (in translation from the Latin) as follows: "I do not mean the force of inertia of Kepler, by which bodies tend toward rest, but the force of remaining in the same state or resting or of moving." In Latin: "Non intelligo vim inertiae Kepleri qua corpora ad quietem tendunt sed vim manendi in eodem quiescendi vel movendi statu."

Newton's Second Law and the Concept of Force in the Principia

⁷ That is, a change in motion (as in the case of a falling body) shows that such a force is acting; but the force may act (as on a body resting on a table) without producing a change in motion. Dr. J. E. McGuire, in commenting on my paper, quite correctly drew attention to this feature of the change in motion being a *sufficient* but not a *necessary* condition for the action of a force.

⁸ The examples in Law III are of both types of force: (1) continuous—a person pressing a stone with a finger, a horse drawing a stone tied to a rope; (2) impulsive (quasi-instantaneous) —a body impinging upon another. Newton says, "This law takes place also in attractions."

⁹ The two versions are as follows:

[*Comment on Law II*]	[*Corol. 1 to Laws of Motion*]
"And this motion (being always directed the same way with the generating force), if the body moved before, is added to or subtracted from the former motion, according as they directly conspire with or are directly contrary to each other; or obliquely joined, when they are oblique, so as to produce a new motion compounded from the determination of both."	"A body, acted on by two forces simultaneously, will describe the diagonal of a parallelogram in the same time as it would describe the sides by those forces separately."

¹⁰ I know of no evidence whatsoever that Newton had read the *Discorsi* prior to writing the *Principia*, although he did obtain and read a copy of the Latin translation in the 1690s; I have elsewhere assembled a considerable number of indications that he was unfamiliar with the *Discorsi*. Newton did not read Italian, but it may be argued that this is not a decisive factor since he might have picked out the bits of Latin. I may add that probabilities are very great against Newton's having had access to that very rare work, Salusbury's English version of the *Discorsi*. On this topic, see I. B. Cohen: Newton's attribution of the first two laws of motion to Galileo, *Atti del Symposium Internazionale di Storia, Metodologia. Logica e Filosofia della Scienza: "Galileo nella Storia e nella Filosofia della Scienza,"* 1967, pp. XXV–XLIV.

¹¹ The optical lectures were alleged to have been read in four installments, starting in Jan. 1669/70, then in Oct. 1670, in Oct. 1671, in Oct. 1672, according to marginal notes in one of the MS versions.

¹² I have referred in this article again and again to the importance of the form of Newton's Second Law: the "force" (impact) being set equal to the "motion" it produces. In the "Waste Book," among the early "Axiomes" we find a statement of the Law of Inertia much as Descartes had written it ("A quantity will always move on in the same straight line [not changing the determination nor celerity of its motion] unless some externall cause divert it"), followed immediately by these two axioms:

"3. There is exactly required so much and noe more force to reduce a body to rest as there was to put it upon motion: *et e contra*.

"4. Soe much force as is required to destroy any quantity of motion in a body, soe much is required to generate it; and soe much as is required to generate it soe much is alsoe required to destroy it."

¹³ I have assumed here that a "force" is measured by its "effect," that is, the "quantity of motion" in a body is a measure of the action of the "force" producing that motion. But, as is well known, in Newton's day there was a considerable controversy as to whether the "force" of a body's motion is to be measured by "mv" or by "mv²." Of course, the "force" might be measured by allowing the moving body to strike a pan of an equal-arm balance, or a dish of soft clay, etc. Colin Maclaurin, who followed Newton closely in all physical questions, and who therefore held that the "force" of a body's motion was (according to Law II) "mv," wrote: "Sir *Isaac Newton,* in his second law of motion, points out to us that the impressed force being considered as the cause, the change of motion produced by it is the effect that measures the cause; and not the space described by it against the action of an uniform gravity, nor the hollows produced by the body falling into clay. This law of motion is the surest guide we can follow, in determining effects from their causes, or conversely the causes from their effects."

154   I. BERNARD COHEN

An account of Sir Isaac Newton's philosophical discoveries (London: printed for the author's children, 1748), p. 137.

[14] The reason is of course that if the same (or an equal) factor appears in weight (providing the motive force) and in the "vis insita" or "vis inertiae" (determining the resistance to a change of state), then the Second Law for continuous forces gives a constant acceleration independent of the mass of the falling body. Def. VIII says that

$$\text{"motive force"} = k_1 \cdot \Delta \, (m\vec{V})/\Delta t$$

but if "motive force" is weight W, proportional to the mass m

$$W = k_2 m$$

it follows that

$$k_2 m = k_1 m \cdot \Delta V/\Delta t$$

and hence if it is the same m on both sides of the equation, the acceleration ($\Delta V/\Delta t$) is constant.

[15] "Hence it is, that near the surface of the earth, where the accelerative gravity, or force productive of gravity, in all bodies is the same, the motive gravity or the weight is as the body; but if we should ascend to higher regions, where the accelerative gravity is less, the weight would be equally diminished, and would always be as the product of the body, by the accelerative gravity. So, in those regions, where the accelerative gravity is diminished into one-half, the weight of a body. . . ."

[16] Newton's famous experiments on mass and weight were made with pendulums of identical length (11 feet) and identical hollow bobs (so as to have the same air resistance). At the center of the bobs, he then placed the same weight of a number of different substances. Any difference in the ratio of mass to weight would have manifested itself by a variation in the period of the pendulum, which did not occur. Newton says (*Principia,* Book III, Prop. VI), "By these experiments, in bodies of the same weight, I could manifestly have discovered a difference of matter less than the thousandth part of the whole, had any such been."

It is interesting to note that in this discussion Newton refers the reader to Section VI of Book II, on pendulums, in particular to Prop. XXIV. Here Newton states the Second Law of Motion for continuous forces, much as we would today: "For the velocity which a given force can generate in a given matter in a given time is directly as the force and the time, and inversely as the matter. The greater the force or the time is, or the less the matter, the greater the velocity generated." Newton does not say that this *is* the Second Law, but rather, that "this is manifest from the second Law of Motion." In other words, the Second Law for continuous forces follows from the Second Law as stated for impulses.

[17] I omit in this abbreviated version the bulk of the present section.

[18] This point is well made by R. G. A. Dolby: A note on Dijksterhuis' criticism of Newton's axiomatization of mechanics, *Isis* 1966, *57*:108–115.

[19] The statements of Lex II and Lex III show no variation in the three editions, but Lex I is altered in the third edition (1726). In all three editions, Lex I begins: "Corpus omne perseverare in statu suo quiescendi vel movendi uniformiter in directum, ..." The concluding portion reads as follows in the two forms:

First (1687) & Second (1713) Editions   *Third (1726) Edition*

". . . nisi quatenus a viribus impressis cogitur statum *illum* mutare."   ". . . nisi quatenus *illud* a viribus impressis cogitur statum *suum* mutare."

I have italicized the words that differ in the two editions.

I have not found this change anywhere among Newton's MS alterations, and it may possibly have originated with Dr. Henry Pemberton, who edited the third edition of the *Principia* under Newton's direction.

[20] At first encounter, it may seem that Newton's statement about "a free body" was a necessary addition and should not have been deleted, after having been added. For clearly a "force" may not generate any motion at all if the body on which it acts is constrained, as might be the case if a blow were to be given to a block of wood nailed to the floor. And even if the blow were of sufficient magnitude to rupture the fastening, some of the "force" would have to be

expended in getting the block loose, and the resulting "motion generated" would thus not be proportional to the "force." On reflection, however, this is seen not to be the case at all. What is in question is not "force" in general but a "force impressed" ("Vis omnis impressa"), which is defined specifically as that "force" which alters a body's state of motion or of rest—and, hence, *only* that part of any "force" in general which *does* so alter a body's state.

21 "And this motion [generated by the force] (being always directed the same way with the generating force), if the body moved before, is added to or subtracted from the former motion, according as they directly conspire with or are directly contrary to each other; or obliquely joined, when they are oblique, so as to produce a new motion compounded from the determination of both."

22 This comment was cancelled by Newton in two stages, first the portion within square brackets (these are Newton's square brackets), then the remainder. An initial sentence, not cancelled, is identical to the initial sentence in the printed version, which reads (in translation): "If some force generate any motion, a double force will generate a double motion, a triple one a triple, whether it has been impressed at one time and at once or by degrees and successively." The second sentence (of the two) in the printed version has been given in n.21 *supra*.

23 But if the motion of the space in the direction $\overrightarrow{AB}$ were accelerated, while the motion $\overrightarrow{AC}$ were uniform, then the curved path would be concave toward AB rather than convex, as Newton has drawn it. Of course, if by some chance the motion of the space were decelerated, the curve would then have the shape indicated by Newton, that is, be convex toward the line AB. Needless to say, there is no indication given in the text to justify the assumption that the motion of the space along the direction of the line AB would be either accelerated or decelerated, since Newton says explicitly, ". . . with the motion which it had in A uniformly continued. . . ."

24 At this time, Newton was also considering a different way of beginning Book One, having a wholly new Prop. I (Theorem I), stating that a body revolving uniformly in a circle will be drawn off from its rectilinear motion and brought back continually to the circular path by a force which is directed to the center, and which will be to the force of gravity as the square of the arc described in a given time to the rectangle (product) of the circle and the space which in that same time that body would describe by falling. Following a number of corollaries, some taken from the old Prop. IV, Theorem IV, Newton would add a note that these results were found by Huygens, who showed in his *Horologium oscillatorium* how to compare the force of gravity with centrifugal forces. Descartes, Borelli, and others, Newton said, had discussed these forces, but it was Huygens who first displayed their quantitative measure.

Prop. II, Theorem II, would then state that if bodies moved uniformly along unequal circles, the forces by which they would be drawn off from rectilinear paths and brought back continually to the circles would be as the arcs described in the same times divided by the squares of the radii. The above discussion of Descartes, Borelli, and Huygens now was to appear in Corol. 2 to a new Prop. V.

25 Newton's complete misrepresentation of Galileo's procedure seems to me strong evidence that he had never read the *Discorsi*, before writing the *Principia*. See n. 10, *supra*.

It is interesting to note that Newton's attribution of the Second Law, as well as the First Law, to Galileo appears in his tract on the motion of bodies in uniformly resisting media. "By means of these two laws," he wrote, "Galileo discovered that projectiles under a uniform gravity acting along parallel lines described parabolas in a non-resisting medium."

26 Newton wants to prove that if "the spaces described [by a freely falling body] in proportional times are as the product of the velocities and the times. . . . And if a body be projected in any direction, the motion arising from its projection is compounded with the motion arising from its gravity. Thus, if the body A by its motion of projection alone could describe in a given time the right line AB, and with its motion of falling alone could describe in the same time the altitude AC; complete the parallelogram ABCD, and the body by that compounded motion will at the end of the time be found in the place D; and the curved line AED, which that body describes, will be a parabola, to which the right line AB will be a tangent at A; and whose ordinate BD will be as the square of the line AB."

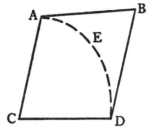

[27] The Motte-Cajori version reads: "When a body is falling, the uniform force of its gravity acting equally, impresses, in equal intervals of time, equal forces upon that body, and therefore generates equal velocities; and in the whole time impresses a whole force, and generates a whole velocity proportional to the time." Since Newton uses the word "vis" only one time less than Motte and Cajori, this passage can hardly illustrate how "Newton uses 'force' in nicely contrasting ways." But see n. 28, *infra*.

[28] Univ. Lib. Cambr. MS Add. 3968, §41, fol. 10. This document is printed *in extenso* in A. Koyré & I. B. Cohen: Newton & the Leibniz-Clarke correspondence, *op. cit.*, n. 5, *supra*, p. 118.

In another MS, never published by Newton, he *did* use the word "force" in English in contrasting senses. It is printed *ibid.*, p. 119 (from MS Add. 3968, §41, fol. 44), and reads in part: "The Theory of Projectiles invented by Galilaeo is founded upon the Hypothesis of uniform gravity, & is generally approved by Mathematicians. Now uniform gravity is that which acts with an uniform force & in equal times by acting with equal force upon the body communicates equal forces to it." This statement is a rough draft, not a final version approved for publication.

BJHS, 1984, **17**

The physical interpretation of the wave theory of light

Frank A. J. L. James*

THERE existed essentially two theories of light during the early nineteenth century: the particulate theory and the wave theory. This we realise today is a gross over-simplification, since there were many varieties of each theory.[1] But to the supporters of one theory the other theory had faults so fundamental that no distinction between varieties of the same theory was sufficient to placate opposition to that theory. This meant that opponents of either the wave or the particulate theory seldom, in their attacks, distinguished between different varieties of either theory.

The conflict, composed of many individual disputes, between the supporters of the wave theory and of the particulate theory has been discussed within a number of contexts in recent years. The change of scientific methodology associated with the acceptance of the wave theory has been discussed by Cantor;[2] the political and institutional pressures on the debate have been discussed by Fox and by Frankel.[3] The work of individual optical researchers in the early ninteenth century has also been studied: Young by Cantor and by Latchford,[4] Fresnel by Silliman,[5] Brewster by Morse.[6]

What I wish to do in this paper is to examine one of the major areas of disagreement between the two groups of supporters which has hitherto received little attention, namely the physical interpretation associated with a theory of light. I shall use this to argue that an additional reason for the general lack of enthusiasm about wave theories until Fresnel's work was that no sustainable physical interpretation was available until then. To

* Royal Institution Centre for the History of Science and Technology, The Royal Institution, 21 Albemarle Street, London, W1X 4BS.

[1] For a taxonomy of the varieties of particulate theories that were advanced see M. L. Schagrin, 'Early Observations and Calculations of Light Pressure', *Am. J. Phys.*, 1974, **42**, 927–40. See below for accounts of three different wave theories.

[2] G. N. Cantor, 'The Reception of the Wave Theory of Light in Britain: A Case Study Illustrating the Role of Methodology in Scientific Debate', *Hist. Stud. Phys. Sci.*, 1975, **6**, 109–132.

[3] R. Fox, 'The Rise and Fall of Laplacian Physics', *Hist. Study. Phys. Sci.*, 1974, **4**, 89–136. E. Frankel, 'Corpusular Optics and the Wave Theory of Light: The Science and Politics of a Revolution in Physics', *Social Stud. Sci.*, 1976, **6**, 141–84.

[4] K. A. Latchford, 'Thomas Young and the Evolution of the Interference Principle', University of London (Imperial College) Ph.D. thesis, 1974. G. N. Cantor, 'The Changing Role of Young's Ether', *B.J.H.S.*, 1970, **5**, 44–62.

[5] R. H. Silliman, 'Fresnel and the Emergence of Physics as a Discipline', *Hist. Stud. Phys. Sci.*, 1974, **4**, 137–62 and 'Augustin Jean Fresnel', *D.S.B.*, V, 165–71. See also Frankel, op. cit. (3).

[6] E. W. Morse, 'Natural Philosophy, Hypotheses and Impiety: Sir David Brewster Confronts the Undulatory Theory of Light', University of California (Berkeley) PhD thesis, 1972. See also R.Olson, *Scottish Philosophy and British Physics 1750–1880*, Princeton, 1975, especially pp. 177–88.

48   *Frank A. J. L. James*

show this I shall discuss the physical interpretations of the wave-theories proposed by Euler, Young and Fresnel and some of the adverse reactions that these theories provoked.

Physical interpretation was not a term employed by scientists in the late eighteenth and early nineteenth centuries; it is, nevertheless, an appropriate expression by which to describe the links between explanations of optical phenomena, by whatever theory, and the behaviour of ponderable matter. The theory of light largely determined the relationship between matter and light that was thought to exist. If there existed an appropriate physical interpretation light could be used to study the structure of matter. The supporters of the particulate theory of light, such as Joseph Priestley, J.-B. Biot and David Brewster, claimed that by the study of light they were directly studying matter.[7] For the wave theory to be used in this way, a theory of how the fluid luminiferous aether (required to transmit light waves) interacted with ponderable matter needed to be developed. Supporters of the wave theory therefore had to argue for a sustainable physical interpretation of the wave theory.

In the early 1760s Leonhard Euler[8] argued against the particulate theory urging for example that the sun, if it emitted light particles, would rapidly diminish in size.[9] According to the wave theory Euler argued that the sun behaved as a 'bell' propagating light waves into the aether.[10] Light was caused by the self-vibrating molecules of self-luminous matter (e.g. flames, the sun, etc) transmitting their motion to the aether. If a molecule of ponderable matter immersed in the aether vibrated quickly enough, the disturbance so caused would be transmitted through the aether to be perceived by the eye as light.[11]

The colour of a particular ray of light was determined by its wave-length. It could easily be imagined that the molecules of self-luminous bodies could only vibrate at certain frequencies thus accounting for their colour. But the theory did not immediately account for the colour of bodies which were not self-luminous. To overcome this problem Euler outlined a theory of resonance to account for the colour of such bodies.[12] He argued that there existed particles on the surface of a body which were not able to vibrate at a particular frequency, and the body could not emit light

[7] J. Priestley, *The History and Present State of Electricity with Original Experiments*, London, 1767, xiii. J.-B. Biot, *Traite de Physiqué*, 3 vols., Paris, 1816, III, 148–9. D. Brewster, 'Report on the recent progress of optics', *Rep. Brit. Ass.*, 1832, 308–22, p. 321.

[8] Leonhard Euler (1707–1783). Most of Euler's writings are published in *Leonhardi Euleri Opera Omnia*, 3 series, Berlin, Gottingen, Leipzig, Heidelberg, Zurich, 1911– . This will be cited as *Opera Omnia*.

[9] L. Euler, *Lettres a une Princesse d'Allemagne*, *Opera Omnia*, 3rd series, volumes 11 & 12. Originally published 3 vols., St Petersburg, 1768–72. Translated into English by Henry Hunter as *Letters of Euler to a German Princess*, 2 vols., London, 1795. Vol. 1, letter 17.

[10] Ibid., Letter 19.

[11] Ibid., Letter 20.

[12] Ibid., Letters 25–6. See also L. Euler, *Nova Theoria Lucis et Colorum*, *Opera Omnia*, 3rd series, Volume 5, *Commentationes Opticae*, Volume 1, pp. 1–45. Originally published in L. Euler, *Opuscula Varii Argumenti*, 1746, **1**, 169–144, art 113.

The Physical Interpretation of the Wave Theory of Light   49

of that wave-length. The colour of a body was therefore the combination of the wave-lengths that the body emitted. This theory of resonance, which explained the colour of bodies, illustrates the idea, not explicitly expressed by Euler, that the study of optical phenomena interpreted according to the wave theory of light could be used to examine the nature of matter. Those optical phenomena which were affected by the nature of ponderable matter, such as the colour of bodies, could be used to elucidate properties of matter which could not otherwise be deduced.

While Euler was confident that there existed a physical interpretation of the wave theory, evinced by his resonance theory, this was not shared by some, if not most of his contemporaries. In 1752 Thomas Melvill[13] argued that there was no optical phenomenon which required Euler's resonance theory; secondly, he suggested that according to this theory 'a body of one colour, placed in homogeneous light of another, ought not to appear of the colour of the light, but of a middle one between that and its own natural colour; which is contrary to experience'.[14] This betrays a distinct lack of understanding by Melvill of Euler's theory. There was nothing in the theory which suggested that such a phenomenon as that which Melvill had proposed could occur; Euler had stated quite explicitly that the potential vibrations of the particles of the body had to be in resonance with the wave-lengths of the light. Instead of Euler's theory, Melvill contended that Newton's theory advanced in the *Opticks*, where the colour of bodies was due to the reflection of light from their surfaces, was perfectly adequate to account for the phenomena.[15]

Melvill's criticism of Euler was approvingly reported by Priestley in 1772.[16] Priestley was committed to upholding what he viewed to be the Newtonian particulate theory of light. He had written in 1767 that light could be used to investigate the internal structure of matter, but he did not refer to this use again as a specific issue;[17] to him such a use was implicit in the assumption that light was particulate. Euler's wave theory could not do this; according to the wave theory the study of matter would always be at one remove from the study of light.

The wave theory enjoyed limited acceptance during the eighteenth century; the dominent theory remained, as the work of Priestley indicates, the particulate theory. Euler's theory was acceptable to its author, and, perhaps, a few others;[18] it did not, however, even for later proponents of the

[13] Thomas Melvill (1726–1753). For biographical information see P. Wilson, 'Biographical Account of Alexander Wilson, M.D., late Professor of practical astronomy in Glasgow', *Edinb. J. Sci.*, 1829, **10**, 1–17, pp. 5–8.
[14] T. Melvill, 'Observations on Light and Colours', *Essays and Observations, Physical and Literary*, 1756, **2**, 12–90. This was read to the Edinburgh Philosophical Society in 1752. p. 38.
[15] Ibid. I. Newton, *Opticks*, 4th ed., London, 1730, reprinted London, 1952, Book I, part II, prop. X.
[16] J. Priestley, *The History and Present State of Discoveries Relating to Vision, Light and Colours*, London, 1772, pp. 357–60.
[17] Priestley, op. cit. (7), xiii.
[18] For some discussion of the limited acceptance of wave-theories in the eighteenth century see A. W. Badcock, 'Physical Optics at the Royal Society 1660–1800', *B.J.H.S.*, 1962, **1**, 99–116. But both this

50   *Frank A. J. L. James*

wave theory, become an acceptable theory, even to signal that a great mathematician had previously believed that light was undulatory. Thomas Young[19] was quite well aware of the adverse criticism that Euler's theory had received and was cautious, in his optical writings at least, to point out where he thought Euler had made mistakes.[20]

In Young's work on the wave theory of light and on the principle of interference we shall see, over a period of about seven years, a distinct shift in his attitude about what constituted a satisfactory physical interpretation of the wave theory. In 1800 he started with an aether which had a specific structure; by 1807, he only required that it exist.

Young made his first public commitment to the wave theory in 1800 in a paper which dealt mainly with acoustical phenomena but had one section devoted to investigating optical phenomena.[21] He could not conceive how light particles could be impelled to the same velocity no matter whether they came from the sun or from the friction of two pebbles;[22] this he said militated against the particulate theory. Without making any attempt to answer this or other objections to the particulate theory Young continued by saying that the phenomenon of electrical discharge demanded the existence of the aether to carry the spark.[23]

He next speculated on the physical conditions that would need to be met by the aether to explain optical phenomena such as refraction and reflection. He assumed that these phenomena were caused by aether being denser in the media than outside them[24] which happened because of an 'attraction'[25] of the media to the aether. At this time the structure of aether was important to Young not only because of its luminiferous properties, but also because it was, as Cantor has shown, central to his natural philosophy.[26] It was natural therefore that Young chose to develop this part of his work, and that this in turn influenced his optical work.

By 1801 Young had given much more thought to the nature of both light and the aether.[27] The wave theory of light, he said, could be defined in four hypotheses. The first three simply stated that a luminiferous aether existed in which waves were propagated when a body became luminous, and the different frequencies of these waves caused different colours to be perceived in the retina.[28] Following each of these three hypotheses Young

and Schagrin, op. cit. (1), demonstrate the dominance of the particulate theory in the eighteenth century.

[19] Thomas Young (1773–1829). Most of Young's writings are published in G. Peacock and J. Leitch (editors), *Miscellaneous Works of the Late Thomas Young*, 3 Volumes, London, 1855. This will be cited as *Works*. The best life of Young remains G. Peacock, *Life of Thomas Young*, London, 1855.

[20] Just to take examples from Young's first optical paper 'Outlines of Experiments and Enquiries Respecting Sound and Light', *Phil. Trans.*, 1800, 106–50, *Works* I, 64–98. On pp. 78, 80 & 81 he mentioned Euler's optical work with adverse criticisms.

[21] Ibid., 78–83.   [22] Ibid., 79.

[23] Ibid.   [24] Ibid., 80.

[25] Ibid.   [26] Cantor, op. cit. (4).

[27] T. Young, 'On the Theory of Light and Colours', *Phil. Trans.*, 1802, 12–48, *Works*, I, 140–69. This was his second Bakerian lecture delivered in November 1801.

[28] 1st hypothesis, Ibid., 142, 2nd, Ibid., 143, 3rd, Ibid., 144.

The Physical Interpretation of the Wave Theory of Light   51

cited and quoted passages taken from various optical writings of Newton (most of them completely out of context as has been noted elsewhere[29]) to show that they did not contradict Newton's opinions. His fourth hypothesis returned to the structure of the aether:

> All material Bodies have an Attraction for the etherial Medium, by means of which it is accumulated within their Substance, and for a small Distance around them, in a State of greater Density but not of greater Elasticity.[30]

This he added was in complete opposition to Newton's view that the aether might have a greater density the further it was away from ponderable matter,[31] but, he continued, 'both [hypotheses] being in themselves equally probable, the opposition is merely accidental, and it is only to be inquired which is the best capable of explaining the phenomena'.[32] Young in his optical work needed this hypothesis to connect ponderable matter and the aether so that he could explain phenomena such as refraction[33] and reflection[34] according to the wave theory. The density of aether surrounding a molecule (different for different molecules) would change the velocity of light passing through it, thus causing refraction.

Within the context of his four hypotheses and his conviction expressed in his previous paper of the close analogy between sound and light,[35] Young enunciated the principle of interference, discovered, he said in May 1801;[36] it has been argued,[37] that he did this by considering the principle of coalescence of musical sounds, discussed in his first paper.[38] His first statement of the principle read

> When two undulations, from different origins, coincide either perfectly or very nearly in Direction, their joint effect is a combination of the motions belonging to each.[39]

He elucidated the principle by explaining how waves behaved when they coincided. From this principle he advanced explanations for such diverse optical phenomena as 'The Colours of Striated Surfaces', 'The Colours of Thin Plates', 'The Colours of Thick Plates', 'Blackness' and 'Inflection'.[40] At this point Young explicitly explained all these phenomena in terms of

---

[29] Cantor, op. cit. (4), 46–7.

[30] Young, *Works*, op. cit. (27), 147.

[31] Ibid. Young presumably had in mind query 21 of Newton's *Opticks*, op. cit. (15). For a discussion of this part of Newton's work and its context see P. M. Heimann 'Ether and Imponderables' in G. N. Cantor and M. J. S. Hodge (eds), *Conceptions of Ether*, Cambridge, 1981, pp. 61–84, especially pp. 64–7. See also pp. 19–24 of the editors' introduction.

[32] Young, *Works*, op. cit. (27), 147–8.

[33] Ibid., 155.

[34] Ibid., 155–6.

[35] Young, *Works*, op. cit. (20), 78–83.

[36] T. Young, *Dr Young's reply to the Animadversions of the Edinburgh Reviewers, on some papers published in the Philosophical Transactions*, London, 1804, *Works*, I, 192–215, p. 202.

[37] See Latchford, op. cit. (4), especially pp. 130–3 & 169–70.

[38] Young, *Works*, op. cit. (20), 83–5. He referred to this passage from *Works*, op. cit. (27), 157 where he enunciated the principle of interference.

[39] Ibid.

[40] Ibid., 158–166. For a discussion of Young's work on inflection, see Latchford, op. cit. (4), p. 180–5.

52   *Frank A. J. L. James*

his wave theory using the principle of interference and to a greater or lesser extent (depending on the phenomenon) his fourth hypothesis of the structure of aether.

In his next paper on light published in late 1802, Young reversed the argument.[41] His new statement of the principle of interference reveals his change of attitude:

> Whenever two portions of the same light arrive at the eye by different routes, either exactly or very nearly in the same direction, the light becomes most intense when the difference of the routes is any multiple of a certain length, and least intense in the intermediate state of interfering portions; and this length is different for light of different colours.[42]

There is no mention here of the luminiferous aether or indeed of light being undulatory. By 'the same light' Young could only have meant, as he had explicitly stated in his first enunciation of the principle, light of the same wavelength. But it would appear that Young wanted this principle established on experimental grounds as a universal law of nature. If this was the case then it was necessary for him to remove all theory loading from his new statement of the principle. To help demonstrate this Young added that he had already shown the sufficiency of the principle to explain numerous optical phenomena;[43] in this he omitted to mention that the explanations of some of these phenomena, especially that of blackness, intimately involved the detailed physical consideration of the behaviour of his structured aether. Using the principle of interference, without referring to the wave theory or the aether, Young offered explanations for the colour of fibres and of mixed plates.[44] He then, fairly tentatively, suggested that the principle of interference, and hence the explanations of these phenomena, would be most sensibly interpreted if light was undulatory.[45] That is, after deducing the principle of interference and the explanations of these phenomena from his original assumption that light was undulatory, Young had reversed the argument and deduced the idea that light was undulatory from his explanations of the phenomena. And there was no mention of the luminiferous aether needed to transmit the waves. This, I think, is significant of Young's inability to maintain a consistent physical conception of the aether,[46] which surely drove him to concentrate on explaining phenomena according to the interference principle. Young could now avoid serious consideration of the structure of the aether—a position which he expressed explicitly in his fourth paper on light where he

[41] T. Young, 'An Account of some cases of the Production of Colours not hitherto described', *Phil. Trans.*, 1802, 387–379, *Works*, I, 170–78.

[42] Ibid., 170.

[43] Ibid. Presumably he was referring to Young, op. cit. (27).

[44] Young, *Works*, op. cit. (41), 171–3.

[45] Ibid., 174–5.

[46] Young's inability to maintain his aether did not go unnoticed. See [H. Brougham], 'The Bakerian Lecture. Experiments and Calculations relative to Physical Optics', *Edinb. Rev.*, 1804, **5,** 97–103, p. 102–3.

The Physical Interpretation of the Wave Theory of Light   53

commented that he no longer thought that ponderable matter affected the density of the aether in its neighbourhood as he had formerly supposed.[47] He could now explain optical phenomena without having to consider the structure of the aether as he had originally done; he now only required it to exist.

Young confirmed that this was indeed his attitude towards the aether in 1807 when he reprinted revised versions of his papers on light.[48] His fourth hypothesis on the nature of light now reads:

> All material bodies are to be considered, with respect to the phenomena of light, as consisting of particles so remote from each other as to allow the ethereal medium to pervade them with perfect freedom, and either to retain it in a state of greater density and of equal elasticity, or to constitute, together with the medium, an aggregate, which may be considered as denser but not more elastic.[49]

Gone from this is Young's concern to explain how the aether accumulates in ponderable matter. For Young the explanation of optical phenomena such as refraction or reflection required the aether to be denser in the refracting and reflecting material than outside them; he did not attempt to explain why. Young had reversed his earlier attitude towards the aether; previously he had assumed that the explanation of optical phenomena must follow from the structure of the aether which he had posited. Now he said that the aether must behave in the way required to explain the phenomena.

Despite Young's work the bulk of scientists continued to support the particulate theory over the wave theory. Fox, Frankel and Cantor have shown how the Newtonian and Laplacian methodological orthodoxies, institutionally entrenched in Britain and France respectively, tended to counter the wave-theory.[50] Important though this is, the lack of a sustainable physical interpretation of the luminiferous aether made it difficult for scientists to accept the wave theory. They were being asked, in effect, to accept an aether without a structure; this made it impossible to use light, interpreted according to the wave theory, for the study of matter. This part of the debate was about whether the study of light was a study of waves or a study of matter. If the former then some structure for the aether was needed to study matter using light; if the latter then the very act of studying light became the study of matter. The wave theory ultimately developed by Young meant that light could not be used to examine the nature of matter.

---

[47] T. Young, 'Experiments and Calculations Relative to Physical Optics', *Phil. Trans.*, 1804, 1–16, *Works*, I, 179–91, p. 188.
[48] T. Young, *A Course of Lectures on Natural Philosophy and the Mechanical Arts*, 2 vols., London, 1807. The papers were reprinted at the end of volume II.
[49] Ibid., II, 618.
[50] On Newtonian orthodoxy see Cantor, op. cit. (2). On Laplacian orthodoxy see Fox, op. cit. (3) and Frankel, op. cit. (3) especially pp. 143–4.

54   *Frank A. J. L. James*

In the work of Augustin Fresnel[51] we shall see a process very much like the reverse of what happened to Young's aether; from beginning his work with an unstructured aether, Fresnel ultimately developed a highly structured aether. He argued in 1814 that the particulate theory could not explain the uniformity of the speed of light, or why the sun had illuminated the earth for a long period without diminishing in volume.[52] It is doubtful whether Fresnel knew of the work of Euler at this time and almost certain that he did not know of Young's work.[53] This however is irrelevant to the argument that there were certain phenomena which were apparently inexplicable according to the particulate theory and that these could be used to detract from the theory. However, to condemn the theory solely on these grounds was less than fair; during the 1800s a considerable amount of effort had been put into successfully explaining phenomena such as polarisation according to the theory.[54]

Fresnel commenced his experimental and theoretical study of the wave theory by considering the nature of diffraction.[55] He presented his detailed results in a paper,[56] written for a prize of the Académie des Sciences, which he was awarded in March 1819.[57] In this Fresnel asserted that he had confirmed experimentally the observation of Berthollet and Malus that diffraction phenomena were independent of the nature of the body which diffracted the light.[58] The only thing on which diffraction did depend was the dimensions of the diffractor. The problem of diffraction should therefore be soluble by considering only the behaviour of waves passing by an obstacle without reference to the nature of the obstacle. Fresnel showed that the velocity of a particle of aether was functionally related to its distance from its centre of vibration at a given time, and the intensity and length of the wave it transmitted.[59] Using this relation Fresnel was able to describe diffraction phenomena with a considerable degree of precision. One of the judges of the prize, Poisson,[60] an opponent of the wave theory, calculated, using Fresnel's equations, that a bright spot of light

[51] Augustin Jean Fresnel (1788–1827). Most of Fresnel's writings are published in H. de Senarmont, E. Verdet and L. Fresnel (editors), *Oeuvres Complètes D'Augustin Fresnel*, 3 volumes, Paris, 1866–1870, reprinted New York, 1965. This will be cited as *Oeuvres*.

[52] A. J. Fresnel to L. Fresnel. 5 July 1814, *Oeuvres*, II, 820–4.

[53] See Frankel, op. cit. (3), 157–8 on Fresnel's general lack of knowledge on optical subjects at this time.

[54] See Ibid. 145–54.

[55] For detailed discussion of this work see Ibid., 159–62 and Silliman, *Hist. Stud. Phys. Sci.*, op. cit. (5) 150–5.

[56] A. J. Fresnel, 'Mémoire sur la Diffraction de la Lumière', *Mém. Acad. Sci.*, 1821–2 [published 1826], **5**, 339–475, *Oeuvres*, I, 247–382.

[57] For an account of the award of the prize and of the publication see Fox, op. cit. (3), 112–4 and the notes contained therein.

[58] Fresnel, *Oeuvres*, op. cit. (56), 278–80. I have been unable to trace the reference to Berthollet and Malus. But see M. Crosland, *The Society of Arcueil*, London, 1967, p. 296.

[59] Ibid., 287–8.

[60] Siméon-Denis Poisson (1781–1840). See Fox, op. cit. (3), 113–4 for Poisson's possible rôle as a judge of Fresnel's paper.

The Physical Interpretation of the Wave Theory of Light   55

would occur in the centre of the diffraction pattern of a small round diffractor. This he thought was a phenomenon which could not occur and thus falsified the theory; however, another judge of the prize, Arago[61] experimentally verified Poisson's prediction.[62] From the outset Fresnel's theory displayed both a remarkably accurate explanatory power, and a considerable predictive ability. These two aspects of the theory led Fresnel to assume that the luminiferous aether must exist as a physical entity. But Fresnel had not, at this time, postulated a structured aether by which the nature of matter could be investigated.

Fresnel next extended the wave theory to explain polarisation and double refraction. In this he discovered that it was necessary to posit a structure of matter for media which possessed these optical properties and also to develop a structured aether. Fresnel had been led to investigate these phenomena after he and Arago had observed that 'two rays of light *polarised at right angles* do not produce any effect upon each other under the same circumstances in which two rays of ordinary light produce destructive interference'.[63] In the paper in which they reported this observation neither Fresnel nor Arago offered any explanation for it according to the wave theory. The implication was, as Fresnel stated later, that light waves were transmitted transversely to their direction of propagation, whereas it had been previously assumed that light waves were propagated longitudinally.[64] This posed considerable conceptual difficulties for Fresnel since no fluid was known to transmit this type of wave and from his account he took some time to accept the implication of his and Arago's work. However, he thought that 'the facts which already furnish so many probabilities in favour of the wave system, and so many objections against that of emission, compel us to recognise this character in the luminous

[61] Dominique François Jean Arago (1786–1853). Most of Arago's papers were collected in J. A. Barral (editor), *Oeuvres Complétes de François Arago*, 17 volumes, Paris, 1854–62. This will be cited as *Oeuvres*.

[62] D. F. J. Arago, 'Rapport fait par M. Arago à l'Académie des Sciences, au nom de la Commission qui avait été chargée d'examiner les Mémoirs envoyés au concours pour le prix de le diffraction', *Ann. Chim. Phys.*, 1819, **11**, 5–30, Fresnel, *Oeuvres*, I, 229–37. This is Arago's report on Fresnel's paper for the prize commission of which he was a member. On p. 236 he reports Poisson's prediction and his own experimental verification.

[63] A. J. Fresnel and D. F. J. Arago, 'Mémoire sur l'Action que les rayons de lumière polarisée exercent les uns sur les autres', *Ann. Chim. Phys.*, 1819, **10**, 288–306, Fresnel, *Oeuvres*, I, 509–22, p. 521. Dans les mêmes circonstances où deux rayons de lumière ordinaire paraissent mutuellement se détruire, deux rayons *polarisés en sens contraires* n'exercent l'un sur l'autre aucune action appréciable. Fresnel's and Arago's emphasis.

[64] A. J. Fresnel, 'Second Mémoire sur la Double Réfraction', (written 1822), *Mém. Acad. Sci.*, 1827, **7**, 45–176, *Oeuvres*, II, 479–596. Translated into English (by A. W. Hobson) as 'Memoir on Double Refraction', *Taylor's Scientific Memoirs*, 1852, **5**, 238–333, p. 244. Fresnel had privately speculated on the possibility of the existence of transverse light waves as early as 1816. See 'Mémoire sur l'influence de la polarization', *Oeuvres*, I, 387–409, p. 394n. Young had also privately proposed in 1817 that light had a 'minute' transverse component which would account for polarisation. See Young to Arago, 12 January 1817, *Works*, I, 380–4, where he further commented that 'in a physical sense, it [the transverse component] is almost an evanescent quantity, although not in a mathematical one'. Young published this suggestion in his article 'Chromatics' in *Supplement to the fourth, fifth and sixth editions of the Encyclopaedia Britannica*, 6 volumes, Edinburgh, 1824, volume III, 141–63, *Works*, I, 279–342, p. 332–3. This was written in September 1817 (*Works*, I, 279).

56   *Frank A. J. L. James*

vibrations'.[65] To Fresnel, a theory which had already explained and predicted so many phenomena must be true, even if the consequences led to peculiar inferences. After demonstrating mathematically that polarised light 'cannot have any vibration normal to the waves'[66] he had to 'suppose that neither does this mode of vibration exist in ordinary light'[67] since otherwise there would be nothing to prevent the light behaving longitudinally once it left the polarising medium.

To explain why some media polarised and doubly refracted light Fresnel argued that it was 'relative to the nature of the . . . constitution of the media possessing the property of double refraction'.[68] He meant here that the 'molecules of doubly-refracting media do not exhibit the same mutual dependence in all directions; so that their relative displacements will give rise to different elasticities according to their directions'.[69] With these hypotheses—of the structure of matter and of aether—Fresnel was able to explain the phenomena of double refraction and polarisation with as much accuracy as he had accorded previously to diffraction phenomena.

This power of explanation alone seems to have been sufficient to convince Fresnel of the physical reality of an aether which transmitted light transversely. However, not surprisingly, he felt compelled to give an account of the 'possibility of the propagation of transversal vibrations in an elastic fluid'.[70] Such a wave propagation might be imagined he said if 'we may suppose that the resistance of the aether to compression is much greater than the force opposed by it to the small displacements of these layers along their own planes'.[71] The particles could only move in the plane of the wave, since according to this hypothesis any normal movement to the plane of the wave was insignificant compared to the force which existed between the planes of aether to overcome it. From this Fresnel was prepared to assert:

> I think I have sufficiently proved that there is no mechanical absurdity in the definition of luminous vibrations which the properties of polarized rays have compelled me to adopt, and which has led me to the discovery of the true laws of double refraction. If the equations of motion of fluids imagined by geometers are not reconcilable with this hypothesis, it is because they are founded on a mathematical abstraction. . . There would be very few philosophers who would reject an hypothesis to which the phenomena of optics so naturally lead, for no other reason than because it does not agree with these equations.[72]

Fresnel thought that even though the equations of fluid mechanics did not at that moment allow for the transmission of transverse waves through a fluid, this was no reason to reject the possibility that such a fluid could not

[65] Fresnel, *Sci. Mem.*, op. cit. (64), 243–4.   [66] Ibid., 249.
[67] Ibid.   [68] Ibid., 243.
[69] Ibid.   [70] Ibid., 258.
[71] Ibid., 261.   [72] Ibid., 262. I have altered the translation of the final sentence.

exist. It would appear that it had not occurred to Fresnel that an aether which would transmit transverse waves would have to be solid; he preferred to think that the mathematical equations of such a fluid had yet to be formulated. Nevertheless, because he had accurately explained certain phenomena according to his hypothesis of transverse light waves, then such a fluid must exist as a physical reality. Fresnel's hesitation in accepting the implication of his and Arago's experimental work on the interference of polarised light indicates that Fresnel spent considerable time in coming to this conclusion; it is reasonable to suppose that he was not simply offering a rationale for accepting transverse waves, but had come to believe in their physical reality, and the consequent reality of the luminiferous aether required to carry them.

With this new work on light and the physical implications contained therein, Fresnel had turned the wave theory from a theory which had not been greatly concerned with the nature of matter to one intimately concerned with it. There were, as he noted in his elementary essay on light published in 1822,[73] other optical phenomena 'such as relate to the absorption of light, for instance, in the reflection of metallic surfaces, and in black bodies; the passage of light through bodies imperfectly transparent, and in the proper colour of bodies'[74] that remained to be reconciled with the wave theory. But he thought that by providing solutions to these problems, other, more fundamental problems would also be solved:

> If anything can contribute very essentially to the advance of this great discovery [of the principles of atomic mechanics], and to assist us in penetrating the secrets of the internal constitution of bodies, it must be by the minute and indefatigable study of the phenomena of light.[75]

This was new so far as the wave theory was concerned. Up to that time the relationship between matter and light waves had mainly been incidental; this had been a serious drawback to the theory. Fresnel in his work on polarisation and double refraction, had made the relationship between matter and light waves explicit. Fresnel thought that by using the wave theory to study optical phenomena a new understanding of the structure of ponderable matter would be achieved. Fresnel's willingness to make such a strong statement about the role of the wave theory in ascertaining the internal structure of matter is a measure of the confidence which he had in the theory's physical validity; by devising suitable structures of matter all optical phenomena would eventually be explained in terms of the wave

---

[73] This essay appeared in the supplement to the French translation of Thomas Thomson, *A System of Chemistry*, 5th edition, 4 volumes, London, 1817: *Système de Chimie*, 5 volumes, Paris, 1818–1822. Fresnel's essay was 'De la Lumière', *Oeuvres*, II, 3–146. This was translated into English (by Thomas Young; see footnote to Fresnel, *Sci. Mem.*, op. cit. (64), 264) as 'Elementary view of the Undulatory Theory of Light', *Quart. J. Sci.*, 1827, **23**, 127–141, 441–54; **24**, 113–35, 431–48; 1828, **25**, 198–215; **26**, 168–91, 389–407; 1829, **27**, 159–65.
[74] Ibid., 1829, **27**, 161.
[75] Ibid., 162.

58   *Frank A. J. L. James*

theory. He did not live long enough to implement this research programme.

Fresnel's theory demanded an aether that possessed some very curious
properties as Young pointed out in 1824:[76]

> This hypothesis of Mr. Fresnel [of transverse waves] is at least very ingenious,
> and may lead to some satisfactory computations: but it is attended by one
> circumstance which is perfectly *appalling* in its consequences. . . [This is] that
> the luminiferous ether pervading all space, and penetrating almost all
> substances, is not only highly elastic, but absolutely solid!!![77]

Young was too committed to the wave theory to let this difficulty deflect
him from supporting Fresnel. He suggested instead that the present
definition of solidity did not exclude the possibility that there existed fluids
which could transmit transverse waves[78] and agreed with Fresnel that in
future the mathematical equations of fluids might be developed to allow for
the transmission of transverse waves.

In the following years Fresnel's theory of light, in spite of the peculiar
properties of his aether, was extended in the way he had envisaged it would
be. During the 1830s several scientists reconciled absorption with the
theory;[79] Cauchy did the same with dispersion[80] and Stokes worked on
stellar aberration which incidentally helped define the structure of the
aether more specifically.[81] In addition the predictive ability of the theory
was confirmed by W. R. Hamilton's discovery of conical refraction.[82]

This is not to say that the wave theory was immediately accepted by
the entire scientific community. One of the few opponents was Brewster
who disliked the way that scientists had totally accepted Fresnel's theory as
a physically valid representation of optical phenomena; they had not,
Brewster contended, shown that there existed a satisfactory physical
interpretation of the wave theory. As evidence for this he cited phenomena
such as the absorption and dispersion of light which in 1833 had not been
reconciled with the wave theory.[83] Brewster had reversed one of the
arguments against the wave theory. In the past opponents of the wave

[76] T. Young 'Theoretical Investigations Intended to Illustrate the Phenomena of Polarisation',
Supplement to the ... Encyclopaedia Britannica, op. cit. (64) VI, 860–3, *Works*, I, 412–7.
[77] Ibid., 415. Young's emphasis.
[78] Ibid., 416–7.
[79] See J. Morrell and A. Thackray, *Gentlemen of Science*, Oxford, 1981, especially pp. 466–72.
[80] A. L. Cauchy, *Noveaux Exercises de Mathématiques (Mémoire sur la Dispersion de la Lumière)*, Prauge,
1835. See also C.-A. Valson, *Le Vie et les Travaux du Baron Cauchy*, Paris, 1868, reprinted Paris, 1970, pp.
147–51.
[81] D. B. Wilson, 'George Gabriel Stokes on Stellar Aberration and the Luminiferous Ether',
B.J.H.S., 1972, **6**, 57–72.
[82] See H. Lloyd, 'On the Phenomena presented by Light in its passage along the Axes of Biaxial
Crystals', *Phil. Mag.*, 1833, III, **2**, 112–20, 207–10, where he reports Hamilton's prediction and his own
experimental verification of conical refraction. Hamilton made this prediction by considering
mathematically how a ray of light obeying Fresnel's equations would behave on entering various
crystals. For details see R. P. Graves, *Life of Sir William Rowan Hamilton*, 3 volumes, Dublin, 1882–9, vol.
I, 623–38, and T. L. Hankins, *Sir William Rowan Hamilton*, Baltimore, 1980, chapter 6.
[83] D. Brewster, 'Observations on the Absorption of Specific Rays, in reference to the Undulatory
Theory of Light', *Philosophical Magazine*, 1833, III, **2**, 360–3.

theory had argued that the theory had little, if any, ability to probe the
nature of matter compared with the particulate theory; this was not the
only argument advanced against the wave theory; but it was one. Brewster
was now complaining that a structured aether had not explained a number
of phenomena, and that he at least, could not accept it as a physical
representation of optical phenomena:

> That the undulatory theory [of light] is defective as a *physical* representation
> of the phenomena of light has been admitted by the more candid of its
> supporters.[84]

Brewster's specific objection to the wave theory was to the aether. He had
no objection to the wave theory as an hypothesis and indeed later wrote
that the theory 'must contain among its assumptions (though, as a physical
theory, it may be false) some principle which is inherent in, and
inseparable from, the real producing cause of the phenomena of light; and
to this extent it is worthy of our adoption of as valuable instrument of
discovery'.[85] To Brewster the wave theory was a heurestic rather than a
physically satisfactory theory; it was a vast calculating device with the
ability to predict new facts. But he thought that 'The power of a
theory, . . . to explain and predict facts, is by no means a test of its truth;
and in support of this observation we have only to appeal to the Newtonian
Theory of Fits and to Biot's beautiful and profound Theory of the
Oscillation of Luminous Molecules'.[86] Brewster did not here suggest that
either Newton's or Biot's theories were true; they were alternative theories
which could explain the phenomena. His methodological objection to the
wave theory was that its physical validity had not been proven. Its
proponents on the other hand claimed that because the theory explained so
many optical phenomena it must be true.[87] Brewster contended that this
was a *necessary* condition for a theory to be true, but not a *sufficient* one since,
he maintained, other theories could also account for optical phenomena.

Brewster thought that only by experimental verification of facts could
true knowledge about the natural world in general and about light in
particular be ascertained; this latter he had attempted to demonstrate in
his *Treatise on Optics* of 1831 in which he hardly discussed any theoretical
issues at all.[88] Brewster when he suggested that the wave theory was
defective as a physical representation of optical phenomena had touched
on what he thought was a weak point of the theory; whereas in fact it was

[84] Ibid., 361, Brewster's emphasis. Brewster was specifically referring to some of John F. W.
Herschel's ambiguous attitudes towards the wave theory in his 'Light', *Encyclopaedia Metropolitana*,
1828, **2**, 341–586.
[85] [D. Brewster], 'Cours de Philosophie Positive. Par M. Auguste Comte', *Edinb. Rev.*, 1838, **67**,
271–308, p. 306.
[86] Brewster, op. cit. (83), 361.
[87] For an extreme statement of this position see B. Powell, 'Remarks on Mr. Barton's reply,
respecting the Inflection of Light', *Phil. Mag.*, 1833, III, **3**, 412–7.
[88] D. Brewster, *A Treatise on Optics*, London, 1831.

60   *Frank A. J. L. James*

one of the strongest parts of Fresnel's theory. Until Fresnel's work the aether did not have any sustainable structure. But Fresnel had devised a theory which was both mathematically extremely accurate and also required the aether to have a quite specific structure. Neither of these conditions had previously existed before. Within this new context the wave theory did become generally accepted; Brewster was the only scientist of note (in Britain at least) who continued to reject the physical validity of the theory although he was slowly forced to concede more and more to the theory.[89]

From all this evidence it can be concluded that what I have termed the physical interpretation of the wave theory was a substantive issue in the debates between the proponents of the two theories of light. Within this context the wave theory was attacked as either not providing an aether by which the structure of matter could be studied (Euler and Young) or in Fresnel's case it was attacked for providing a highly structured aether which did not immediately explain certain phenomena. In the former case the lack of a sustainable physical interpretation contributed to the minimal acceptance of the wave theory; the opponents of Euler and Young pointed out this weakness. In the latter case Fresnel's mathematical theory by demanding a specific structure for the aether, which could be used to study matter, helped initiate a research programme in optics which the earlier wave theories proposed by Euler and by Young had not done.

[89] See Morrell & Thackray, op. cit. (79), 469.

Speculation and Experiment in the Background of Oersted's Discovery of Electromagnetism

By Robert C. Stauffer *

IN the intellectual life of Hans Christian Oersted (1777–1851), philosophy and experimental science were both abiding interests, and his discovery of electromagnetism is best understood as arising from a fertile union of speculation and experiment. As Oersted wrote about himself: "Throughout his literary career, he adhered to the opinion, that the magnetical effects are produced by the same powers as the electrical. He was not so much led to this, by the reasons commonly alleged for this opinion, as by the philosophical principle, that all phenomena are produced by the same original power." [1] This belief in the unity of the powers of nature was characteristic of the German Romantic school of *Naturphilosophie*. Both Oersted's metaphysical speculations with their echoes of *Naturphilosophie* and his extensive laboratory experience deserve more general recognition; for, contrary to the interpretation of many writers who have touched on Oersted and electromagnetism, his discovery was no mere accident. [2]

Oersted's career as a whole has been surveyed in the excellent monograph with which Dr. Kirstine Meyer introduced her noteworthy edition of his scientific papers. [3] Here Mrs. Meyer has given a sound and thorough analysis of all the varied facets of Oersted's scientific life. The range of his interests included studies on Chladni's sound figures, the compressibility of water, electromagnetism, diamagnetism, thermoelectricity, the first isolation of the element aluminum, [4] and proposals for a new Scandinavian chemical nomenclature.

---

* University of Wisconsin.

[1] H. C. Oersted, *Naturvidenskabelige Skrifter . . . Scientific Papers*, ed. Kirstine Meyer, 3 vols. (Copenhagen, 1920), II, p. 356. This collected edition will hereafter be referred to as *Skrifter*. For this passage see also Appendix 1 of this paper.

[2] The "accident version" of the history of Oersted's discovery was first originated by Ludwig Wilhelm Gilbert in his *Annalen der Physik*, 1820, *66*: 292; see R. C. Stauffer, "Persistent Errors Regarding Oersted's Discovery of Electromagnetism," *Isis*, 1953, *44*: 307–310. Gilbert was an enemy of the speculative approach to natural science, typical of *Naturphilosophie*, and his *Annalen* was the organ for a number of attacks on *Naturphilosophie*; see Ludwig Choulant, "Versuch über Ludwig Wilhelm Gilbert's Leben und Wirken," *Ann. Phys.*, Lpz., 1824, *76*: 468–469; and Johann Eduard Erdmann, *Grundriss der Geschichte der Philosophie* (4th ed., Berlin, 1896), II, p. 530.

[3] "The Scientific Life and Works of H. C. Oersted," in Oersted, *Skrifter*, I, pp. xiii–clxvi; also published separately. Dr. Meyer's monograph in Danish on Oersted's varied activities in the Danish community is in *Skrifter*, III, pp. xi–clxvi. For an excellent summary of Dr. Meyer's monographs, see J. Rud Nielsen's article, "Hans Christian Oersted — Scientist, Humanist and Teacher," *American Physics Teacher*, 1939, 7: 10–22. For another biographical sketch, see Rollo Appleyard, *Pioneers of Electrical Communication* (London, 1930), pp. 143–176, but note that Appleyard ignores Meyer's work.

[4] Cf. J[ohan B.] Fogh, "Ueber die Entdeckung des Aluminiums durch Oersted im Jahre

34   ROBERT C. STAUFFER

Mrs. Meyer's work is invaluable for any full understanding of Oersted as a scientist, teacher, and intellectual leader in Denmark.

This present article will concentrate especially upon speculative and experimental factors involved in Oersted's greatest discovery. My purpose is to bring together into one account relevant material which previously could only be found scattered through a considerable number of older books and articles, and which has often been ignored in recent discussions of Oersted. Besides presenting new material, I hope to promote wider recognition for Dr. Meyer's discoveries, which I have found most valuable. The first part of this article will sketch briefly Oersted's philosophical interests and their setting, and then will proceed to illustrate his views on the unity of the powers of nature. To Oersted, his belief in the existence of a physical relation between electricity and magnetism seemed a logical corollary to this general principle of unity. His metaphysical faith in this relationship afforded a motive for his persistent experimentation in the field of electricity and magnetism. This metaphysical factor in Oersted's motivation illustrates the general possibility that the intellectual environment can influence the evolution of science along with the basic influence of the internal logic of scientific ideas, and the influence of social, political, technological, and economic factors.[5] Oersted's experiments culminated in the discovery of a clear and simple demonstration of the physical effect of current electricity on the magnetic needle. The history of this experimental work will be the subject of the second part of this paper.

&

In the early part of his career, Oersted was almost more a "nature philosopher" than a natural scientist. While a student at the University of Copenhagen, he became interested in Kant. In a Danish journal he published a critical account of Kant's *Metaphysische Anfangsgründe der Naturwissenschaft*, and, later in 1799, he continued this subject in his doctoral dissertation.[6] The influence of Kant's view that "a rational doctrine of nature deserves the name of natural science only when the natural laws at its foundation are cognised *a priori*, and are not mere laws of experience"[7] is evident throughout Oersted's intellectual life, as the following pages will illustrate. These philosophical interests were probably strengthened in 1801 while he was travelling in Germany on a fellowship. At Berlin he attended lectures by Fichte and made

---

1825," *K. Danske Videnskabernes Selskab, Copenhagen. Mathematiskfysiske Meddelelser,* 1921, 3: no. 14, 1–17.

[5] Cf. L. Rosenfeld, "On the Method of History of Science," *Archives internationales d'histoire des sciences,* 1947, *1:* 126–129; and introduction to "Essays on the Social History of Science," *Centaurus,* 1953, *3:* 7–11; S. Lilley, "Social Aspects of the History of Science," *Arch. int. hist. sci.,* 1949, *2:* 382–386; and "Cause and Effect in the History of Science," *Centaurus,* 1953, *3:* 58–72; and Richard H. Shryock, "The Interplay of Social and Internal Factors in Modern Medicine: An Historical Analysis,"

Centaurus, 1953, *3:* 107–109, 119.

[6] *Skrifter,* I, pp. 33–105, i.e., "Grundtraekkene af Naturmetaphysiken tildeels efter en nye Plan," separate number of *Philosophisk Repertorium* (Copenhagen, 1799), and *Dissertatio de forma metaphysices elementaris naturae externae* (Copenhagen, 1799).

[7] Immanuel Kant, in *Vorrede* to *Metaphysische Anfangsgründe der Naturwissenschaft* (Riga, 1786), p. vi. I found that I could not improve upon Belfort Bax's translation, so I have quoted it from his edition of Kant's *Prolegomena* and *Metaphysical Foundations of Natural Science* (London, 1903), 138.

his personal acquaintance as well. At Jena he became friends with Friedrich Schlegel and heard August Schlegel's lectures. In his autobiography Oersted mentions many other leading German thinkers encountered at this time.[8]

Oersted's faith in the unity of the forces of nature links his philosophic views with the Romantic "philosophy of nature" which was so prominent in German thought at the beginning of the nineteenth century. Although Oersted did not accept blindly the more extreme views of Friedrich W. J. Schelling, the leading exponent of this *Naturphilosophie*, he certainly responded to the aesthetic appeal of Schelling's writings, and it is reported that Oersted, a few years before his death, ascribed his discovery of electromagnetism to the stimulus received from Schelling.[9]

For this reason a brief glance at some of Schelling's views on nature will not be an irrelevant digression, particularly since Schelling's speculations on the unity of nature specifically stressed electricity and magnetism. In connection with the problem of understanding Oersted, our interest in Schelling and in *Naturphilosophie* in general lies in the acknowledged influence on Oersted's philosophical thought and in the specific emphasis placed upon the sciences of electricity and of magnetism because of the supposed metaphysical significance of the forces there made manifest. We can therefore ignore the complex problems of the derivation and development of Schelling's thought and the successive changes in this point of view and his intellectual interests.

Schelling's Romantic style of expression makes it easy for us to agree with Mme. de Staël, who concluded her discussion about him and about German Romantic philosophy in general with the remark that "systems which aspire to the explanation of the universe cannot be analysed at all clearly by any discourse: words are not appropriate to ideas of this kind and the result is that, in order to make them serve, one spreads over all things the darkness which preceded Creation but not the light which followed." [10] Schelling's mode of thought in statements such as "Nature is only the visible organism of our understanding" [11] would seem alien and almost meaningless to most scientists today. Nor would the following possibly distorted echo of Kant sound much better:

> The assertion is, that all phenomena are correlated in one absolute and necessary law, from which they can all be deduced; in short, that in natural science all that we know, we know absolutely *a priori*. Now, that experiment never leads to such a knowing, is plainly manifest, from the fact that it can never get beyond

[8] Meyer in Oersted, *Skrifter*, I, p. xxv; Oersted's autobiography in Hans Ancher Kofod's *Conversations-Lexicon* .... XXVIII (Copenhagen, 1828), pp. 521–522.

[9] W[ilhelm] Beetz in *Der Antheil der k. bayerischen Akademie der Wissenschaften an der Entwickelung der Electricitätslehre* (Munich, 1873), p. 17, stated that Oersted "mich wenige Jahre vor seinem Tode versicherte, der von Schelling erhaltenen Anregung verdanke er diese Entdeckung, welche seinem Namen verewigt hat."

[10] Anne Louise Germaine (Necker) Staël-Holstein, *De l'Allemagne* (1810), in *Oeuvres*

complètes XI (Paris, 1820), part IV, chap. vii, p. 273.

[11] Friedrich Wilhelm Joseph Schelling, *Sämmtliche Werke*, ed. K. F. A. Schelling (14 vols., Stuttgart and Augsburg, 1856–1861), III, p. 272, hereafter cited as *Werke*. The sentence translated is in "Einleitung zu dem Entwurf eines Systems der Naturphilosophie. Oder über den Begriff der Speculativen Physik und die innere Organisation eines Systems dieser Wissenschaft" (1799). This work presents a good brief account of Schelling's approach to science. A valuable translation by Tom Davidson was published in the *Journal of Speculative Philosophy*, 1867; *1*: 193–220.

36   ROBERT C. STAUFFER

the forces of Nature, of which itself makes use as means. . . . By this deduction of all natural phenomena from an absolute hypothesis, our knowing is changed into a construction of Nature itself, that is, into a science of Nature *a priori*. If, therefore, such deduction itself is possible, a thing which can be proved only by the fact, then also a doctrine of Nature is possible as a science of Nature; a system of purely speculative physics is possible, which was the point to be proved.[12]

On the other hand, Schelling's reaction against the thinkers revered by most of the previous generation, that of the Age of Reason and of the great French Encyclopedia, is clear enough: "Coming after the purblind mode of investigating nature which became generally established beginning with the ruination of philosophy at the hands of Bacon and of physics at the hands of Newton and Boyle, a higher perception of nature begins with *Naturphilosophie*; it forms a new organ of intuition for understanding nature." [13]

Thus in general, *Naturphilosophie* placed the highest value upon intuition as the pathway to the understanding of nature, and Schelling held that "the concept of an empirical science is a mongrel notion. . . . That which is pure empiricism is not science; and, conversely, that which is science is not empiricism." [14]

Schelling seemed fascinated by the underlying identity of apparent opposites. Physical phenomena exhibiting polarity aroused his especial interest, and he asserted, "The first principle of a philosophical system of science is *to go in search of polarity and dualism throughout all of nature*." [15] The polar forces of electricity and of magnetism were treated at length in half a dozen of his works.[16] Schelling fully shared the contemporary enthusiastic interest in Galvani's mysterious animal electricity, and he seemed entranced by the potential significance of this new-found force of nature.[17] At one time he even asserted that the processes of magnetism, electricity and galvanism "are thus as it were the prime numbers of nature," and further that "galvanism governs all of organic nature and is the true border-phenomenon of both [organic and inorganic] natures." [18] Later he stated that galvanism was essentially identical with "the chemical process." [19] A little earlier he had written: "For a long

[12] Schelling, *Werke*, III, pp. 276–278; translation quoted from Davidson, *J. Specul. Philos.* 1867, I: 196.

[13] Schelling, *Werke*, II, p. 70; i.e., "Ideen zu einer Philosophie der Natur . . . ," 2nd ed. (1803), in "Zusatz zur Einleitung."

[14] *Werke*, III, p. 282; i.e., "Einleitung zu dem Entwurf . . . ."

[15] Werke, II, p. 459; i.e., "Von der Weltseele, eine Hypothese der höheren Physik zur Erklärung des allgemeinen Organismus" (1798).

[16] "Ideen zu einer Philosophie der Natur" (1797). "Von der Weltseele . . ." (1798). "Erster Entwurf eines Systems der Naturphilosophie" (1799). "Einleitung zu dem Entwurf . . ." (1799). "Allgemeine Deduktion des dynamischen Processes oder der Kategorien der Physik," originally published in Schelling's *Zeitschrift für speculative Physik*, 1800, vol. 1. "Darstellung meines Systems der Philosophie,"

orig. pub. in *Zeitschrift für spec. Physik*, 1801, vol. 2. These works are to be found in vols. II, III, and IV of the *Werke*, where the specific sections can be located through the tables of contents. Since there is no index to this edition of Schelling, scholars may find useful the index in volume III of Otto Weiss's edition: F. W. J. v. Schelling. *Werke. Auswahl in Drei Bänden* (Leipzig, 1907).

[17] Cf. Friedrich Schlegel's remark to Schleiermacher in a letter (undated, but the editor places it in 1799): "Hier gehts ziemlich bunt und störend durcheinander — Religion und Holberg, Galvanismus und Poesie." *Aus Schleiermacher's Leben. In Briefen*, ed. Ludwig Jonas and Wilhelm Dilthey, III (Berlin, 1861), p. 133.

[18] *Werke*, IV, pp. 73, 75, i.e., "Allgemeine Deduktion . . . ." (1800).

[19] *Werke*, IV, p. 185, i.e., "Darstellung meines Systems . . . ." (1801).

time it has been said that magnetic, electrical, chemical, and finally even organic phenomena would be interwoven into one great association. . . . This great association, which a scientific physics must set forth, extends over the whole of Nature." [20] There are volumes of Schelling's prose which defy any attempt at summary quotation, but Kuno Fischer has epitomized much of his approach to science in saying, "his views in *Naturphilosophie* were all determined by those ideas of a universal unity of all the phenomena of nature. . . . In inorganic nature, Schelling's point of view was directed toward the *unity of physical forces*. . . ." [21]

In similar style, Oersted expressed this theme of the unity of nature repeatedly in his lectures and writings. In the published version of a lecture on the history of chemistry given early in the winter semester of 1805–6, he said, "I see the transfigured line of heroes in the history of the human spirit, quietly looking down on this confusion. They say to us, 'We have sacrificed more time and greater powers than our brothers, to fathom the depths of nature and of reason. Only part of the way did we encounter darkness and doubt; the deeper we penetrated the more all before us became light and unity.' " [22] A few years later, he wrote that "a more deeply penetrating search into nature shows us a wondrous unity in all this infinite variety" and "thus all natural laws form together a unity which, when their operation is pondered, constitutes the essence of the whole world." [23] Even in the last years of his life, Oersted worked over these same essays to prepare them in the form in which they appeared in the Danish edition of his collected literary works.[24]

These philosophical predilections help to explain Oersted's youthful enthusiasm for the fantastic system of the Hungarian chemist Jakob Joseph Winterl. By the publication of his *Prolusiones ad Chemiam Saeculi Decimi Noni* (Buda, 1800), Winterl tried to replace Lavoisier's antiphlogistic system of chemistry by one founded on the hypothetical substances *Andronia*, the principle of acidity, and *Thelycke*, the principle of alkalinity. The metaphysical appeal of the unity of the forces of nature implicit in Winterl's thesis that the positive and the negative principles of electricity formed the common basis of the phenomena

[20] *Werke*, III, p. 320 *n*, i.e., "Einleitung zu dem Entwurf" (1799).

[21] *Schellings Leben, Werke und Lehre*, vol. VII in Fischer's *Geschichte der neuern Philosophie* (3rd ed., Heidelberg, 1902), pp. 324–325. Fischer gives the fullest account I have seen of Schelling and *Naturphilosophie*. Readers who wish to pursue the will-o'-the-wisp of *Naturphilosophie* farther than I have here, may find the following useful: Franz Schnabel, *Deutsche Geschichte im Neunzehnten Jahrhundert*, III (Freiburg im Breisgau, 1934), pp. 171–175, gives a most informative brief account. Hinrich Knittermeyer, *Schelling und die Romantische Schule* (Munich, 1929), i.e., vol. 30/31 of Gustav Kafka's collection, *Geschichte der Philosophie in Einzeldarstellungen* is briefer than Fischer, but contains more recent bibliography. For sources of *Naturphilosophie* in Kant, see *Kants Naturphilosophie als Grundlage seines Systems* (Berlin, 1894), by Arthur Drews, who held that "die Naturphilosophie Schellings ist die Vollendung der Kantischen Naturphilosophie . . ." (p. 494).

For discussions of *Naturphilosophie* available in English, see the histories of philosophy by Johann Eduard Erdmann, Harold Höffding, and Friedrich Ueberweg.

[22] *Skrifter*, I, p. 318, i.e., "Betrachtungen ueber die Geschichte der Chemie," [Gehlen's] *Journal fuer die Chemie, Physik und Mineralogie*, 1807, 3: 194–231; republished in Oersted, *The Soul in Nature*, trans. Leonora and Joanna B. Horner (London, 1852), p. 303.

[23] *Skrifter*, III, pp. 156, 158; also in *Soul in Nature*, pp. 448, 450. Cf. Oersted, *Recherches sur l'identité des forces chimiques et électriques*, trans. Marcel de Serres (Paris, 1813), p. 2.

[24] *Samlede og efterladte Skrifter* (8 vols. Copenhagen, 1849–1852). The passages translated are to be found in vol. 5, pp. 7, 111, and 113.

of heat and light, acids and bases, electricity and magnetism, blinded Oersted
to the flimsiness of Winterl's claims to have achieved experimental verification
of his views. Oersted made himself an enthusiastic missionary for the spread
of Winterl's system, and for this purpose he wrote an exposition of it entitled
Materialien zu einer Chemie des Neunzehnten Jahrhunderts (Regensburg,
1803). At the end of this work we come to Oersted's own statements correlating
electricity and magnetism:

> The constituent principles of heat which play their role in the alkalis and
> acids, in electricity, and in light are also the principles of magnetism, and thus
> we have the unity of all forces which, working on each other, govern the whole
> cosmic system, and the former physical sciences thus combine into one united
> physics (we do not yet have one complete natural philosophy); for do not friction
> and percussion produce heat as well as electricity, and do not dynamics and
> mechanics interlock completely? (If it were necessary, this would become still
> more evident, if we could survey at one glance all of Ritter's beautiful discoveries
> related to this subject, in part made long ago.) Our physics would thus be no
> longer a collection of fragments on motion, on heat, on air, on light, on electricity,
> on magnetism, and who knows what else, but we would include the whole universe
> in one system.[25]

Oersted's continuing interest in this theme is reflected in his formulation
of the prize question posed in 1809 by the Royal Danish Society of Sciences:
"What is the connection between the variation and inclination of the magnetic
needle and physical forces, both in their usual, mild modes of action such as
wind, atmospheric electricity, northern lights, etc. and in their unusual, more
violent modes of action, lightning, earthquakes, hurricanes, etc. — The Society
wishes this subject to be investigated in its historical as well as its experimental
and speculative aspects." [26]

By the age of thirty-five, Oersted had become dissatisfied with the obscurity
of Winterl's views and disillusioned by his errors of fact,[27] but his faith in the
unity of the forces of nature was basically unchanged. He acknowledged the
influence of *Naturphilosophie* on his general views of science, although he had
turned away from its fantastic extremes.[28] His next notable statements regarding electricity and magnetism show a significant recognition of the necessity of new experimental evidence. This appears in the account of his own
electrochemical system first presented under the title of *Ansicht der chemischen
Naturgesetze durch die neueren Entdeckungen gewonnen* (Berlin, 1812). An
extensively revised version was published in Paris in 1813 under the more apt
title of *Recherches sur l'identité des forces chimiques et électriques*. Excerpts
from chapter eight, on magnetism, will make Oersted's views clear:

> Men have always been inclined to compare magnetic forces with electrical
> forces. The great resemblance between the electrical and magnetic attractions and

[25] *Skrifter*, I, pp. 209–10.
[26] K. Danske Videnskabernes Selskab, *Det
Kongelige Danske Videnskabernes Selskab 1742–
1942 Samlinger til Selskabets Historie* (Copenhagen, 1942), I, p. 540.
[27] *Skrifter*, II, pp. 175–176, i.e., *Recherches
sur l'identité*, p. 18.

[28] "Nous ajouterons enfin que la philosophie
de la nature [i.e., *Naturphilosophie*, cf. p. xv],
qu'on a cultivée depuis vingt ans en Allemagne,
pourrait aussi réclamer ses droits sur quelques
vues que nous allons proposer." *Recherches*, p.
11. Cf. *Skrifter*, II, pp. 41–42, i.e., *Ansicht der
Chemischen Naturgesetze durch die neueren
Entdeckungen gewonnen* (Berlin, 1812).

repulsions and the similarity of their laws must necessarily lead to this comparison. . . .

There is one phenomenon, however, which seems adverse to this opinion; electrical bodies act upon magnetic bodies as if they were not animated by any particular force whatsoever. To remove this difficulty completely would be very interesting for science; but, since the present state of physics has not yet furnished facts sufficient for that, we shall show at least that this involves merely a difficulty, not a fact absolutely contrary to the identity of the electrical and magnetic forces. . . . The galvanic mode of activity lies midway between the magnetic mode and the electrical. There the forces are more latent than in electricity and less than in magnetism.

. . . Magnetism exists in all the bodies of nature, as Bruckmann and Coulomb have proved. For this reason it is felt that magnetic forces are as general as electrical forces. One should test whether electricity in its most latent form has any action on the magnet as such. This experiment would offer some difficulty because electrical effects are always likely to be involved, making the observations very complicated.[29]

Although Oersted was susceptible to the aesthetic and emotional appeal of *Naturphilosophie*, his training as a pharmacist, begun in his father's shop and continued at the University of Copenhagen, gave him a start in the accumulation of practical experience in experimental science, and this is reflected in his increasing appreciation of the necessity of the experimental verification of speculative ideas. Even as a student, while noting "beautiful and great ideas" in Schelling's early works, Oersted had also observed that "the empirical propositions adduced are often utterly false."[30] A little later during his *Wanderjahr* in Germany he wrote about Schelling that "he wants to give us a complete philosophical system of physics, but without any knowledge of nature except from textbooks. . . ."[31]

In his teaching at the University of Copenhagen he warned against uncontrolled speculation. A pupil and friend reports Oersted's telling him around 1810: "It is also my firm conviction, and my lectures bear witness thereof, that a great fundamental unity pervades the whole of nature; but just when one has become convinced of this, it becomes doubly necessary to direct one's whole attention to the world of the manifold, wherein this truth above all finds its confirmation. If one does not do this, unity itself remains an unfruitful and empty idea which leads to no true insight."[32]

Thus, however close he was to *Naturphilosophie* in regard to faith in the unity of the forces of nature, Oersted differed sharply from Schelling in his acceptance of the fundamental importance of careful observation. I have dealt first with his speculative views, for they have no necessary logical connection with his experimental work, but it is now time to examine Oersted's own experience with observations on electricity and magnetism.

৵

[29] *Recherches* . . . , pp. 234–238.

[30] *Skrifter*, I, p. xviii, i.e., "Grundtraekkene af Naturmetaphysiken," *Philosophisk Repertorium* (1799).

[31] Meyer in Oersted, *Skrifter*, I, p. xxv, who cites *Breve fra og til Hans Christian Ørsted udgivne af Mathilde Ørsted* (Copenhagen, 1870), I, pp. 81–82.

[32] Johannes Carsten Hauch, *H. C. Oersted's Leben. Zwei Denkschriften von Hauch und Forchhammer*, trans. H. Sebald (Spandau, 1853), p. 13.

40   ROBERT C. STAUFFER

Like many other scientists in 1800 Oersted was stimulated by Volta's great
invention of the galvanic battery. While managing the Lion Pharmacy in
Copenhagen, he had the opportunity of making galvanic experiments. In 1801
he published an account of a new form of Voltaic apparatus which he had
developed, and he also developed an instrument for measuring galvanism by
the rate of evolution of gas during the electrolysis of water. In Germany, later
in 1801, where he was travelling on a fellowship, he was often asked to demonstrate his new galvanic battery.[33]

By the beginning of the nineteenth century, the debate over the relations
between electricity and magnetism had already had a long history. The analogy
between electrostatic and magnetic attractions and repulsions had been noted.
Mariners had observed an occasional reversal of polarity in compass needles
aboard ships struck by lightning.[34] Franklin had magnetised needles by dis-
charging through them a battery of Leyden jars.[35] Similar experiments were
repeated by Johann Karl Wilcke, Martin van Marum and Jan Hendrik van
Swinden among others, but the results were seemingly contradictory. Nor did
the essay prize offered in 1776 and 1777 by the Bavarian Academy of Sciences
bring forth any conclusive settlement of the question.[36]

Whatever may have first attracted Oersted's attention to the experimental
work on this subject, it seems clear that his major stimulus came from Johann
Wilhelm Ritter. Oersted met Ritter when he was in the region of Weimar in
September 1801.[37] Reminiscing a quarter of a century later, Oersted wrote
in his autobiography:

> Of all his acquaintanceships, the one he formed with J. W. Ritter, then at the
> height of his fame, became especially important to him. After a few days'
> acquaintance Oersted even took part in the conclusion of the long series of experi-
> ments on magnetism which one year later were published in Ritter's *Beiträge zur
> näheren Kenntniss des Galvanismus*. Those of the experiments in which Oersted
> took part were perhaps not suited to give him full confidence in the results
> derived from them, but he relied all the more on the whole series of repeated
> experiments which Ritter had made earlier; so that it was not until several years
> later that he convinced himself of their inaccuracy by repeated experiments by
> himself and others. The more proofs he saw of Ritter's diligence in his experi-
> ments and of his acute sense of observation, the more reluctant he was to believe
> that Ritter had so completely allowed himself to be led astray by his own biased
> imagination — something which later on happened all too often to this otherwise
> so highly gifted man.[38]

[33] Meyer in Oersted, *Skrifter*, I, pp. xxi, xxii,
xxiv.

[34] See, e.g., anon., ". . . . Narrative of a
strange effect of thunder upon a magnetic sea-
card . . . ," *Royal Society London, Phil. Trans.*,
1676, 11: 647–648.

[35] Benjamin Franklin, *Experiments and Observations on Electricity*, ed. I. B. Cohen (Cambridge, Mass., 1941), pp. 242–243 (Letter VI).

[36] See Oersted's judicious historical account
in his article on electromagnetism (under the
heading "Thermoelectricity") in the *Edinburgh*

Encyclopaedia; also *Skrifter*, II, pp. 353–354.
John J. Fahie, *A History of Electric Telegraphy
to the Year 1837* (London, 1884), chap. 9; Beetz,
Antheil der bayerischen Akademie, p. 15. For
papers submitted, see *Bayerische Akademie der
Wissenschaften, Neue philosophische Abhandlungen*, 1780, vol. 2, and Jan Hendrik van Swinden,
*Recueil de mémoires sur l'analogie de l'électricité
et du magnétisme* . . . . (3 vols., The Hague,
1784).

[37] Meyer in Oersted, *Skrifter*, I, pp. xxiv, xxvi,
xxix.

[38] Oersted autobiography, pp. 522–523.

OERSTED'S DISCOVERY OF ELECTROMAGNETISM   41

Ritter's activities merit a brief review; they present a Romantic combination of real and imaginary empirical discoveries with the most fantastic of speculations. His training in pharmacy at Jena served as the introduction to a scientific career of wide-ranging interests. In February 1801, Ritter discovered ultraviolet rays through their darkening of silver chloride.[39] In his history of electricity, Hoppe considered over half a dozen of Ritter's discoveries worthy of note.[40] Among these were his galvanic decomposition of water in 1800, independent of the work of Nicholson and Carlisle, and his invention of the storage column or secondary cell.

Extensive as Ritter's experimental work was, his imagination ranged even farther. In philosophy, his views often approached those of Schelling, particularly in regard to the unity of the forces of nature and the tendency to exaggerate the significance of analogies in science.[41] As to the value of empirical observation, their views differed. Schelling is reported to have reproached Ritter for his empirical philistinism,[42] while in a letter to Goethe, Ritter observed about Schelling that "Nature has grounds to be dissatisfied with his material procedure in physics." [43] Yet Ritter's empiricism was not the sort to preclude an interest in animal magnetism or to prevent his experimenting with divining rods.[44] His search for relationships in nature even tended toward astrology when he tried to correlate periods of maximum inclination of the ecliptic with outstanding discoveries in electricity: Kleist's invention of the Leyden jar in 1745, Wilcke's electrophorus in 1764, Volta's condenser in 1783, and Volta's pile in 1800. He went on to prophesy that the next epoch-making discovery would come in "1819⅔ or 1820." This, strangely enough, was written in a letter to Oersted, the very man who was to fulfill the prophecy.[45]

Oersted was a close friend and admirer of Ritter's by the time he proceeded to Paris in 1802. There at the Philomathic Society he reported on Ritter's electrical research, and Biot asked that Oersted persuade Ritter to publish his recent discoveries and submit them for the prize offered by the Institute.[46] In his autobiography Oersted tells us:

> Napoleon as First Consul had established an annual prize of 3000 francs to be awarded by the French Institute for the most important electrical or galvanic discovery made there which could be considered as important as Franklin's or Volta's. Ritter had at that time invented his storage column which might well deserve the annual prize. He wrote a paper on the subject in his customary

[39] [Gilbert's] *Annalen der Physik*, 1801, 7: 527; 1802, 12: 409–415.

[40] Edmund Hoppe, *Geschichte der Elektrizität* (Leipzig, 1884), pp. 138–139, 141, 152, 164–165, 171–172, 174, 178. Wherever I have checked Hoppe, I have found him outstanding in thoroughness and reliability.

[41] Carl v. Klinckowstroem, "Johann Wilhelm Ritter und der Elektromagnetismus," *Archiv für die Geschichte der Naturwiss. u. d. Tech.*, 1922, 9: 70, 75, 79.

[42] Klinckowstroem, *ibid.*, p. 72, cites N. F. Link, *Ueber Naturphilosophie* (Leipzig & Rostock, 1806), p. 122.

[43] "Ich verkenne Schellings grosse Tendenz nicht; ich bin ihn früh gefolgt und ehre ihn, —

was kann ich aber dafür, wenn die Natur mit dem Materiellen seines Verfahrens in der Physik Ursach hat, unzufrieden zu seyn!" Quoted in Klinckowstroem, *ibid.*, p. 71; also in Klinckowstroem, "Goethe und Ritter (mit Ritters Briefen an Goethe)," *Goethe-Gesellschaft, Weimar, Jahrbuch*, 1921, 8: 145.

[44] See letters by Ritter in *Correspondance de H. C. Örsted avec divers savants*, ed. M. C. Harding (Copenhagen, 1920), II, pp. 184, 193–198, 210.

[45] Oersted, *Correspondance*, II, pp. 35–36.

[46] Meyer in Oersted, *Skrifter*, I, p. xxxi, cites Oersted, *Breve*, I, p. 137; see also Harding in Oersted, *Correspondance*, II, pp. 4–5.

42   ROBERT C. STAUFFER

obscure style, and requested Oersted to translate it. Word for word, it was impossible. He rewrote it entirely as a French dissertation, which Ritter later declared he understood better than his own. No one received the prize that year, since the French Institute believed that Ritter's experiments were not any more significant than many older ones. Oersted thereupon announced several experiments designed to contradict this opinion, but this time without effect. Nevertheless, recent years seem completely to have awarded Ritter's experiments the importance earlier denied them. Ritter had also informed Oersted of the experiment whereby a galvanic charged wire assumed the position of a magnetic needle. and accordingly claimed the great prize; but the experiment was pronounced unsound [p. 524].

Acting thus as Ritter's spokesman in Paris, Oersted wrote a number of articles. He discussed the discovery of ultra-violet rays.[47] In the paper which was presented to the scientists of the National Institute, Oersted described Ritter's storage column or secondary pile. In a postscript he mentioned Ritter's claim "that the earth has electrical poles just as it has magnetic poles." [48] In the article concerning experiments on magnetism Oersted reported Ritter's claims to have discovered a difference in oxidizability between the north and the south poles of a magnet.[49] Here, as in the case of Ritter's later claim to have observed that a suspended needle with one end of silver and the other half zinc would turn to the direction of the magnetic meridian,[50] Oersted eventually recognized that Ritter had been misled by experimental errors.[51]

It must have been embarrassing for Oersted to find that he had sponsored Ritter's claims to experiments which no one in Paris could repeat. This whole experience seems to have done much to sharpen Oersted's critical judgment in regard to the validity of experimental claims and the interpretation of evidence. On the other hand Ritter's enthusiastic faith and his tireless experimentation apparently stimulated and strengthened Oersted's own faith in the unity of the forces of electricity and magnetism.[52]

After Oersted's return to Copenhagen in 1804 and his appointment in 1806

---

[47] *Skrifter*, I, pp. 245–248, i.e., "Experiences sur la lumière par M. Ritter, à Jéna, communiquées par Orsted, docteur à l'université de Copenhague," *Journal de physique, de chimie, d'histoire naturelle et des arts*, 1804, *57*: 409–411. Cf. "Expériences sur les rayons invisibles du spectre solaire, par M. Ritter de Jena" (Note communiquée par M. Vicktred [sic], docteur à l'université de Copenhague)," *Société philomatique de Paris, Bulletin des sciences*, 1803, *3*: 197–198.

[48] *Skrifter*, I, p. 232, i.e., "Expériences sur un appareil à charger d'électricité par la colonne électrique de Volta par M. Ritter, à Jena," *Journal de physique* . . . , 1804, *57*: 363.

[49] *Skrifter*, I, pp. 242–245, i.e., *Journal de physique* . . . , 1804, *57*: 406–409.

[50] See M. C. S. Weiss, "Extrait d'un ouvrage allemand de M. J. W. Ritter, intitulé: Das electrische System der Koerper; Le systême électrique des corps," *Annales de chimie et de physique*, 1807, *64*: 80.

[51] *Skrifter*, II, pp. 354–355, i.e., article "Thermoelectricity." See *Annales de chimie* . . . ,

1808, *65*: 211–215, for experimental evidence against Ritter's claims.

[52] In view of the close relations between Ritter and Oersted it seems unnecessary to speculate about the possibility of any influence on Oersted from the obscure experiment made by Gian Domenico Romagnosi in 1802. For the text of Romagnosi's article and a new source reference, see *Epistolario di Alessandro Volta*, IV (Bologna, 1953), pp. 540–541. For a translation and critical discussion, see Appleyard, *Pioneers of Electrical Communication*, pp. 157–161. See also Paul Fleury Mottelay, *Bibliographical History of Electricity & Magnetism* (London, 1922), pp. 365–367.

In the reprint of Romagnosi's article in *Epistolario di Alessandro Volta*, a name is supplied for the "Trent gazette"; the original source is cited as *Ristretto dei foglietti universali di Trento*, n. 52, of 3 August 1802. Here also is mentioned *Gian Domenico Romagnosi — Letter edite a cura di Stefano Fermi*, under the auspices of the R. Istituto Lombardo di Scienze e Lettere, in Milano, 1935.

OERSTED'S DISCOVERY OF ELECTROMAGNETISM   43

as professor extraordinary at the University, his experimental interests continued to develop along with growing sureness of technique, although his teaching load, which in some winters reached five hours of lecturing a day, together with outside responsibilities such as those starting in 1815 as secretary of the Royal Danish Society of Sciences often left him little time for the laboratory. He continued to touch on electricity and magnetism in his papers. As we shall see, his theory of the "conflict of electricities" and the undulatory propagation of electric current, published in 1806,[53] was to play a role in his formulation of the famous experiment which demonstrated the existence of electromagnetism. His experiments made jointly with Lauritz Esmarch led in 1816 to the development of a new galvanic cell using a copper trough instead of a wooden one; this improved apparatus generated a strong current and was frequently employed thereafter.[54]

In the decades immediately preceding Oersted's decisive discovery, which was made in the spring of 1820, scientists were far from agreement as to the existence of any physical relation between electricity and magnetism. In 1802, Ampère had announced that he would "DEMONSTRATE that the electrical and magnetic phenomena are due to *two different fluids* which act independently of each other." [55] In 1807, Thomas Young, in his *Lectures on Natural Philosophy*, had held that "there is no reason to imagine any immediate connexion between magnetism and electricity. . . ." [56] In 1818, John Bostock remarked in regard to galvanism that "although it may be somewhat hazardous to form predictions respecting the progress of science . . . . it does not appear that we are at the present in the way of making any important additions to our knowledge of its effects. . . ." [57] In 1819, David Brewster published volume XIII of his *Edinburgh Encyclopaedia*. Here J. B. Biot, at the end of his article on magnetism, after stating that "there exists the most complete, the most perfect, and the most intimate analogy between the laws of the two magnetic principles, and those of the two electrical principles," concluded in regard to the magnetic principles that "the independence which exists between their actions and the electric actions does not allow us to suppose them to be of the same nature as electricity" (p. 277). In Germany, *Naturphilosophie* still appeared in association with the opposite views, such as those which Julius von Yelin set forth in a public address to the Bavarian Academy of Sciences in 1818. There Yelin said, "I speak of magnetism and electricity as identical yet individualised fundamental forces of nature." [58] And in Denmark, "Oersted *was*

[53] *Skrifter*, I, pp. 267–273, i.e., "Ueber die Art, wie sich die Electricität fortpflanzt," [Gehlen's] *Neues allgemeines Journal der Chemie*, 1806, 6: 292–302.

[54] *Skrifter*, II, pp. 206 f.; Hoppe, *Geschichte*, p. 172.

[55] François Arago, *Oeuvres complètes*, II (Paris, 1854), p. 50.

[56] *A Course of Lectures on Natural Philosophy* . . . (London, 1807), I, p. 694.

[57] *An Account of the History and Present State of Galvanism* (London, 1818), pp. 101–102. In this book, Bostock revised and amplified his article in the *Edinburgh Encyclopaedia*.

[58] *Ueber Magnetismus und Electricität als identische und Urkräfte* (Munich, 1818), pp. 20–21: "Ich spreche von Magnetismus und Electrizität als identischen, jedoch individualisirten Grundkräften der Natur.

"Ich habe darum für nötig gefunden, in der Einleitung aus naturphilosophischer Betrachtung einer Natur überhaupt, und in so weit es für meine Absicht gehören konnte, eine solche Identität und Individualisirung als möglich, ja als erforderlich für eine Manchfachheit derselben nachzuweisen, und daraus umgekehrt, die Pflicht des Physikers erkennbar zu machen: bei Erklärung der Naturphänomene, so wenig

44   ROBERT C. STAUFFER

searching for the connection between those two great forces of nature. His previous writings bear witness to this, and I, who associated with him daily in the years 1818 and 1819, can state from my own experience that the thought of discovering this still mysterious connection constantly filled his mind," Johan Georg Forchammer recalled in a commemorative address.[59] But as Oersted himself explained a few years later, "For a long time he imagined that it would be more difficult to confirm this idea by experiments than it later turned out to be." [60]

୬ई

In the spring of 1820, Oersted might well have repeated the comment he had made at the beginning of his teaching career, "To a certain extent, my investigations in physics run parallel to my lectures," [61] for this explains the occasion of the discovery of electromagnetism. As Dr. Sarton and Dr. Cohen have remarked, this is the one case known in the history of science when a major scientific discovery has been made before an audience of students during a classroom lecture.

Of the three accounts of the circumstances which Oersted himself has left us, the earliest was written the following spring,[62] some years before the other two, and therefore will be quoted as the primary source:

On Electromagnetism

(A.) The History of my previous Researches
on this Subject.[63]

When I began to investigate the nature of electricity, I conceived the idea that the propagation of electricity consisted of an incessant disturbance and restoration of equilibrium and thus included an abundance of activity which would not

---

Hülfsmittel, als nur möglich anzuwenden, und besonders vorsichtig und sparsam in der Annahme nicht darstellbarer Materien zu seyn." For further evidence of the influence of *Naturphilosophie*, see pp. [7]–9, 18–20, 46–49, and passim.

[59] *Hans Christian Örsted. Et Mindeskrift, laest i det Kongelige danske Videnskabernes Selskabs Möde den 7de November 1851* (Copenhagen, 1852), p. 13; also *Oersted's Leben. Denkschriften von Hauch und Forchhammer*, p. 101. For Oersted's own position in regard to general differences between the approach to science in Germany and in France, consider the following statement written to his wife in 1823: "If in Germany I am often tempted to protest against Nature Philosophy when I see how it is misapplied, in France I feel so much the more called upon to defend it, or rather I feel a fundamental difference in scientific thought which I should not have imagined to be so great if I had not so often felt its vital presence." *Skrifter*, I, p. cxiv; Meyer cites *Breve*, II, p. 53 as source.

Soon after, with the rise of the generation of Liebig and Schleiden, the reaction of German scientists against *Naturphilosophie* was to be-

come general. As James F. W. Johnston reported opinion at the 1830 Versammlung deutscher Naturforscher: "Oersted's experimental are far more valuable than his theoretical memoirs. 'Il fait des belles expériences,' said a German doctor to me, 'c'est ce qu'il fait bien — mais quand il écrit — nons [sic] ne trouvons ordinairement que des phantasies.' This expression is no doubt much too strong; but it shows the general opinion of his tendency to speculation." *Edinburgh Journal of Science*, 1831, 4: 229.

[60] *Skrifter*, II, p. 447, i.e., "Meddelelse om Electromagnetismens Opdagelse," *Videnskabernes Selskabs Oversigter* (1820–1821), pp. 12–21.

[61] *Skrifter*, I, p. 211, i.e., "Correspondenz," [Gehlen's] *Neues allgemeines Journal der Chemie*, 1804, 3: 322.

[62] Ludwig Hartmann, "Unveröffentlichte Briefe und Documente des Physikers Hans Christian Oersted," *Archiv für Geschichte der Mathematik, der Naturwiss. u. d. Tech.*, 1931, 13: 331–332; cf. extract of letter from Oersted to Blainville, *Journal de physique . . .*, 1821, 92: 316.

[63] *Skrifter*, II, pp. 223–225, i.e., "Betrachtungen ueber den Electromagnetismus," [Schweig-

OERSTED'S DISCOVERY OF ELECTROMAGNETISM   45

be foreseen in regarding it as a uniform current.* Hence I regarded the transmission of electricity as an electric conflict and found myself, particularly in my investigations of the heat produced by the electric discharge, induced to show that the two opposite electrical forces in the conductor heated by their action are there combined so as to escape all observation, without, however, having reached perfect equilibrium; † so that they might still show great activity, although under a form entirely different from that which can properly be termed electrical. In spite of my efforts to justify my idea, this complete disappearance of electrical forces indicated by the electrometer, accompanied by very considerable action of another kind, seemed very improbable to most physicists. Perhaps this feeling may be attributed partly to the obscurity of the subject and partly to the imperfect manner in which I explained my theory; for it must be admitted that new ideas rarely present themselves with perfect clarity even to their author. Nevertheless, an inward feeling of the agreement of my theory with the facts inspired in me such a strong persuasion of its truth that upon this basis I ventured to construct my theory of heat and of light, and to attribute to these seemingly neutralized forces a radiating action capable of penetrating to the greatest distances.

Since for a long time I had regarded the forces which manifest themselves in electricity as the general forces of nature, ‡ I had to derive the magnetic effects from them also.** As proof that I accepted this consequence completely, I can cite the following passage from my *Recherches sur l'identité des forces chimiques et électriques*, printed at Paris, 1813. "It must be tested whether electricity in its most latent state has any action on the magnet as such." †† I wrote this during a journey, so that I could not easily undertake the experiments; not to mention that the way to make them was not at all clear to me at that time, all my attention being applied to the development of a system of chemistry. I still remember that, somewhat inconsistently, I expected the predicted effect particularly from the discharge of a large electric battery and moreover only hoped for a weak magnetic effect. Therefore I did not pursue with proper zeal the thoughts I had conceived; I was brought back to them through my lectures on electricity, galvanism, and magnetism in the spring of 1820. The auditors were mostly men already considerably advanced in science; so these lectures and the preparatory reflections led me on to deeper investigations than those which are admissible in ordinary lectures. Thus my former conviction of the identity of electrical and magnetic forces developed with new clarity, and I resolved to test my opinion by experiment. The preparations for this were made on a day in which I had to give a lecture the same evening. I there showed Canton's experiment on the influence of chemical effects on the magnetic state of iron. I called attention to the variations of the magnetic needle during a thunderstorm, and at the same time I set forth the conjecture that an electric discharge could act on a magnetic needle placed outside the galvanic circuit. I then resolved to make the experiment. Since I expected the greatest effect from a discharge associated with incandescence, I inserted in the circuit a very fine platinum wire above the place where the needle was located. The effect was certainly unmistakable, but still it seemed to me so confused that I postponed further investigation to a time when I hoped to have more leisure.‡‡ At the beginning of July these experiments were resumed and continued without interruption until I arrived at the results which have been published.

---

ger's] *Journal fuer Chemie und Physik*, 1821, *32*: 199 f.; "Considérations sur l'électro-magnétisme," *Journal de physique* . . . , 1821, *93*: 161–163; minor differences indicate Oersted wrote both the German and French versions of this article, and I have used details from each for the new translation given. For evidence that the account of the discovery by Christopher Hansteen (published in Henry Bence Jones, *Life and Letters of Faraday* (1st ed., London, 1870), II, pp. 395–396) was based on hearsay, see Stauffer, *Isis*, 1953, *44*: 309.

46   ROBERT C. STAUFFER

* My article on this subject is in Gehlen's *Neues Journal der Chemie*, 1806, and in the *Journal de physique* for the same year.

† See my *Ansicht der chemischen Naturgesetze* (Berlin, 1812), pp. 133–284.

‡ See the letter added at the end of my *Materialien zu einer Chemie des neunzehnten Jahrhunderts* (Regensburg, 1803), also *Ansicht der chemischen Naturgesetze*, p. 135, *Recherches sur l'identité des forces*, p. 127 and many other places.

** *Ansicht der chemischen Naturgesetze*, pp. 246–251; *Recherches.* pp. 234–238.

†† *Ansicht*, p. 251; *Recherches*, p. 238.

‡‡ All my auditors are witnesses that I mentioned the result of the experiment beforehand. The discovery was therefore not made by accident, as Professor Gilbert has wished to conclude from the expressions I used in my first announcement.

Both the other two accounts by Oersted were probably written in 1827.[64] The one from the article on Thermo-electricity in Brewster's *Edinburgh Encyclopaedia* (given in appendix 1 of this paper) is the fullest, and here Oersted added the point that he anticipated his discovery that the electric current exerts a force transverse to the direction of the current. Here he also added the statement that three months elapsed between the initial discovery and his resumption of the experiments at the beginning of July. The account (appendix 2 of this paper) included in the autobiography he wrote for Kofod's *Conversationslexikon* is less technical but serves to confirm the other two.

The four-page article announcing the discovery of electromagnetism is well known,[65] but from it the reader can get only hints of the series of over sixty experiments [66] which lay behind it. From Oersted's posthumous papers, Dr. Meyer has published the collection of laboratory notes involved, together with facsimiles of nearly one hundred diagrams. Some are dated July 1820, and they are all clearly related. Here we can observe Oersted proceeding methodically from his initial observation that the influence of a galvanic current can deflect a magnetic needle to his fundamental law "that the magnetical effect of the current has a circular motion around it." [67]

The first account was too brief and condensed to be perfectly clear on all points. From these notes we can see that Oersted had already discovered that the conductor need not be incandescent to deflect the needle.[68] The notes and diagrams also make more clear the latter part of the paper treating the experiments with the conducting wire bent back upon itself. This forms one turn of a solenoid, with a north and a south side, and as Dr. Meyer points out, "Oersted saw already at this period that a closed electrical circuit acts like a magnet, but he has not sufficiently defined its position to the needle." [69]

Further laboratory notes published by Dr. Meyer cover the continuation of the first series of experiments and Oersted's discovery that only a single cell was necessary to obtain a very marked deflection of the needle.[70] Thus Oersted

[64] Oersted, Autobiography, pp. 537, 540.
[65] For a facsimile of original Latin version and of English translation see *Isis*, 1928, *10*: [437]–[444].
[66] Oersted, *Skrifter*, II, p. 358, i.e., art. "Thermoelectricity."

[67] Meyer in Oersted, *Skrifter*, I, pp. lxxii-lxxxviii.
[68] Meyer, *ibid.*, p. lxxvi.
[69] Meyer, *ibid.*, p. lxxxi.
[70] Meyer, *ibid.*, pp. xciv-xcvi.

OERSTED'S DISCOVERY OF ELECTROMAGNETISM   47

proceeded to the important new discoveries announced in his second paper on electromagnetism, which forms a logical continuation of the first, "an appendix published two months later (in Schweigger's Journal)." [71] Hoppe and Meyer both rightly stress the historical importance of this often forgotten second paper.[72] Here Oersted stated: "The magnetic effects do not seem to depend upon the intensity of the electricity, but solely on its quantity. The discharge of a strong electric battery [i.e., of Leyden jars] transmitted through a metallic wire produces no alteration in the position of the needle." [73] Suspending a single light but powerful galvanic cell with the circuit closed by a conducting wire in the form of a narrow loop, he demonstrated, as he expected, that "one of the poles of an energetic magnet . . . will set the galvanic apparatus in motion." He concludes: "Here we have an apparatus whose extremities act like the poles of a magnet. But it must be admitted that only the faces of the two extremities, and not the intermediate parts, have this analogy. . . . I have not yet found the way to make a galvanic apparatus capable of directing itself toward the poles of the earth. For this object it would be necessary above all to have a much more movable apparatus." [74]

The first paper, in Latin, was really in the form of a preliminary announcement giving results without much experimental detail; the second paper was more detailed and the diagrams published make the form of Oersted's apparatus clear. Especially now that we can study the original laboratory notes, if we consider the two papers together, we can see that Oersted had accumulated abundant experimental evidence and had presented a well-rounded account of the basic facts of electromagnetism.

Oersted's second paper was received in Paris on 29 September 1820,[75] just twenty-five days after Arago had announced Oersted's discovery to the Academy of Sciences.[76] Already Ampère had begun to build upon the foundations established by Oersted, and on September 25 had announced to the Academy his discovery of the mutual forces between two parallel electric currents.[77] Many other scientists soon entered the field. Oersted had started a new era in the study of electricity.

~ఢ

As now can be seen from the evidence presented, the experimental demonstration of the existence of electromagnetic forces was the vindication of a faith which Oersted had already held for many years before 1820. For Oersted the foundation of this faith was the conviction that in reality there exists an underlying unity of the forces of nature. Although this conviction was primarily metaphysical rather than strictly scientific in origin, it nevertheless helped to

[71] Oersted, *Skrifter*, II, p. 364, i.e., art. "Thermoelectricity."

[72] Hoppe, *Geschichte*, p. 197; Meyer in Oersted, *Skrifter*, I, p. xcvii.

[73] "Nouvelles expériences électro-magnétiques," *Journal de physique* . . . , 1820, *91*: 78. This July number, containing both Oersted's first and second papers on electromagnetism, was not published until 30 October 1820; see

Oersted, *Correspondance*, II, p. 272. The German version in Schweigger's *Journal* was translated by J. L. G. Meinecke from the French; see Oersted, *Correspondance*, II, p. 431.

[74] *J. de physique* . . . , 1820, *91*: 80.

[75] *J. de physique* . . . , 1820, *91*: 78.

[76] *Académie des sciences, Paris. Procès-verbaux des séances* (1820–1823), 7: 83.

[77] *Ibid.*, p. 95.

48   ROBERT C. STAUFFER

motivate a continued search for conclusive experimental evidence. And it was only after decades of extensive experience in the laboratory that Oersted arrived at the actual design and execution of the decisive experiments.

In the history of science, any speculation, no matter how fantastic, which leads to important experimental discovery, deserves some notice. Thus the stimulus Oersted received from Schelling's "beautiful and great ideas" and from *Naturphilosophie* in general and also the influence of both the speculations and the experiments of Ritter should be recognized as factors involved in a major discovery in physics. This exemplifies the significance of intellectual factors outside the realm of science as potential influences upon the development of science.

Thus in order to have a full account of the history of science, although we start with the internal logic of scientific ideas, we must also consider not only social and political history, economic history, and the history of technology but general intellectual history as well.[78]

APPENDIX 1
Oersted's Account of the Discovery of Electromagnetism Published in his Article on "Thermo-Electricity" in *The Edinburgh Encyclopaedia*.[79]

Electromagnetism itself, was discovered in the year 1820, by Professor *Hans Christian Oersted*, of the university of Copenhagen. Throughout his literary career, he adhered to the opinion, that the magnetical effects are produced by the same powers as the electrical. He was not so much led to this, by the reasons commonly alleged for his opinion, as by the philosophical principle, that all phenomena are produced by the same original power. In a treatise upon the chemical law of nature, published in Germany in 1812, under the title *Ansichten der chemischen Naturgesetze*, and translated into French, under the title of *Recherches sur l'identité des forces électriques et chymiques*, 1813, he endeavoured to establish a general chemical theory, in harmony with this principle. In this work, he proved that not only chemical affinities, but also heat and light are produced by the same two powers, which probably might be only two different forms of one primordial power. He stated also, that the magnetical effects were produced by the same powers; but he was well aware, that nothing in the whole work was less satisfactory, than the reasons he alleged for this. His researches upon this subject, were still fruitless, until the year 1820. In the winter of 1819–20, he delivered a course of lectures upon electricity, galvanism, and magnetism, before an audience that had been previously acquainted with the principles of natural philosophy. In composing the lecture, in which he was to treat of the analogy between magnetism and electricity, he conjectured, that if

[78] Support from the research funds of the Graduate School of the University of Wisconsin provided for the basic translation of Oersted's autobiography made by Miss Martha K. Trytten. It is a pleasure to acknowledge the assistance of Miss Trytten and the generous cooperation of Professor Einar I. Haugen in regard to translations from the Danish used in this paper.

[79] XVIII (Edinburgh, 1830), also in *Skrifter*, II, pp. 356–358. Aspects of the problems with the English language which Oersted faced in writing this article may be glimpsed in the following excerpts from an undated draft, written in English, of a letter from Oersted to David Brewster: "I have now finished the article upon thermoelectricity . . . I have spend much time in writing it, partly because for the language partly because I desired to make the article at the same time short and complete. . . . Still the language whereof I am not master, may often have hindred my from the perfection I aimed at. I have not been able to find any scientific Brittish Gentleman here, of whom I could ask that he should take the pains of perusing my article . . . . I hope that You will be so kind as to mend the faults that still remain." *Correspondance*, II, pp. 279–280.

OERSTED'S DISCOVERY OF ELECTROMAGNETISM   **49**

it were possible to produce any magnetical effect by electricity, this could not
be in the direction of the current, since this had been so often tried in vain, but
that it must be produced by a lateral action. This was strictly connected with
his other ideas; for he did not consider the transmission of electricity through a
conductor as an uniform stream, but as a succession of interruptions and re-
establishments of equilibrium, in such a manner, that the electrical powers in
the current were not in quiet equilibrium, but in a state of continual conflict. As
the luminous and heating effect of the electrical current, goes out in all directions
from a conductor, which transmits a great quantity of electricity; so he thought
it possible that the magnetical effect could likewise eradiate. The observations
above recorded, of magnetical effects produced by lightning, in steel-needles not
immediately struck, confirmed him in his opinion. He was nevertheless far from
expecting a great magnetical effect of the galvanical pile; and still he supposed
that a power, sufficient to make the conducting wire glowing, might be required.
The plan of the first experiment was, to make the current of a little galvanic
trough apparatus, commonly used in his lectures, pass through a very thin
platina wire, which was placed over a compass covered with glass. The preparations for the experiments were made, but some accident having hindered him from
trying it before the lecture, he intended to defer it to another opportunity; yet
during the lecture, the probability of its success appeared stronger, so that he
made the first experiment in the presence of the audience. The magnetical needle,
though included in a box, was disturbed; but the effect was very feeble, and
must, before its law was discovered, seem very irregular, the experiment made no
strong impression on the audience. It may appear strange, that the discoverer
made no further experiments upon the subject during three months; he himself
finds it difficult enough to conceive it; but the extreme feebleness and seeming
confusion of the phenomena in the first experiment, the remembrance of the
numerous errors committed upon this subject by earlier philosophers, and particularly by his friend *Ritter*, the claim such a matter has to be treated with
earnest attention, may have determined him to delay his researches to a more
convenient time. In the month of July 1820, he again resumed the experiment,
making use of a much more considerable galvanical apparatus. The success was
now evident, yet the effects were still feeble in the first repetitions of the experiment, because he employed only very thin wires, supposing that the magnetical
effect would not take place, when heat and light were not produced by the
galvanical current; but he soon found that conductors of a greater diameter give
much more effect; and he then discovered, by continued experiments during a
few days, the fundamental law of electromagnetism, viz., *that the magnetical
effect of the electrical current has a circular motion round it.*

When he had discovered this fundamental law, he thought it proper to publish
the discovery, in order that it might be as soon as possible perfected by the
co-operation of other philosophers. Apprehending that others might lay claim
to this discovery, he sent a short Latin description of his experiments to the most
distinguished philosophers and learned bodies; and though, by this means, he
has not avoided the pretensions which have been made to his discovery by
others, still he has rendered them ineffectual. It deserves, perhaps, to be noticed,
that the above-mentioned Latin description, consisting of four pages in 4to., of
which the first gives the introduction and the description of the apparatus, the last
the conclusions, contains upon the two intermediate pages, the results of more than
60 distinct experiments. From this brevity, it has happened, that some philosophers have thought that he had treated his subject in a superficial manner.

APPENDIX 2
Oersted's Account of his Discovery Published in his Autobiography.[80]

The year 1820 was the happiest in Oersted's scientific life. It was during this

50   ROBERT C. STAUFFER

year that he discovered the magnetic effect of electricity. In consequence of the
over-all unity of things he had, even in his earliest writings, assumed that magnetism and electricity were produced by the same forces. This opinion, incidentally,
was not new; quite the contrary, it had alternately been accepted and rejected
throughout more than two centuries; but heretofore no one who accepted the
connection had been able to find decisive proof. The investigators had expected
to find magnetism in the direction of the electric current, so that the north and
south poles would act either just like or just the reverse of positive and negative
electricity. All investigations had shown that nothing was to be found along this
path. Oersted therefore concluded that just as a body charged with a very strong
electric current emits light and heat at all times, so it might also similarly emit
the magnetic effect he assumed to exist. The experiences of the past century, in
which lightning had reversed the poles in a magnetic needle without striking it,
confirmed his belief. The idea first occurred to him in the beginning of 1820
while he was preparing to treat the subject in a series of lectures on electricity,
galvanism, and magnetism. He had set up his apparatus for the experiment
before the lecture hour, but did not get around to carrying it out. During the
lecture, the conviction so grew upon him that he offered his listeners an immediate
test. The results corresponded to expectations, but only a very weak effect was
obtained, and no particular law could immediately be observed from it. It was
only observed that the electric current, like other magnetic effects, penetrated
glass. So long as the experiments were not more conclusive he feared that he,
like Franklin, Wilcke, Ritter, and others, would be deceived by a mere coincidence. Burdened by daily routine for several months, he did not venture further
experiments in the meantime. To what extent a certain tendency to procrastinate
and to utilize his free moments to live in the world of thoughts contributed, he
himself could not easily determine. In July he renewed the experiments, with a
very large galvanic circuit of copper containers, zinc plates, and dilute acid. As
witnesses and assistants, two friends were present: Commander and Director of
Navigation Wleugel and Councillor Esmarch. A very strong effect was immediately obtained and tested under varying conditions. Nevertheless, many days of
experimenting were required before he could find the law governing the effect.
As soon as he had discovered it, he rushed to publish his work. This was done in
a very short Latin article filling two closely printed quarter-sheets, in which he
compressed the description of his experiments to such an extent that the two
middle pages contain as many suggested experiments as they do lines.

This half-sheet was now sent on the same post day to the most important
scientific centers in Europe. The experiments were soon repeated in all countries
in which there were friends of science. And the greatest reward an inventor can
enjoy, that of seeing his invention become the object of the most industrious
investigation, seeing it expanded and fructifying, was his to the fullest. The
number of those who have written on electromagnetism amounts to well over a
hundred. Therefore, with the united efforts of so many, this knowledge too has
been expanded and enriched in content far beyond what one would expect in the
short period of seven years. In addition, the discoverer received honors from
various places abroad. Many learned societies elected him to membership. The
Royal Society of London sent him the Copley Medal; the French Institute, which
no longer offered the prize for galvanism, gave him as an extraordinary award
the prize of the Mathematical Class, a medal worth three thousand francs. During all these testimonials of approval, the inventor could not but feel deeply that
even those discoveries which we have sought industriously are not the fruit of
our efforts alone, but are conditioned by a series of events and a situation in the
scientific world which are subject to a law higher than all those he is seeking.

* Han Ancher Kofod's *Conversations-Lexicon* . . . XXVIII (Copenhagen, 1828), pp. 536–8.

The Mutual Embrace of Electricity and Magnetism

In the development of his electromagnetic field theory Maxwell relied heavily on this suggestive image.

M. Norton Wise

Scientific imagination has often been guided by visual images. Notable in the physical sciences among those who claimed to think in pictures, images, or embodied mathematics are Bohr, Einstein, and Maxwell. As historically important as visual images have been, however, the process of concept formation through imagery has not often been easy to unravel. Therein lies the intended significance of the present article for, once uncovered, the creative role of Maxwell's imagery is relatively transparent. Moreover, it explains much that has been obscure in the emergence of his electromagnetic theory—for example, how his ideas were related to Faraday's and how the well-known reciprocal symmetry of electric and magnetic fields first appeared. Before considering electromagnetism, however, a few remarks on the general function of images may serve at the outset to focus attention on those aspects which seem most clearly exemplified in Maxwell's work.

Without attempting to distinguish sharply between visual images, physical analogies, and models, I will use the phrase visual image to refer to those pictorial representations of natural phenomena which function primarily as symbols and which often have metaphorical connotations. Kepler's representation of the cosmos by a sun-centered sphere, in metaphorical analogy to the Christian trinity as a sphere with God the Father at the center, was such a symbol. So was Bohr's "formal representation" of electron stationary states by definite orbits, particularly when coupled with his metaphor of "free will" of the electron in transitions between states. Visual images as symbols, or hieroglyphs, depict more or less abstractly those characteristics of a phenomenon seen to be core characteristics, while merely evoking or standing for much that is not depicted, sometimes a physical model or analogy, sometimes a mathematical structure. The metaphorical connotations of visual images, on the other hand, often reveal strong commitments—psychological, religious, and philosophical—which help to explain their power in concept formation and their continual reappearance in the work of a single individual.

Partly because of these symbolic and metaphorical aspects, powerful images suggest relations and concepts that extend far beyond the empirical subject matter they are taken to represent. Much as models and analogies, but with less concreteness, images serve as heuristic devices, as guides to what one hopes to discover, as well as symbols for what one supposedly knows. All such heuristic devices share the property of generating problems whose answers force reformulation of the representations from which they arose. But images conceived specifically as symbols have a special characteristic, not appropriate to models and analogies. A symbol for physical reality may remain the same while the content symbolized changes radically. In this sense, the function of visual images is like that of words and of mathematical symbols, which also shape perceptions and are reshaped in meaning by the perceptions they produce. But visual images are easier to grasp as symbols than are words and they possess more immediate perceptual significance than abstract mathematical symbols. For that reason an historical examination of the role of imagery in the formation of a powerful concept may display quite sharply the process of shaping and reshaping of content.

With the latter goal in mind I present here an analysis of a striking example, an image of mutually embracing curves that was present throughout the formative stages of Maxwell's electromagnetic theory, roughly 1855 to 1870. Through the metaphor of mutual embrace, symbolized as two interlocked rings, Maxwell first conceived the reciprocal dynamics of electric currents and magnetic forces. His original conception was physically and mathematically incomplete, yet it acted as both motivation and guide for completion. The immediate result was a symmetrical set of relations between steady-state currents and magnetic forces. This initial success, however, more nearly opened the problem than solved it and led to a thoroughgoing reformulation that we now recognize as Maxwell's equations. In order to see this process clearly, it will be helpful first to survey the 19th-century prelude to Maxwell's transformation of electricity and magnetism.

Replacing Laplacian Imagery

Following the examples of Newton's universal gravitation between atoms of matter and his corpuscular theory of light, a coherent and fruitful program for all of physics emerged in France in the early 19th century. This program—which has been called Laplacian, after its most illustrious exponent—attempted to describe each of the areas of physical phenomena, such as gravity, electricity, magnetism, and heat, in terms of a separate "matter" whose action was the sum of the independent actions of its supposed constituent particles. Two such actions were possible: either the particles exerted forces, such as the gravitational force, on points at a distance, or the particles themselves were transmitted to points at a distance, as the corpuscles of light. Thus there were electric and magnetic "fluids" between whose respective particles inverse square forces acted; heat was also identified with a fluid, the self-repulsive caloric fluid, but conduction of heat through normal matter took place as radiation of particles of caloric between molecules of matter. In each area, then, the controlling image was one of direct action at a distance between independent particles, whether through forces or secondary particle transmission. This single image for many phenomena represented a coherent explanatory program for physics. It did not, however, unify the various branches on a shared physical basis,

The author is an assistant professor in the Department of History, University of California, Los Angeles 90024.

0036-8075/79/0330-1310$02.00/0 Copyright © 1979 AAAS

a demand that would soon be widely felt and that Maxwell's theory would partially satisfy.

The Laplacian type of mechanical reduction was not conceived by its supporters as merely a heuristic device to guide research; Poisson, Biot, Laplace, and others insisted that it was the only acceptable mode of explanation (*1, 2*). As a mathematically sophisticated and very successful program, therefore, it demanded an equally powerful opposition before its associated imagery could be replaced. The most serious challenges came after 1815 in the form of highly articulated mathematical alternatives, presented by Fresnel in his wave theory of light and by Fourier in his macroscopic description of heat conduction as a flow process. Beneath the mathematics of their descriptions lay new and very simple imagery. Both Fresnel and Fourier substituted indirect transmission of effects by propagation through a largely unspecified medium for direct transmission of particles or forces to a distance. Fourier in particular highlighted the positive, descriptive character of his heat conduction analysis, which would elevate it beyond any hypothesis concerning the true nature of heat. He provided neither a new model nor (directly) a new analogy, but rather a new technique of analysis and a new image. In a picture consisting only of flow across constant-temperature surfaces, he stressed the simple linear relation between the amount of heat crossing any unit surface per unit time (flux) and the rate of change of temperature across the surface (temperature gradient). In vector notation, this is $\mathbf{Q} =_, k\mathbf{I}$, where $\mathbf{Q}$ is flux of heat ("quantity," in the language of Faraday and Maxwell), $\mathbf{I}$ is temperature gradient ("intensity"), and k is relative conductivity of the medium.

Fourier's description effectively made temperature gradient the cause of heat flow through a medium. That would later be seen to have raised the concept of the gradient to independent and fundamental status, analogous to the steepness of a hill, whereas in the Laplacian picture the gradient had been only a secondary, purely mathematical, expression. Similarly, the wave theory of light substituted two concepts, displacement and tension in a medium, for the former flux of corpuscles. Both new propagation schemes, in retrospect, characterized propagation by two parameters, one a quantity and the other an intensity.

Fourier's treatment of heat transformed the Laplacian picture of direct transmission to a distance into an in-

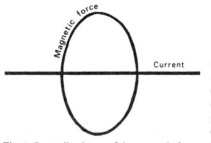

Fig. 1. Oersted's picture of the magnetic force in the vicinity of a current-carrying wire. For a complete representation Oersted imagined magnetic circles drawn at all radii from the wire.

direct propagation view. This generalized statement, however, contains an implication that Fourier never intended. Because the older picture applied not only to the transfer of material corpuscles, such as caloric, but also to nonmaterial electric or magnetic action at a distance, the transformed picture could potentially be applied to the transmission of forces as well as to heat conduction. That potential was recognized to varying degrees by several mathematically oriented physical scientists, notably Green, Gauss, Whewell, and especially William Thomson (later Lord Kelvin). Thomson began to develop the analogy explicitly and extensively in the 1840's, but primarily as a mathematical analogy. He stopped short of a physical analogy, which would have involved a quantity-intensity distinction. When Maxwell learned the mathematics from Thomson in the early 1850's, however, he learned it as what may be called physical geometry. He gave the mathematical form itself a certain reality. The full flow picture, therefore, with a quantity-intensity distinction displayed, would come to play a major role in his new synthesis of electricity and magnetism. The quantity-intensity distinction will be seen below to be one of the basic sources for his dual-field description of electric and magnetic forces, with two fields for each force ($\mathbf{E}$, $\mathbf{D}$ and $\mathbf{H}$, $\mathbf{B}$ in modern notation).

Interconvertibility and Faraday's Imagery

Although Fourier and Fresnel upset the coherence of Laplacian physics, they did not mount the fatal challenge. That challenge derived from several discoveries of the interconversion of powers or forces in nature: chemical to electric in the voltaic cell (Volta, 1800); electric to magnetic in electromagnetism (Oersted, 1820); and magnetic to electric in electro-

magnetic induction (Faraday, 1831). Such interconversions could not occur between independently conserved fluids. At the very least all aspects of electricity and magnetism had to be reduced to a common basis. Today we recognize the demand for unity through conservation of energy, which was just emerging as a principle in the 1840's (*3*). During that emergence unity was sought in at least three different directions. One could attempt to salvage the action at a distance view by reducing many phenomena to complex forces between only electric particles. Among continental analysts this approach was widely adopted, and with remarkable success. Wilhelm Weber's velocity- and acceleration-dependent law of force was the classic attainment of the 1840's. A second alternative, which attempted more but initially achieved less, was to reduce electricity and magnetism, along with heat and light, to mechanical processes in the luminiferous ether that underlay the wave theory of light. The all-subsuming ether was originally most popular among British natural philosophers, such as Whewell, Herschel, and Challis. It led directly to the mathematical work of Thomson and Maxwell, although with a strong influence from a third alternative.

The third alternative was pursued by Michael Faraday. Instead of subtle fluids and ethers, Faraday spoke of natural powers and forces and attempted to investigate them experimentally. Focusing on the relations evident between powers in interconversion, he hoped to uncover the unity of all natural powers. But describing relations between powers, while avoiding concrete models, is not a straightforward task, and therein lies the importance of imagery to Faraday's discussion. All of Faraday's descriptions and analyses of electric and magnetic phenomena were couched in graphic images of lines of force, which represented, ostensibly without prejudice, whatever action might actually have been occurring. If we look at these pictures closely, they will take us a long way toward understanding the theoretical structure that Maxwell extracted from them.

The 19th-century image of electromagnetism derived from Oersted's discovery that the magnetic power of an electric current is directed in circles around, and is perpendicular to, a wire carrying the current, as shown in Fig. 1. These two "axes of power," the current line and the magnetic line, became Faraday's fundamental lines of force, although each line had other important representations. More complex pictures, even models, emerged from Ampère's suggestion that all magnetism derives from currents. The lodestone, for example, would consist of molecules of iron surrounded by tiny currents. Any small magnet, similarly, could be replaced by a current loop, and the well-known attractive and repulsive effects of magnets would apply equally to currents. Figure 2 shows typical resulting configurations. Working in the Laplacian tradition, Ampère reduced these interactions to an action at a distance law of force between infinitesimal sections, or elements, of current. In simplified form, adjacent parallel elements of current always attract, while adjacent antiparallel elements repel.

Faraday accepted neither the reality of action at a distance nor the propriety of arbitrarily defined current elements, but he did agree that two adjacent lines of electric force, representing parallel currents and taken as a whole, would attract laterally. Any single closed line, furthermore, would tend to elongate, as though antiparallel elements on opposite sides of the loop repelled each other. These lateral and longitudinal relations of lines of force, represented in Fig. 3, Faraday took to be the proper descriptive basis for understanding the nature of the electric power as it appeared in currents: electric current lines attract laterally and extend longitudinally.

Considering the closely associated magnetic lines of force, Faraday thought of them as a series of little magnets placed end to end, with the net effect that magnetic lines contract longitudinally and repel laterally, as represented for the lines of a bar magnet in Fig. 4a. Thus magnetic lines and electric current lines each have their own lateral and longitudinal relations, but if the bar magnet is replaced by a bundle of current loops, as shown in Fig. 4b, those two sets of relations are seen to be reciprocal. Lateral attraction between current lines has the same effect as longitudinal contraction in magnetic lines, and longitudinal extension in current lines has the same effect as lateral repulsion between magnetic lines. With a simplicity that would soon rivet Maxwell's attention, Faraday symbolized this mutual relation of the two axes of power as two linked rings, perpendicular to each other, but unified in their reciprocity. That image is reproduced in Fig. 5.

The perpendicular rings of electricity and magnetism, Faraday observed, exhibit a certain unity, a "oneness of condition of that which is apparently two powers or forms of power, electric and magnetic" (4). This statement is from a

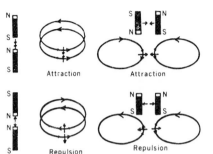

Fig. 2. The forces between current loops conceived as magnets led Ampère to reduce their actions to forces between elements of current (short sections along the currents). In simplified form, adjacent parallel elements attract and adjacent antiparallel elements repel.

paper written in 1851, after Faraday had struggled for at least 20 years with the nature of lines of force. He had by then developed a large body of associated concepts, of which the most immediately important for Maxwell were the notion of conduction of lines of force, and an associated distinction, similar to Fourier's, between quantity and intensity of the power represented by the lines.

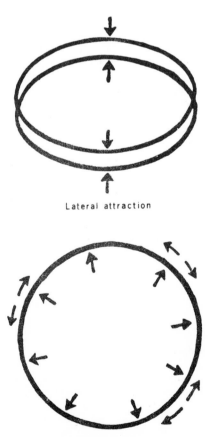

Lateral attraction

Longitudinal extension

Fig. 3. Faraday replaced Ampère's forces between current elements by a dynamics of lines of force: adjacent electric current lines attract laterally, but any particular line tends to extend longitudinally.

All of these ideas developed originally out of Faraday's analysis of electro-chemical processes, such as those in voltaic cells. He imagined the electric tension between the two plates of a voltaic cell to inhere in chains of polarized water molecules produced by differential affinities of the two plates for the oxygen and hydrogen of water. Each such chain represented a line of electric force. If the plates were connected by a conductor—a material that could not support electric tension—the entire chain would momentarily break up, relieving its tension through the conductor, only to immediately recombine and break up again. Continuation of the process would constitute a current of electricity through the circuit.

In this picture it is apparent that the quantity of the current would depend on how many lines of tension were undergoing breakup and recombination; that is, on the size of the plates or on the number of lines through any lateral section cutting all of them. Similarly, the power of the current to overcome any resistance to conduction in the connecting circuit, its intensity, would depend on the power of the cell to develop longitudinal tension in the lines of force, or on the relative affinities among its components. Every current, then, had two characteristics, quantity and intensity, the first a lateral measure of power and the second a longitudinal one. In any resisting circuit, the actual quantity of the current would be proportional to its intensity and to the conducting power of the circuit, yielding a description very similar to Fourier's earlier conduction equation [although to Faraday the meaning was quite different (2, pp. 142–148)]. It is understandable, then, that the mathematically inclined Maxwell would read the two versions as one, $Q = kI$, particularly when he had learned the conduction equations of Ohm and Kirchhoff.

Faraday applied his schema for conduction of lines of force between contiguous particles successively to electrostatics and magnetism, developing by 1850 a highly articulated version for magnetic lines conducted through magnetic materials and through space To the continental notion of force at a point arising from the action of point particles at a distance, he specifically opposed the notion of conduction of powers through an intervening medium, or through space. Ponderomotive forces, he reasoned, were exerted on objects only as a result of their participation in the conduction process, or in the field of force. When Maxwell came to Faraday's work in 1854 the lateral and longitudinal properties of quantities and intensities in conduction were an integral corollary to the reciprocal lateral and longitudinal dynamics of electric and magnetic lines, although Faraday admitted to an only imperfect understanding of what the exact relation might be. This, I will argue, was the primary problem Maxwell sought to resolve in his first assault on electromagnetism.

Maxwell's Original Conception

Maxwell set out in 1854 to develop further the relation of electric currents to magnetic forces in the manner of Faraday's conception of a field of force. He reported his earliest ideas to Thomson (5):

I have heard you speak of "magnetic lines of force" and Faraday seems to make great use of them, but others seem to prefer the notion of attractions of elements of currents directly [Ampère]. Now I thought that as every current generated magnetic lines and was acted on in a manner determined by the lines through which it passed that something might be done by considering "magnetic polarization" as a property of a "magnetic field" or space and developing the geometrical ideas according to this view.

Maxwell went on to sketch out a geometrical description of the magnetic field in relation to currents. As would soon be typical of him, he attempted to distill the entire conception into a simple set of basic theorems. The two theorems presented below were a direct attempt to express Faraday's reciprocal dynamics of the linked rings in the language of quantities and intensities.

Here and throughout the remainder of this article, I have added in brackets modernized equations for the benefit of mathematical readers. Table 1 shows the stages of development of these equations and compares them with present-day Maxwell's equations. In translating the mathematics I have consistently used the terms quantity and intensity with the notations $\mathbf{Q}$ and $\mathbf{I}$ in order to maintain a focus on symmetry (and missing symmetry) between electric and magnetic descriptions. This corresponds to Maxwell's first image and to the continuing geometrical significance of his variously denoted field variables, which he ultimately grouped under fluxes (quantities) and forces (intensities) (6, 7). Although the evolution of Maxwell's equations can be conveniently expressed mathematically, nothing in the present discussion depends on the bracketed equations, and they should not be allowed to obscure the primary power of Maxwell's imagery. In his initial letter to Thomson, Maxwell himself provided no equations for

(a)

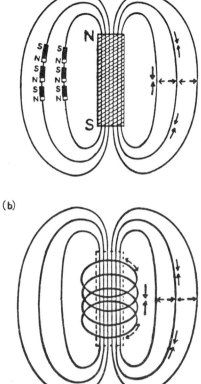
(b)

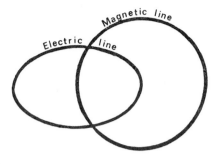

Fig. 4 (left). (a) Dynamics of Faraday's magnetic lines of force. Each line tends to contract along its length, but adjacent lines repel laterally. The result is a dynamic balance. (b) Bundle of current loops and associated lines of magnetic force. When the two sets of lines are taken as a system it is apparent that electric and magnetic effects are reciprocal. Longitudinal contraction in the magnetic lines has the same effect as lateral attraction between current lines, and lateral repulsion between magnetic lines has the same effect as longitudinal extension of current lines.　　Fig. 5 (right). Faraday's symbol of oneness between electric and magnetic axes of power. This would soon become the central image of Maxwell's first paper on electromagnetism, where he would label it mutually embracing curves.

his theorems, which is an essential part of the story. That it is essential provides a general clue, I believe, to the way in which Maxwell developed his later arguments. With this in mind, we may return to the original dynamics of electric and magnetic lines.

A natural measure of lateral attraction of contiguous lines of magnetic force was the quantity or number of the lines crossing a unit area, $\mathbf{Q}_m$, and a natural measure of the total tendency of a line to contract longitudinally was the sum of intensity $\mathbf{I}_m$ along the line. Through a translation of this kind, I suggest, Faraday's identity between the longitudinal contraction of a magnetic line and the lateral attraction $\mathbf{Q}_e$ of any current lines linked through it became Maxwell's first theorem, which I express in the slightly simplified and retrospective form:

1) The magnetic intensity summed around a closed loop is measured by the total quantity of current through the loop (8).

$$\left[\oint \mathbf{I}_m \cdot d\lambda \propto \iint \mathbf{Q}_e \cdot d\sigma \right]$$

where $d\lambda$ is a line element and $d\sigma$ is a surface element.

That much might be seen as merely a result of Ampère's work on the relation of magnetic forces to currents. Now

called Ampère's law, the relation had been utilized and extended by both Gauss and Thomson well before Maxwell. However, they employed only the single theorem, for force to them had only a single aspect. Maxwell's use of a second theorem has no precedent in any work but Faraday's. If I am correct, the second theorem was a loose attempt, following the associations of theorem 1, to express an identity between lateral attraction of magnetic lines passing through any area and longitudinal extension of any current line surrounding the area, thereby completing the symmetry of Faraday's linked rings (again simplified):

2) The total magnetic quantity through any surface is measured by the current around its edge.

$$\left[\oint (\text{current?}) \cdot d\lambda \propto \iint \mathbf{Q}_m \cdot d\sigma \right]$$

To the degree that these two theorems were the foundation of Maxwell's electromagnetic theory, and it was a very large degree, his deepest insight was to have made two laws out of what had been one. That move was more than merely reminiscent of Fourier's transformation of heat conduction; it was much the same transformation, here arrived at by reinserting Faraday's quantity-in-

Table 1. Forms of Maxwell's equations (abbreviated set) as they evolved from his initial letter to Thomson in 1854 to the "Electromagnetic theory of light" in 1868. The problem of energy and its relation to moving forces is omitted. The last column translates quantities and intensities into present-day symbols.

1854 (reconstruction of verbal description)		1855 (here limited to the differential forms)	1868 (for source-free space)	Present (including sources)
		$Q_e' = k_e'I_e$	$D = \epsilon E$	
Q_m, I_m (distinguished geometrically)	Law II	$Q_m = k_mI_m$	$Q_m = k_mI_m$	$B = \mu H$
$Q_e = k_eI_e$ (simplified)	Law IV	$Q_e = k_eI_e$		$j = \sigma E$
$\nabla \cdot Q_m = 0$ (implicit)		$\nabla \cdot Q_m = 0$	$\nabla \cdot Q_m = 0$	$\nabla \cdot B = 0$
$\nabla \cdot Q_e = 0$ (closed currents)		$\nabla \cdot Q_e = 0$	$\nabla \cdot Q_e' = 0$	$\nabla \cdot D = 4\pi e$
$\oint I_m \cdot d\lambda = \iint Q_e \cdot d\sigma$	Law III	$\nabla \times I_m = Q_e$	$\nabla \times I_m = 4\pi \frac{\partial Q_e'}{\partial t}$	$\nabla \times H = 4\pi j + 4\pi \frac{\partial D}{\partial t}$
$\oint (current?) \cdot d\lambda = \iint Q_m \cdot d\sigma$	Law I	$\nabla \times I_0 = Q_m$		$\nabla \times A = B$
$\oint I_e \cdot d\lambda = -\frac{\partial}{\partial t}\iint Q_m \cdot d\sigma$	Law VI	$I_e = -\frac{\partial I_0}{\partial t}$	$\nabla \times I_e = -\frac{\partial Q_m}{\partial t}$	$\nabla \times E = -\frac{\partial B}{\partial t}$

tensity physics into Thomson's mathematical heat analogy for electric and magnetic forces.

Now two observations can quickly be made regarding the second theorem: first, it is wrong as stated, since the magnetic quantity would depend on the size and shape of the loop and not only on the current; second, the usage of "current around the edge" cannot be quite "intensity of the current, I_e, summed around the edge" (as it would have to be for symmetry with theorem 1), because magnetic quantity has no direct relation to current intensity, or tension. Maxwell was therefore either not following Faraday's dynamic reciprocity, or he was in something of a predicament concerning its exact mathematical expression. All the evidence points to the latter explanation. I suggest that we are seeing here, in the loose verbal statements of the theorems, the initial metaphorical usage of an image which, if stated precisely, Maxwell would immediately have seen to be false. Once used loosely, however, as a means for visualizing a complex physical situation, it suggested creative new ways of treating the old problem.

A Formalized Mutual Embrace

Only a little more than 1 year after his theoretical sketch, Maxwell had generated, in his first paper on electricity and magnetism, a new structure. He did so by inventing a new mathematical description to fulfill Faraday's unifying image of linked rings. In fact, he now raised Faraday's captivating but rather dry description of the linked axes of power to a completely new level of metaphorical appeal with the term "*mutually embracing curves*" (9, 10). The process of Maxwell's creation is of considerable interest for the history of science because it illustrates simply and clearly the importance of pure *imag*ination. Maxwell was at this time neither a sophisticated mathematician (although his talent was great) nor an experimentally or theoretically learned electrician. The mathematics he employed came entirely from Thomson's recent work in electricity and magnetism—to such a degree that Maxwell hesitated to continue his own efforts, thinking Thomson must have "the whole draught of the thing lying about in loose papers" (*11*).

We recall that the mutually embracing curves symbolized two sorts of relations between the curves: a conduction description for each, involving lateral quantities and longitudinal intensities, and dynamic reciprocity, interrelating lateral and longitudinal forces. Maxwell's seminal paper "On Faraday's lines of force," was in two parts, which mirror these two relations. In part I he developed at length an analogy for lines of force in terms of fluid flow through a resistive medium, to make the conduction picture for forces as intuitively lucid as the traditional action at a distance description. In part II he redeveloped the reciprocity of electric currents and magnetic lines; and here, at the crucial juncture where his former theorem 2 had been inadequate, he showed how a new intensity could be defined, consistent with the mathematics of continuous flow, that would complete the missing symmetry of his two former theorems.

Theorem 1, Ampère's law, stated that the sum of current quantity through any surface is equal to the sum of magnetic intensity around the bounding edge. Mathematically, Maxwell now realized, that relation depended only on the current forming a continuous closed loop [$\nabla \cdot Q_e = 0$]. Theorem 2 was an attempt at a reciprocal equation for magnetic quantity through any surface, which should have been related to a longitudinal property of current around the edge, but it could not be current intensity as usually defined. Maxwell simply postulated that an intensity of the required kind should exist, since the only mathematical condition for it was that the magnetic quantity behaved as a continuous flow [$\nabla \cdot Q_m = 0$]. The analogical conduction description of forces in part I, therefore, as applied to closed flow circuits of both magnetic force and electric current [$\nabla \cdot Q_m = 0$; $\nabla \cdot Q_e = 0$], guaranteed the possibility of reciprocity in the mutually embracing curves of part II. Adopting a long-standing label of Faraday's for a supposed electromagnetic condition of matter, an electrotonic state, Maxwell called his new invention the electrotonic intensity. It was not yet a physical state, but it could nevertheless be imagined physically (*9*, p. 205):

We may conceive of the electro-tonic state at any point of space as a quantity determinate in magnitude and direction, and we may represent the electro-tonic condition of a portion of space by any mechanical system which has at every point some quantity, which may be a velocity, a displacement, or a force, whose direction and magnitude correspond to those of the supposed electro-tonic state. This representation involves no physical theory, it is only a kind of artificial notation.

With the seemingly small addition of this new intensity, Maxwell completed mathematically his earlier attempt at reciprocal description of current lines and magnetic lines in electromagnetic space.

(We would say today that he had invented the vector potential, as Maxwell himself later called the electrotonic state.) In order to capture the accomplishment, he returned again to a set of simple laws, with two laws now for conduction symmetry and two for reciprocity (9, p. 206):

Law I. The entire electro-tonic intensity [I₀] round the boundary of [any] surface measures the quantity of magnetic induction which passes through that surface, or, in other words, the number of lines of magnetic force which pass through that surface. [Compare the former theorem 2.]

$$\left[\oint \mathbf{I}_0 \cdot d\boldsymbol{\lambda} = \iint \mathbf{Q}_m \cdot d\boldsymbol{\sigma} \text{ or } \nabla \times \mathbf{I}_0 = \mathbf{Q}_m \right]$$

Law II. The magnetic intensity at any point is connected with the quantity of magnetic induction by a set of linear equations, called the equations of conduction [$\mathbf{Q}_m = k_m \mathbf{I}_m$].

Law III. The entire magnetic intensity round the boundary of any surface measures the quantity of electric current which passes through that surface. [Compare the former theorem 1.]

$$\left[\oint \mathbf{I}_m \cdot d\boldsymbol{\lambda} = \iint \mathbf{Q}_e \cdot d\boldsymbol{\sigma} \text{ or } \nabla \times \mathbf{I}_m = \mathbf{Q}_e \right]$$

Law IV. The quantity and intensity of electric currents are connected by a system of equations of conduction [$\mathbf{Q}_e = k_e \mathbf{I}_e$].

Although these four laws express the embracing curves, they are completely general and do not depend on following lines of force themselves around closed loops. Any embracing loops imagined in a space of currents and magnetic lines will yield the same relations as a physical current loop and a definite magnetic line. The theorems represent the structure of a constant electromagnetic field in any region, no matter how small.

Note that the conduction equations and the reciprocity equations express two different sorts of symmetry. The electrotonic intensity stands by itself in law I without relation to either conduction equation. In that sense the descriptions of electric and magnetic lines are not entirely symmetrical. The reader knowledgeable in electromagnetic theory will note further that the symmetries expressed here are not at all those familiar from the present-day Maxwell's equations, and since they apply only to the steady state, they have nothing to do with electromagnetic induction (12). Induction became the centerpiece of Maxwell's theory only at a later stage of development. It did have a prominent place even here, however, although not in the symmetries.

The phenomenon of a current being produced in a closed conductor when the magnetic quantity through the loop changed—electromagnetic induction— was Faraday's first major discovery and

has always been his most famous one. By 1850 no credible electrical theory could fail to subsume it. Maxwell, therefore, from the beginning of his work in 1854, had been concerned to describe current induction along with the general relations of electric and magnetic curves. Loosely following Faraday, he had expressed the effect originally as follows (13):

The electromotive force along any line [the driving intensity tending to produce a current along the line] is measured by the number of lines of [magnetic force] which that line cuts in unit of time. Hence the electromotive force round a given circuit depends on the decrease of the number of lines which pass through it in unit of time.

$$\left[\oint \mathbf{I}_e \cdot d\boldsymbol{\lambda} \propto -\frac{\partial}{\partial t} \iint \mathbf{Q}_m \cdot d\boldsymbol{\sigma} \right]$$

There was a problem with this description, however, and one recognized by Faraday long before Maxwell. How could the magnetic quantity merely passing through the loop affect the electromotive force at the loop? Would that not be action at a distance all over again? "It is natural," observed Maxwell in 1855, "to suppose that a force of this kind, which depends on a change in the number of lines, is due to a change of state which is [merely] measured by the number of these lines" (9, p. 187). That was Faraday's reasoning in originally proposing the electrotonic state. A wire in a magnetic field would supposedly be in the peculiar state, and when the field was removed, the collapse of the state would appear as an induced current. Maxwell's new mathematical expression for an electrotonic state fit the requirement precisely. It provided "the means of avoiding the consideration of the quantity of magnetic induction which *passes through* the circuit" (9, p. 203). Law VI, following those quoted above, expressed the relevant relation (14):

Law VI. The electro-motive force on any element of conductor is measured by the instantaneous rate of change of the electro-tonic intensity on that element, whether in magnitude or direction.

$$\left[\mathbf{I}_e = -\frac{\partial \mathbf{I}_0}{\partial t} \right]$$

Applied to a closed circuit, that meant "the electro-motive force in a closed conductor is measured by the rate of change of the entire electrotonic intensity round the circuit referred to unit of time" (9, p. 207).

$$\left[\oint \mathbf{I}_e \cdot d\boldsymbol{\lambda} = -\frac{\partial}{\partial t} \oint \mathbf{I}_0 \cdot d\boldsymbol{\lambda} \right]$$

Only the electrotonic intensity in the circuit itself was now in direct action.

Electromagnetic Propagation and Light

With the introduction of an electrotonic state, Maxwell had attained what seemed to him at the time a unified geometrical, or structural, description of electromagnetism, where the structural emphasis was on the steady-state relations of currents and magnetic forces. But the attainment brought a series of new problems, including the physical nature of the electrotonic state; how it was transmitted from its origin in a magnet or current through nonconducting spaces; and how, if it arrived at a secondary conducting circuit, it induced a current there. For Maxwell, following the trend of British mathematical natural philosophy, these problems suggested an answer based on a mechanical model of the luminiferous ether. That new focus would shift dramatically the direction of his thinking, and would change ultimately even the content symbolized by the mutually embracing curves. The image itself, however, would maintain its symbolic form and guiding role.

To see the problem a little more clearly, and in one of the ways Maxwell saw it (15, 16), consider two conducting loops, one connected to a battery and containing a switch that is to be closed at the beginning of a thought experiment. According to Maxwell's analysis of 1855, developed above, the initial increase of current would produce an increasing electrotonic state everywhere in the surrounding space which, when it reached the second circuit, would appear as an induced current. The process could be described macroscopically as resulting from an increasing linkage of magnetic lines through the two circuits, but that image conveyed no sense at all, not even structurally, either of the physical nature of the electrotonic state or of its mode of propagation through the nonconducting space between two circuits. The image of linkage, furthermore, did not apply to the peculiar case of induction in open circuits, where the effect could only be an electrostatic tension between the open ends of the conductor. How then were electrostatic force and electrotonic intensity related?

Maxwell had been concerned about these problems in his 1855 paper, where he had expressed the hope of providing later a mechanical representation of the electrotonic state. He had even supplied some formal mathematics applicable to open circuits, but had explicitly limited his application to closed conducting loops, with the excuse that "we know little of the magnetic effects of any currents which are not closed." Even while

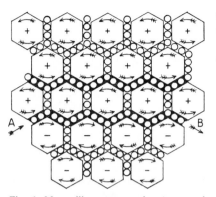

Fig. 6. Maxwell's representation (corrected slightly) of an electromagnetic ether in and surrounding a current-carrying conductor. The section is transverse through the vortex cells but longitudinal through the current, which proceeds from *A* to *B* as an actual translation of idler particles. By tangential action the cells adjacent to the current are set in vortex rotation, which in turn exerts a tangential action on the next adjacent idler particles. If free to translate, the latter particles will roll along between the rows of cells and will constitute an induced current; if not free, they will rotate in place, exerting a tangential action on the next adjacent cells and setting them in vortex rotation. Thus, the magnetic lines (linear vortices) will propagate through space.

thus distinguishing open and closed circuits, he had nevertheless believed for some time, following Faraday, that electrostatic induction (tension) and current conduction were the same, differing only in degree of resistivity of the medium: "Thus the analogy between statical electricity and fluid motion turns out more perfect than we might have supposed, for there [in statics] the induction goes on by conduction just as in current electricity, but the quantity conducted is insensible owing to the great resistance of the dielectrics" (*17*). Could it be, then, that electrostatic induction took place in closed loops; that an incomplete circuit for currents was actually closed by static induction in the open space? Maxwell apparently did not see the problem in quite that explicit form in 1855, or doubtless he would have carried it further, but many of the elements of such a conception were lying in readiness for a burst of new vision, and were surely a powerful motivation for its eventual appearance. First, however, he would have to connect the problem of magnetic propagation between electric circuits with electrostatic action, while simultaneously both distinguishing current conduction from electrostatic induction and equating their magnetic effects. That complex of relations would form a new basis of interpretation for the mutually embracing curves, this time as a structural image for

the physical process of propagation itself.

Maxwell's flurry of fresh insights arose in 1861, when he made good his promised mechanical representation of magnetic lines, currents, and the electrotonic state. In "On physical lines of force" (*18*), he presented his notorious vortex model of the magnetic ether, using the diagram shown in Fig. 6. Each line of force was the axis of a vortex filament. For adjacent vortices to have the same sense of rotation about their axes, thereby representing lines of force in the same direction, they were supposed to be contained in cellular regions separated by "idler" particles, the particles being in rolling contact with the cells. These idler particles constituted electricity. They were free, in all substances, to rotate in place between rotating cells, but in conductors they could also translate between cells, with more or less resistance. A stream of them would constitute a current which, because it exerted a tangential action on adjacent cells, would set them in rotation as vortices, or magnetic lines, surrounding the current. The next adjacent idler particles would then also be set rolling by tangential action of these first vortex cells (constituting electromotive force) and if the particles could not translate (being in a nonconductor) they would set the next adjacent cells in rotation, and so on out into space in the vicinity of the initiating current. Lines of magnetic force would thereby be formed surrounding the current at larger and larger radii. Whenever the propagating effects reached a closed conductor the idler particles would simply translate while rolling against the next adjacent cells, thereby constituting an induced current in the opposite direction from the original. Any resistance to translation, however, would eventually set those next adjacent cells in rotation, and the induced current would be stopped, thus reproducing the observed phenomena of electromagnetic induction.

Considering this strange machinery as a heuristic model, valuable for showing the possibility of a mechanical theory, Maxwell proceeded to show that it reproduced exactly the required relations between currents, magnetic force, and the electrotonic state. And the electrotonic state had now a simple mechanical interpretation as rotational momentum of the vortices. One problem remained, however, in the initial description: what exactly was the mechanism of tangential action that transmitted rotation from idler particles to

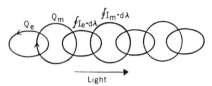

Fig. 7. The mutually embracing curves reconstituted as the basis of propagation of electromagnetic effects. Curve Q_e represents an actual line of current that is increasing in strength and curve Q_m represents an actual magnetic line, also increasing in strength. The remaining curves are imaginary lines around which electric and magnetic intensities are summed.

cells, and from the exterior of cells through their interior? The vortices had to be fluid in order to produce the correct dynamic behavior of magnetic lines, but they could not be perfectly fluid because then no tangential action could occur in their interiors. Neither could the connection between idler particles and cell walls be perfectly rigid in normally nonconducting space, for then the starting and stopping of induced currents would be instantaneous, to say nothing of the problem of motion of normal bodies through such a system. To avoid these problems, apparently, as well as to avoid dissipation of energy in a merely viscous medium, Maxwell proposed in a later installment that the cells were elastic and that they interacted elastically with the idler particles (*19*). Some such mechanism would explain conveniently propagation from point to point throughout the ether. It would do more, however, and that was soon to be the basis for transformation of the imagery we have been discussing. It would provide an explanation of electrostatic effects in nonconductors and in open circuits.

Maxwell by now realized that static induction could not be simply weak conduction, for equally good insulators showed widely varying capacities for static induction. But if static electric charge were merely elastic displacement of idler particles, whereas currents involved extended motion, the difficulty would be resolved. The displacement, Q_e', while it was occurring, would constitute a brief current, a displacement current, $\partial Q_e'/\partial t$. And just as the effect of a changing magnetic field, when propagated to a closed conductor, would be an induced current, so the process of propagation itself would take place by successive induction of displacement currents. Each such induction by a magnetic vortex would reinduce a second vortex and so on through space. With that famous invention, the essentials were complete for Maxwell's remarkable synthesis of all contemporary electricity and magnetism. To set the capstone, he showed that the propagation of magnetism by electric displacement would occur at the known velocity of light. His electromagnetic theory of light was immediate: "We can scarcely avoid the inference that *light consists in the transverse undulations of the same medium which is the cause of electric and magnetic phenomena*" (*16*, p. 500).

The Mutual Embrace Reconstituted

We have seen that Maxwell's initial goal in employing the mutually embracing curves was a macroscopic description, in the steady-state case, of the relation between a magnetic field and its associated electric currents. Only secondarily was electromagnetic induction a consideration. If the constant-field description led naturally to an explanation of current induction, it was, at that level of analysis, satisfactory; and the image of closed linked rings beautifully fit the criterion. Nevertheless, the problem of providing a mechanical vortex description of a changing field, and of the communication of its effects, had refocused Maxwell's attention on induction as the fundamental aspect of a field description. Any change in the magnetic field would propagate by successive, and reciprocal, inductions of displacement currents and magnetic lines.

Use of the term reciprocal introduces the last stage of Maxwell's analysis that will be considered here. Dynamic reciprocity had always been the foundation, for both Faraday and Maxwell, of the mutual embrace of electric currents and magnetic lines, but a new sense of that reciprocity now occurred to Maxwell and has since been taken as the symbol of his system. Any changing magnetic quantity would induce a net electromotive force (summed electric intensity) in any curve surrounding it. The response at any point in the medium to the electric intensity would be a displacement current, or electric quantity $[Q'_e = k'_e I_e]$, obeying continuity of flow $[\nabla \cdot Q'_e = 0]$. That changing electric quantity, similarly, would induce a net magnetic intensity in any curve surrounding it, and the response at any point in the medium would be again a continuously flowing magnetic quantity $[Q_m = k_m I_m; \nabla \cdot Q_m = 0]$. As the process repeated itself the initial disturbance would propagate in all directions in such a way as to preserve the recipro-

cal relation between electric and magnetic sums over arbitrary closed curves. Thus the mathematical relation of mutual embrace would itself propagate through space. This might be represented as the chain of Fig. 7, although it is important to recognize that the links in the chain could represent actual lines of force only in very special circumstances. Since any two linked curves could be imagined as small as one pleased, the mutual embrace provided a structural representation for the overall process of propagation from point to point throughout the field. With that view, the mechanical model became less immediately significant. At the geometrical level mechanisms in the medium gave way to structure in the field, and the electrotonic state concomitantly sacrificed its prominent place to the displacement current.

True to his style, Maxwell encapsulated his new conception in a set of simple theorems, which once again defined the reciprocity of electric and magnetic action. They are most clearly set out in the form of embracing curves in an 1868 "Note on the electromagnetic theory of light" (*20*). The new theorems should be compared directly with the laws stated above in order to capture the transformation in content. Law III appears here virtually unchanged, but for the explicit term embracing:

Theorem A. If a closed curve be drawn embracing an electric current, then the integral of the magnetic intensity taken round the closed curve is equal to the current multiplied by 4π.

$$\left[\oint \mathbf{I}_m \cdot d\boldsymbol{\lambda} = 4\pi \iint \mathbf{Q}_e \cdot d\boldsymbol{\sigma}\right]$$

The intent, however, of the old theorem is radically altered, for Maxwell is now not so much concerned with current as with displacement current.

Theorem D. When the electric displacement increases or diminishes, the effect is equivalent to that of an electric current in the positive or negative direction.

$$\left|\frac{\partial \mathbf{Q}'_e}{\partial t} \text{ replaces } \mathbf{Q}_e\right|$$

In other words, if the total electric displacement through any closed curve varies, the rate of change will measure the total magnetic intensity around the curve.

$$\left[\oint \mathbf{I}_m \cdot d\boldsymbol{\lambda} = 4\pi \frac{\partial}{\partial t} \iint \mathbf{Q}'_e \cdot d\boldsymbol{\sigma}\right]$$

The relation reciprocal to this is no longer the old law I, connecting magnetic quantity and electrotonic intensity in the steady state, but law VI, on induction, and in the circuital form that had been secondary:

Theorem B. If a conducting circuit [following theorem D, any closed curve in a dielectric] embraces a number of lines of magnetic force [magnetic quantity], and if, from any cause whatever, the number of lines is diminished, an electromotive force will act round the circuit, the total amount of which will be equal to the decrement of the number of lines of magnetic force in unit of time.

$$\left[\oint \mathbf{I}_e \cdot d\boldsymbol{\lambda} = -\frac{\partial}{\partial t} \iint \mathbf{Q}_m \cdot d\boldsymbol{\sigma}\right]$$

The mutually embracing curves have been here reconstituted as mutually inducing curves of magnetism and electric displacement. Two additional equations, of conduction, relate magnetic quantity to magnetic intensity, and electric displacement (quantity) to electromotive force (intensity). Thus the reciprocity of the new curve is complete in a way in which the earlier image was not, for the quantities and intensities of the conduction relations are the same as those reciprocating in the induction relations.

Table 2. Comparison of Maxwell's equations in the absence of sources in 1868 and at present.

1868	Present
$\mathbf{Q}'_e = k'_e \mathbf{I}_e$	$\mathbf{D} = \epsilon \mathbf{E}$
$\mathbf{Q}_m = k_m \mathbf{I}_m$	$\mathbf{B} = \mu \mathbf{H}$
$\oint \mathbf{I}_e \cdot d\boldsymbol{\lambda} = -\frac{\partial}{\partial t} \iint \mathbf{Q}_m \cdot d\boldsymbol{\sigma}$	$\oint \mathbf{E} \cdot d\boldsymbol{\lambda} = -\frac{\partial}{\partial t} \iint \mathbf{B} \cdot d\boldsymbol{\sigma}$
or $\nabla \times \mathbf{I}_e = -\frac{\partial \mathbf{Q}_m}{\partial t}$	or $\nabla \times \mathbf{E} = -\frac{\partial \mathbf{B}}{\partial t}$
$\oint \mathbf{I}_m \cdot d\boldsymbol{\lambda} = 4\pi \frac{\partial}{\partial t} \iint \mathbf{Q}'_e \cdot d\boldsymbol{\sigma}$	$\oint \mathbf{H} \cdot d\boldsymbol{\lambda} = 4\pi \frac{\partial}{\partial t} \iint \mathbf{D} \cdot d\boldsymbol{\sigma}$
or $\nabla \times \mathbf{I}_m = 4\pi \frac{\partial \mathbf{Q}'_e}{\partial t}$	or $\nabla \times \mathbf{H} = 4\pi \frac{\partial \mathbf{D}}{\partial t}$
$\nabla \cdot \mathbf{Q}'_e = 0$	$\nabla \cdot \mathbf{D} = 0$
$\nabla \cdot \mathbf{Q}_m = 0$	$\nabla \cdot \mathbf{B} = 0$

That is the tight symmetry familiar in our present Maxwell's equations in the absence of charges or currents, as shown in Table 2.

Conclusion

We have followed the evolution of a single image through mutliple stages in Maxwell's conceptualization of the electromagnetic field. The mutually embracing curves served to anchor his ideas, from start to finish, on a geometrically solid framework to which all speculation could be tied. At the same time the structure itself suggested new paths of investigation—through the flow analogy, for example, and through reciprocal dynamics. A consistent mathematical formulation for the electrotonic state was the first result, with strong hints of more to come. The extensions emerged as the displacement current and the concomitant picture of propagation of electromagnetic effects through a medium, all firm ground for an electromagnetic theory of light. But the new perceptions, reflected back on the old imagery, required its reformulation. With propagation at the center of attention, the mutual embrace of electricity and magnetism was reconstituted as mutual induction—still a mutual embrace, but productive now of offspring.

A somewhat different summary of Maxwell's work arises from reflecting on the usual reconstruction of his insights. It has often been said that Maxwell's contribution to electromagnetic theory consisted fundamentally in adding the displacement current $[4\pi\, \partial\mathbf{D}/\partial t]$ to Ampère's law $[\nabla \times \mathbf{H} = 4\pi\mathbf{j}]$ and that he was motivated either by the need to preserve charge conservation in open circuits or by his perception of a need to complete the symmetry with Faraday's law of induction $[\nabla \times \mathbf{E} = -\partial\mathbf{B}/\partial t]$. There is a sense in which each of these

claims is true, but as the analysis above has shown, they miss entirely the historical core of Maxwell's thinking. He did add the displacement current, but far more fundamental was his expression of Ampère's law with a quantity-intensity distinction built in. He was concerned with charge conservation in open circuits, but before that problem could even arise explicitly he had to distinguish between conduction and static induction. He was deeply concerned with the symmetry between electricity and magnetism, but his original symmetry did not involve electromagnetic induction at all; instead he followed Faraday's dynamic reciprocity between a constant current and its associated constant magnetic force. Completing the symmetry of these mutually embracing curves produced a second form of Ampère's law and an electrotonic state [mathematically our vector potential $\nabla \times \mathbf{A} = \mathbf{B}$]. The displacement current arose not from symmetry considerations but from Maxwell's attempt to elaborate mechanically the meaning of the electrotonic state, especially for open circuits and for the transmission of magnetic effects to produce current induction. Only with the displacement current in full view did Maxwell reformulate his conception of mutual embrace on the basis of electromagnetic induction rather than steady-state currents. Thus the present-day symmetry of Maxwell's equations derived from the role of the displacement current, rather than the displacement current arising from symmetry.

References and Notes

1. For a discussion of the intransigence of the Laplacians see J. Herivel, *Joseph Fourier: The Man and the Physicist* (Clarendon, Oxford, 1975), especially pp. 153–159.
2. The context directly relevant here is discussed in M. N. Wise, thesis, Princeton University (1977); University Microfilms 77-14, 252, pp. 25–32.
3. T. S. Kuhn in *Critical Problems in the History of Science* [M. Clagett, Ed. (Univ. of Wisconsin Press, Madison, 1969), pp. 321–356] has given a lucid survey of interconversion, and other problems suggesting energy conservation.
4. M. Faraday, *Experimental Researches in Electricity* (three volumes, London, 1834, 1844, 1855), vol. 3, paragraph 3268; the diagram accompanies paragraph 3265.
5. From a letter of 13 November 1854 reprinted in "The origins of Clerk Maxwell's electric ideas, as described in familiar letters to William Thomson," *Proc. Cambridge Philos. Soc.* **32**, 702 (1936).
6. J. C. Maxwell, "On the mathematical classification of physical quantities," *Proc. London Math. Soc.* **3** (1870); reprinted in (7), vol. 2, pp. 257–266; see p. 261.
7. _____, *The Scientific Papers of James Clerk Maxwell* (Cambridge Univ. Press, Cambridge, 1890).
8. This simplified version of Maxwell's theorem underestimates the complexity of temporal evolution in his thought. He had not distinguished physically, at this stage, between quantity and intensity. He had distinguished only two geometric aspects, surface and linear, of magnetic force conceived as a condition of space, magnetic polarization. This geometrical mode of viewing physical entities was very deep in Maxwell's perception generally.
9. J. C. Maxwell, "On Faraday's lines of force: Parts I and II," *Trans. Cambridge Philos. Soc.* **10** (1856).
10. See (7), vol. 1, pp. 184, 194, and 194n, where "embracing" is twice italicized. Not only here but repeatedly in later papers Maxwell employed the metaphor of "embrace."
11. Letter of 13 September 1855, in (5), p. 712.
12. Much the same point has been made by C. W. F. Everitt in his excellent article on Maxwell for the *Dictionary of Scientific Biography* (Scribner's, New York, 1974), vol. 9, pp. 206–207; reproduced with additions as *James Clerk Maxwell: Physicist and Natural Philosopher* (Scribner's, New York, 1975), pp. 90–93.
13. Letter of 13 November 1854, in (5), p. 703.
14. I have omitted law V, on the force between currents, in order to avoid an extended discussion of Maxwell's energy considerations, which are not directly essential for the image of embracing curves.
15. See J. C. Maxwell, "On physical lines of force: Parts I and II," *Philos. Mag.* **21** (1861), figure 3; (7), vol. 1, p. 488 and the description beginning on p. 477.
16. Also see "On physical lines of force: Part III," *Philos. Mag.* **23** (1862); (7), vol. 1, pp. 489–491.
17. This statement is from part I of (9), p. 181, where it is largely unexamined; the preceding phrase is from part II of (9), p. 196, which was written several months later. Maxwell may thus have begun in part II to see some difficulties with his earlier thoughts, but having once seen current conduction and static induction as the same, he was all the more likely to invent a new conception to fulfill the old vision.
18. (15), pp. 451–488; the vortex description is on p. 477. Maxwell's diagram requires a slight correction as pointed out in Everitt's book (12), pp. 95–96.
19. (16), pp. 489–502. A fourth part concluded the work.
20. Appended to "On a method of making a direct comparison of electrostatic with electromagnetic force," *Philos. Trans. R. Soc. London* **158** (1868); (7), vol. 2, pp. 125–143, theorems quoted from pp. 138–139.
21. I thank J. G. Burke and E. S. Abers for helpful comments and C.W.F. Everitt for a significant and detailed critique.

Kinetic atom

David B. Wilson

Departments of History and Mechanical Engineering, Iowa State University, Ames, Iowa 50011
(Received 26 June 1980; accepted 18 November 1980)

In the middle of the 19th century, physicists abandoned the idea of heat as a special kind of matter in favor of the idea that heat was a "mode of motion" of ordinary matter. From this "dynamical" view of heat, they proceeded to a more precise "kinetic" theory and, along with it, the idea of the kinetic atom. This paper will survey the research of men like Joule, Kelvin, Maxwell, Clausius, and Boltzmann, as it comments on the basic conceptual issues involved in this area of the history of physics.

The concept of a kinetic atom refers most specifically to the development of the kinetic theory of gases by Clausius, Maxwell, and Boltzmann beginning in the late 1850s. But, in one way or another, the idea involved a great deal of nineteenth-century physics—including physicists' specific scientific theories as well as their methodologies or philosophies of science. And, in the minds of some—perhaps most notably, Lord Kelvin—the real kinetic theory penetrated much more deeply than the kinetic theory of gases. Consequently, in this paper, I shall try to suggest something of the significance of the kinetic theory of gases by surveying physicists' ideas throughout the century and by dicussing Kelvin's own ideas at some length.

In early nineteenth-century Europe, physical research was centered in France. And French physicists developed at least two competing methodologies—one which insisted that theories incorporate statements about the smallest particles of matter, the second insisting that they do not.[1]

Linked with Laplace and Poisson, the first viewpoint has been called the "astronomical" view of nature. Just as Newton had explained the universe in terms of large chunks of matter interconnected by the universal force of gravitation, so also, according to this viewpoint, microscopic nature should be explained in terms of material particles and their action-at-a-distance forces. The forces could be attractive or repulsive, and they acted only over short, "insensible" distances. The material particles included both particles of ordinary, or ponderable, matter and particles of the various imponderable fluids used at the time to explain heat, electricity, magnetism, and light. In any given situation, then, the general problem was to imagine a system of ponderable and/or imponderable particles interconnected by a force or forces which would account for observations.

The second methodology—associated with Fourier and the positivist philosopher August Comte—greatly deemphasized the relevance of all these unobservable particles and forces. At its extreme, this viewpoint insisted that such entities, because unobservable, were simply unknowable and, therefore, ought to be excluded from reliable, positive scientific theories.

By the middle of the century, the center of physics research was shifting— or had shifted—away from France especially to Britain. The leading members of the famous Cambridge "school" of physics—Green, Stokes, Kelvin, Tait, and Maxwell—took their degrees from Cambridge between 1837 and 1854. Largely in reaction against the competing French methodologies, they developed a middle ground—a highly flexible methodology of cautious realism coupled with the use of mechanical analogies or models.[2]

They were *realists* in their rejection of the positivism associated with Fourier and Comte; but they saw themselves as more *cautious* than Laplace and Poisson in incorporating unobservables into their theories. It seemed that man's *ability* to know unobservables demanded an extra responsibility in discussing them. Thus when one lacked sufficient evidence, one either emphasized a phenomenological approach or turned to models to help probe the unseen. Because the microscopic realm behaved in accordance with Newtonian laws of motion, models yielded, by analogy, insights into nature's hidden workings, though without necessarily providing detailed pictures of them.

The approach is evident in Stokes's research on the viscosity of fluids in the 1840s, which led to Stokes's law concerning the resistance of a viscous fluid to the motion of a sphere moving through it. In contrast to positivists, Stokes *was* willing to discuss the "ultimate molecules" of a fluid. In both a preliminary draft of his major paper on friction in fluids and in the paper itself, Stokes considered two different hypotheses regarding the molecules of a fluid. Both hypotheses involved the fact that, as he stated in his paper: "It is an undoubted result of observation that the molecular forces, whether in solids, liquids, or gases, are forces of enormous intensity, but which are sensible at only insensible distances."[3] Whether the molecules were separated from one another or whether they were smooth particles in actual contact, the results, Stokes declared, would be the same.[4] Hence, he felt his work contrasted with the Frenchman Navier's treatment of fluid friction in which repulsive forces between two molecules depended on the relative motion between them "so that," as Stokes explained, "two molecules repel each other less strongly when they are receding, and more strongly when they are approaching, than they do when they are at rest."[5] The strength of Stokes's own method was that, unlike Navier's astronomical approach, it, as Stokes said, "does not necessarily require the consideration of ultimate molecules."[6] That is to say, Stokes did not know enough about molecules to support a specific theory about them, but he did know enough to know that he could legitimately treat the fluid as a continuum. And because his results did not rest upon a specific molecular hypothesis, they would remain valid whatever the truth of the molecular situation might one day turn out to be. Stokes's research on fluid friction also helps illustrate the role of mechanical models, for he later explained his original idea in a letter to Kelvin:

> Suppose you had an Indian rubber bag, in tension, filled with smooth marbles, or at least marbles which were very nearly smooth. Then if you were to give it a

motion of distortion, every now and then a marble would start into a new position among its fellows, and the change would be accompanied by a sudden disturbance among the whole set . . . .[7]

Stokes related the shifting of marbles to fluid viscosity.

Maxwell, using a similar methodology to Stokes's, discussed particles of fluids in the 1860s in his kinetic theory of gases much more specifically than Stokes had in the 1840s. A major reason that he could do so was the discovery of the mechanical equivalent of heat. Supported by the experiments of Joule and others and generalized into the first law of thermodynamics by Kelvin and others, the conclusion that heat had a mechanical equivalent implied that heat was simply, as the phrase had it, a "mode of motion" of ordinary matter.[8] It was natural, therefore, to resurrect the theory that heat was the motion of microscopic particles of matter, and the kinetic theory of gases, then, followed from this basic concept.

A successful kinetic theory of gases resulted mainly from the efforts of Clausius, Maxwell, and Boltzmann from the late 1850s through the 1870s.[9] Clausius's first paper on the topic in 1857 expressed much of the framework within which his and later research was carried out. He assumed that particles of a gas were constantly in motion, that they were very small compared to the distances between them, and that their interactions with one another were of very short duration. Using these assumptions, he correlated the particles' motions with observed relations between the pressure, volume, and temperature of a gas.[10] When Clausius was criticized that a gas whose particles moved as fast as he claimed would diffuse through a given volume much more rapidly than actually observed, he responded with his concept of the mean free path of gas particles.[11] Maxwell employed many of Clausius's ideas, but introduced the notion that the velocities of gas particles were not all the same, but were distributed over a wide range of values.[12] Maxwell reached the unexpected conclusion that the viscosity of a gas ought to be independent of its density,[12] and experimental confirmation of this prediction was important in gaining acceptance for the kinetic theory. Boltzmann's work strengthened Maxwell's concept of the distribution of velocities and Maxwell's statistical interpretation of the second law of thermodynamics in the context of the kinetic theory of gases.[13] As Maxwell's famous demon (so named by Kelvin) illustrated, the dissipation (again Kelvin's term) of energy in a gas was exceedingly probable, but not necessary.[14]

But what about the gas particles themselves? Here Clausius, Maxwell, and Boltzmann were noncommittal. Clausius spoke of the "sphere of action" around a gas particle, but without saying exactly what the particle itself was like and without, for example, explicitly supporting Boscovichean atoms.[15] Maxwell and Boltzmann both thought the same results would follow whether one regarded gas particles as solid spheres or as centers of force. Maxwell, using spheres for convenience in 1860, seems to have viewed his system of many interacting spheres as a mechanical model, for he wrote: "If the properties of such a system of bodies are found to correspond to those of gases, an important physical analogy will be established, which may lead to more accurate knowledge of the properties of matter."[16] In 1866 Maxwell used a molecular force depending on the inverse fifth power of the distance. Although such a force fit more naturally with Boscovichean atoms, even here

Maxwell declared that the particles themselves could be "mere points, or pure centres of force . . . [or] small solid bodies . . . ."[17]

Consequently, as a result of the kinetic theory, one could say a good deal about the sizes and motions of particles of gases, but much less about the nature of the particles themselves. Also, it should be noted that the kinetic theory was not an unqualified success. In spite of the Maxwell–Kelvin–Boltzmann statistical interpretation, some were still bothered by the concept of irreversibility applied to a collection of particles which individually moved in accordance with Newton's laws of motion.[18] One of Kelvin's "clouds" over the dynamical view of nature at the end of the century concerned the failure of the Maxwell–Boltzmann equipartition theorem to deal with gases composed of polyatomic molecules.[19] Such difficulties apparently prevented universal assent to the kinetic theory in the 1890s.

At the same time the well-known kinetic theory was being developed, Kelvin was developing another, and potentially more fundamental, kinetic theory. The theoretical background to it included not only thermodynamics, but the wave theory of light and the Faraday effect as well. The wave theory of light established the existence of the luminiferous ether so securely that, contrary to what is sometimes thought, even the Michelson–Morley experiment of 1887 did not cause physicists to give it up.[20] As Kelvin wrote in the 1904 edition of his famous Baltimore Lectures, ". . . We must not listen to any suggestion that we are to look upon the luminiferous ether as an ideal way of putting the thing. A real matter between us and the remotest stars I believe there is, and that light consists of real motions of the matter, . . . motions in the way of transverse vibrations."[21] Dating from 1845, the Faraday effect, of course, showed that the plane of these transverse vibrations was rotated when light was transmitted through leaded glass within a magnetic field.[22] A final part of the background was Kelvin's own apparent willingness to consider ether and ponderable matter as, ultimately, not two different things at all. In an 1854 paper, for example, he thought that the ether of space was simply an extension of the Earth's atmosphere. As he wrote, "Whether or not this [luminiferous] medium [of space] is (as appears to me most probable) a continuation of our own atmosphere, its existence is a fact that cannot be questioned . . . ."[23] Hence, keeping in mind the well-established luminiferous ether, the Faraday effect, the dynamical theory of heat, and Kelvin's views of the possible identity of ether and matter, we can turn to Kelvin's own kinetic ideas in the late 1850s.

He imagined a perfect, frictionless fluid with solids ("motes" he called them) distributed through it. Though he did not think this conception could ever provide a complete theory of nature, he wrote in his notebook in January 1858 that "as a temporary mechanical illustration of some of the agencies hitherto looked upon as among the most inscrutable phenomena of inorganic physics it seems to me well worth while to work out something of the dynamics of a perfect liquid with motes . . . ."[24] The system depended on two dynamical propositions: (1) that the translational motion of such a mote through such a fluid would be converted to rotational motion of the mote; and (2) that such a rotating mote would transmit a pressure through the fluid repelling other motes. The model could represent an everyday liquid like water by illustrating the dissipation of the liquid's macroscopic motion into microscopic rotations of

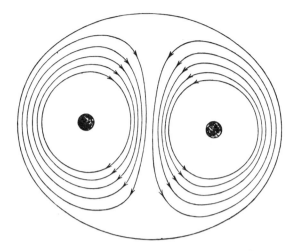

Fig. 1. From Kelvin's *Mathematical and Physical Papers* [28]: "In the smoke-rings which have been actually observed, [the ring's velocity] seems to be always something smaller than the velocity of the fluid along the straight axis through the centre of the ring [see the arrows in the diagram]; for the observer standing beside the line of motion of the ring sees, as its plane passes through the position of his eye, a convex outline of an atmosphere of smoke in front of the ring." Hence, the diagram can represent the cross section of a vortex atom, with the outermost line separating ether within the atom from ether surrounding the atom.

motes—that is, into heat. With the mutual repulsions of large numbers of rotating motes, Kelvin wrote in his notebook, "the elasticity of solids, of real liquids, and of gases, may possibly come to be illustrated."[25] And as he wrote in a letter to Stokes, "An infinite number of such motes all rotating with great angular velocities will repel one another and keep up the kind of stability and relative stiffness, required for luminiferous vibrations."[26] Moreover, if the axes of rotation of the motes were aligned in the same direction, then, he said, "undulations among the system would have Faraday's rotatory property, when the planes of the waves are perp[endicula]r to this direction."[26] Though many difficulties faced the system and though Kelvin never developed it into the precise, quantitative account necessary for acceptance, he clearly regarded it as potentially the basis for a comprehensive, unified, dynamical theory of nature. It would have removed the usual dichotomy between ether and matter, and the various motions within the system would have provided a comprehensive dynamical underpinning for observed phenomena in optics, electricity, magnetism, and heat. This vision formed much of the unstated conceptual background to his concluding remarks in a lecture at the Royal Institution in London in 1860:

It does seem that the marvelous train of discovery, unparalleled in the history of experimental science, which the last years of the world has seen to emanate from experiments within these walls, must lead to a stage of knowledge, in which laws of inorganic nature will be understood in this sense that one will be known as essentially connected with all, and in which unity of plan through an inexhaustibly varied execution will be recognised as a universally manifested result of creative wisdom.[27]

Although mostly unpublished, Kelvin's ideas helped set the stage for a theory he did publish in 1867—one that caught the imagination of many late nineteenth-century physicists. This was his theory of the vortex atom.[28] (See

Fig. 1.) In 1867, Kelvin became aware of a decade-old paper by Helmholtz arguing that whirlpool-like motions in a frictionless fluid were permanent.[29] Under natural circumstances, they could not come into existence within such a fluid, nor, once established, could they pass out of existence. Also in 1867, P. G. Tait showed Kelvin his smoke-ring demonstration. With a smoke-filled box having one flexible side and one side with a circular opening, Tait demonstrated that a blow on the flexible side sent a ring of smoke from the circular opening. If, as the ring floated across the room, it encountered a similar ring coming from the other direction, the two rebounded from one another. The smoke-ring model suggested to Kelvin that permanent, Helmholtz-like whirlpools in a frictionless fluid of an ether would behave similarly. Hence, the solid motes of his earlier conception gave way to smoke-ring-like motions of the fluid, and Kelvin hoped that, as this theory was further developed, it would "become the foundation of the proposed new kinetic theory of gases. [Also] the possibility of founding a theory of elastic solids and liquids on the dynamics of more closely packed vortex atoms may be reasonably anticipated."[30] The vortex atom would *really* be a kinetic atom—not only an atom that moved, but one which itself was merely a "mode of motion" of ether. Physical nature, then, would no longer contain ponderable matter, imponderable ether, and possibly other imponderables, nor would it be divided into the fluid and solids of his previous model; nature would consist only of a fluid ether and its various motions.

How was Kelvin's theory of the vortex atom received? Conveniently for us, about a decade after Kelvin published his theory, Maxwell was asked to write an encyclopedia article entitled "Atom," in which he wrote at length on the vortex atom. In discussing atoms, Maxwell emphasized the growing science of spectroscopy as well as other aspects of physics which we have already noted. By the 1870s it seemed evident that the particles of matter of any one element must vibrate in complicated but definite ways in order to produce the several spectral lines characteristic of that element. Summarizing his views of possible atomic theories, Maxwell wrote:

The small hard body imagined by Lucretius, and adopted by Newton, was invented for the express purpose of accounting for the permanence of the properties of bodies. But it fails to account for the vibrations of a molecule as revealed by the spectroscope. We may indeed suppose the atom elastic, but this is to endow it with the very property for the explanation of which, as exhibited in aggregate bodies, the atomic constitution was originally assumed. The massive centres of force imagined by Boscovich may have more to recommend them to the mathematician, who has no scruple in supposing them to be invested with the power of attracting and repelling according to any law of the distance which it may please him to assign .... [But] it is in questionable scientific taste, after using atoms so freely to get rid of force acting at sensible distances, to make the whole function of atoms an action at insensible distances.

On the other hand, the vortex rings of Helmholtz, imagined as the true form of the atom by Thomson [i.e., Kelvin], satisfies more of the conditions than any atom hitherto imagined.[31]

Moreover, Maxwell explained that whereas the followers

of Lucretius and Boscovich might add one *ad hoc* hypothesis after another to make their theories work, "he who dares to plant his feet in the path opened up by Helmholtz and Thomson has no such resources." Kelvin's fluid had, wrote Maxwell, "no other properties than inertia, invariable density, and perfect mobility," and, in dealing with the fluid, Kelvin and his followers had no other method than "pure mathematical analysis." "The difficulties of this method are enormous," Maxwell observed, "but the glory of surmounting them would be unique."[32]

Unfortunately, such glories did not come to pass. Within a decade or so of Maxwell's optimistic comments, Kelvin had abandoned the theory of vortex atoms. He had been unable to apply it in certain areas, and, more significantly, as he said, "after many years of failure to prove that the motion in the ordinary Helmholtz circular ring is stable, I came to the conclusion that it is essentially unstable, and that its fate must be to become dissipated . . . ."[33] Accordingly, in the 1890s, when Kelvin spoke of atoms, he spoke of spheres or of centers of force, but with no confidence that he was describing the atoms of nature. Indeed, his pessimism regarding man's knowledge of atoms was part of his overall viewpoint reflected in his references to the limited state of scientific knowledge, to the lack of even a "finger post" to point the way toward truth,[34] and to his own "failure" to understand ether, matter, and electricity any better at the end of his career than he had at the beginning.[35] Ironically, even as physics itself progressed, Kelvin found himself in 1890 much less sure than he had been in 1870 that he was within reach of a true theory of the atom.

If we look at others during the 1890s—both British and continental—we can find many who were even more uncertain about atoms than Kelvin. Kelvin's younger countryman, Karl Pearson, for example, wrote in 1892 that atoms and ether "exist only in the human mind, and they are 'shorthand' methods of distinguishing, classifying, and resuming phases of sense impressions. They do not exist in or beyond the world of sense impressions, but are the pure product of our reasoning faculty."[36] J. H. Poynting and Horace Lamb—members of Pearson's generation—echoed these positivistic views, Poynting declaring that "our hypotheses are in terms of ourselves rather than in terms of Nature itself . . ."[37] and Lamb explaining that, unlike the generation of Kelvin and Stokes, modern physicists no longer viewed "the physical world as a mechanism . . . whose most intimate details might possibly some day be guessed."[38] Nevertheless, these men thought such mental images aided scientific research, and Pearson, for instance, proposed that an atom might be a squirt of ether. The idea was that an atom was something like a squirt of water from an underwater tap, except that in the case of an ether squirt there was no tap.[39] In the German-speaking part of Europe, Ernst Mach's positivism was similar to Pearson's, and, indeed, Mach had been an important influence on Pearson. The energetics movement associated with Planck and Ostwald called for the elimination of unobservable entities from science. The Austrian Boltzmann, whose views were akin to those of the older British physicists, was surrounded by such an alien intellectual climate that he felt himself in stark isolation. In France, the prestige of Duhem and Poincaré militated against atomic or molecular theories, and the Frenchman, Jean Perrin, was fighting prevailing views when he tried to prove the reality of molecules in his

work on Brownian motion.[40] To simplify: older British physicists tended to regard atoms as real, but thought they knew little of what they were actually like; many younger British physicists viewed atoms as merely mental constructs, but agreed with their older colleagues that such mental pictures aided physical research; and most continental physicists appear to have wanted to hear as little as possible about atoms and molecules in any form.

Before concluding, I should say something about chemists' discussions of atoms during the nineteenth century. Briefly, the chemical tradition was largely separate from the physical, and chemists gave Daltonian atomism a difficult time. Although Dalton's laws of chemical combinations seemed valid enough, they did not convince everyone of the existence of atoms. One problem was that the atomic theory sometimes seemed to require the truth of highly dubious propositions, such as the curious claim that equal volumes of quite different gases would contain the same number of gas particles. Also, with the atomic theory, there were great obstacles in working out an acceptable system of compound formulas and relative atomic weights. As one British chemist pointed out in 1869, "on the one hand all chemists use the atomic theory and, . . . on the other hand, a considerable number view it with mistrust, some with positive dislike."[41] The chemical and physical traditions probably approached each other most closely when Maxwell claimed that his kinetic theory of gases showed that, "the number of particles in unit of volume, is the same for all gases at the same pressure and temperature. This result agrees with the chemical law that equal volumes of gases are chemically equivalent."[42] Apparently, however, chemists largely ignored such physical support for the chemical atomic theory and equivocation regarding chemical atoms persisted into the 1890s. Indeed, in Britain some of the strongest support for chemical atoms came from physicists like Maxwell and Stokes, both of whom weighed in on the side of atoms in debates at the Chemical Society of London[43] and both of whom espoused the natural theological argument that the sameness of all atoms of an element indicated that they were "manufactured articles," all made the same way by God.[44]

In conclusion, in spite of the unsettled aspects of nineteenth-century science, I would re-emphasize its identifiable patterns and developments and progress. Both Laplace and Fourier, for example, did represent well-defined research traditions—traditions which yielded leadership to British and German research traditions in midcentury. Chemists may still have disagreed whether atoms were real entities or merely heuristic devices, but by the end of the century they had established a system of compound formulas and relative atomic weights within the framework of Mendeleev's periodic table. And the kinetic theory of gases obviously was possible in the 1860s as it simply had not been only a few decades earlier. Moreover, the kinetic theory allowed physicists to say much more about atoms than earlier in the century. Yet, the developments and progress provided no decisive, agreed-upon answer to the question, "What is an atom?"[45] In fact, what one thought about atoms in the 1890s depended largely on whether he was a chemist or a physicist, a continental European or a Briton, and, if British, whether young or old. At the very least, it is certainly mistaken to think that late nineteenth-century physicists placed complacent confidence in a naive view of a world made of billiard-ball-type atoms and that they were rudely jolted from such views by coming events. No doubt the coming events were jolting and surprising, but not because of physicists' complacent acceptance of billiard-ball atoms. Scientific problems and philosophical trends precluded such complacency. Instead, I should guess that the biggest surprise was that the severely intractable, or highly uninteresting, question, "What is an atom?," turned out to have some kind of meaningful and interesting answer after all.

[1] For French views, see Robert Fox, Hist. Stud. Phys. Sci. **4**, 89–136 (1974); and J. W. Herivel, Brit. J. His. Sci. **3**, 109–132 (1966).

[2] For a study emphasizing such a flexible methodology in Maxwell's thought and its relationship to French views, see Robert Kargon, J. Hist. Ideas **30**, 423–436 (1969). For a convenient bibliography of recent publications on Maxwell, see Donald Franklin Moyer, Stud. Hist. Philos. Sci. **8**, 262–263 (1977), Footnote 68. For an examination of the introduction of French physics into British thought, see Maurice Crosland and Crosbie Smith, Hist. Stud. Phys. Sci. **9**, 1–61 (1978).

[3] G. G. Stokes, *Mathematical and Physical Papers*, 5 Vols., Vols. IV and V edited by Joseph Larmor (Cambridge University, Cambridge, 1880–1905), Vol. I, p. 79. The paper was published in 1845. Stokes's law mentioned above appeared in his 1850 paper, in *ibid.*, Vol. III, pp. 55–67.

[4] For the published version, see Ref. 3, Vol. I, pp. 84–86. The unpublished version reads: "To recapitulate: the following are the very simple hypotheses on which it now appears that the equations of motion which have been found rest. That a fluid is composed of molecules, perhaps separated from each other, perhaps, in some cases at least, in contact, but smooth if the latter be the case: that when the average of a great number is taken there is no particular arrangement of the molecules in one direction rather than in another: and that if a single molecule were displaced from its position of relative equilibrium, the accelerating force of restitution would be very large unless the displacement were very small . . . ." (Stokes, manuscript entitled, "Theory of the friction of fluids in motion," Cambridge University Library, Stokes Collection, Add. MS 7656, PA224.)

[5] Reference 3, Vol. I, p. 182.

[6] Reference 3, Vol. I, p. 184.

[7] G. G. Stokes to Lord Kelvin, 10 and 11 March 1888, Cambridge University Library, Kelvin Collection, Add. MS 7342, S481. This and the letter cited in Footnote 26 will be included in my forthcoming edition of the Stokes–Kelvin correspondence, to be published by the Cambridge University Press. I am grateful to the Syndics of the Cambridge University Library for permission to publish here extracts from the letters and from the manuscript cited in Footnote 4.

[8] For studies of this aspect of Joule's and Kelvin's researches, see Thomas S. Kuhn, in *Critical Problems in the History of Science*, edited by Marshall Clagett (University of Wisconsin, Madison, WI, 1959), p. 321–356; P. M. Heimann, Centaurus **18**, 147–161 (1974); Crosbie W. Smith, Arch. Hist. Exact Sci. **16**, 231–288 (1976); and M. Norton Wise, Hist. Stud. Phys. Sci. **10**, 49–83 (1979). See also John Tyndall, *Heat Considered as a Mode of Motion* (Longman, London, 1863), a book referred to by Kelvin below in Footnote 34.

[9] For an extensive treatment of kinetic theory in the nineteenth century, see Stephen G. Brush, *The Kind of Motion We Call Heat: A History of the Kinetic Theory of Gases in the 19th Century*, 2 Vols. (North-Holland, Amsterdam, 1976). He includes a highly detailed bibliography of relevant nineteenth-century works, some of the most significant of which he had already reprinted in *Kinetic Theory*, edited by S. Brush, Vol. I: *The Nature of Gases and Heat* (Oxford University, Oxford, 1965); Vol. II: *Irreversible Processes* (Oxford University, Oxford, 1966).

[10] Rudolf Clausius, in *Kinetic Theory*, Ref. 9, Vol. I, pp. 111–134.

[11] Reference 10, Vol. I, pp. 135–147.

[12] James Clerk Maxwell, in *Kinetic Theory*, Ref. 9, Vol. I, pp. 148–171.

[13] See Ludwig Boltzmann, in *Kinetic Theory*, Ref. 9, Vol. II, pp. 88–175 and 188–193.

[14] See, for example, Kelvin, in *Kinetic Theory*, Ref. 9, Vol. II, pp. 176–187.

[15] R. Clausius, in *Kinetic Theory*, Ref. 9, Vol. I, p. 139.

[16] Maxwell, in *Kinetic Theory*, Ref. 9, Vol. I, p. 150.

[17] Maxwell, in Ref. 16, Vol. II, pp. 32–33. On Maxwell, see also P. M. Heimann, Stud. His. Philos. Sci. **1**, 189–211 (1970), and M. J. Klein, Am. Sci. **58**, 84–97 (1970).

[18] See S. Brush, *The Kind of Motion We Call Heat*, Ref. 9, Vol. II, pp. 627–640.

[19] Kelvin, "Nineteenth Century Clouds over the Dynamical Theory of Heat and Light," reprinted as Appendix B in Kelvin, *Baltimore Lectures on Molecular Dynamics and the Wave Theory of Light* (Clay, London, 1904).

[20] See, for example, Loyd S. Swenson, *The Ethereal Aether: A History of the Michelson–Morley–Miller Aether Drift Experiments, 1880–1930* (University of Texas, Austin, 1972) and my review of it in Hist. Sci. **12**, 220–227 (1974). See also Gerald Holton, Isis **60**, 133–197 (1969).

[21] Kelvin, Ref. 19, pp. 8–9.

[22] See Michael Faraday, Philos. Trans. R. Soc. London **CXXXVI**, 1–20 (1846). See also J. Brookes Spencer, Isis **61**, 34–51 (1970).

[23] Kelvin, *Mathematical and Physical Papers*, 6 Vols., Vols. IV–VI edited by Joseph Larmor (Cambridge University, Cambridge, 1882–1911), Vol. II, p. 28.

[24] Quoted in Ole Knudsen, Centaurus **16**, 48 (1972). For a more detailed account of the material discussed in this and the next paragraph, see the "Introduction" to The Correspondence between Sir George Gabriel Stokes and Sir William Thomson, Baron Kelvin of Largs, edited with an introduction by David B. Wilson (to be published). See also Ole Knudsen, Arch. Hist. Exact Sci. **15**, 235–281 (1976); and Jed Buchwald, *Dictionary of Scientific Biography* (Scribner, New York, 1970). Knudsen, for instance, notes the difference between Clausius's and Kelvin's dynamical theories of heat in the 1850s.)

[25] Knudsen, Centaurus **16**, 48 (1972).

[26] Kelvin to Stokes, 17 June 1857, Stokes Collection, K98.

[27] Kelvin, *Reprint of Papers on Electrostatics and Magnetism*, 2nd ed. (MacMillan, London, 1884), p. 225.

[28] Kelvin, *Mathematical and Physical Papers*, Ref. 23, Vol. IV, pp. 1–12. See also Robert Silliman, Isis, **54**, 461–474 (1963).

[29] H. Helmholtz, Philos. Mag. **33**, 485–512 (1867). This is a translation by P. G. Tait and Helmholtz of the 1858 paper. See the added notes by Tait and Kelvin on pp. 511–512.

[30] Kelvin, Ref 28, pp. 2–3.

[31] James Clerk Maxwell, in *The Scientific Papers of James Clerk Maxwell*, edited by W. D. Niven, 2 Vols. (Herman, Paris, 1927), Vol. II, pp. 470–471. Maxwell clearly expressed his antipathy for the concept of action at a distance in his articles "On Action at a Distance" and "Attraction," favorably quoting, for example, Newton's famous letter to Richard Bentley in which he called the idea "so great an absurdity, that I believe no man, who has in philosophical matters a competent faculty of thinking, can ever fall into it." (Volume II, p. 487; Volume II, pp. 311–323.) Hence, for Maxwell, neither in the case of large distances—as with gravity—nor insensibly small distances—as with Boscovich's atoms—did action at a distance provide a satisfactory account of the true situation in nature. In both articles, for instance, he stressed Faraday's and Kelvin's successes in reducing magnetic attractions and repulsions to actions transmitted through a continuous material medium.

[32] J. Maxwell, Ref. 31, pp. 471–472.

[33] Quoted in Silvanus P. Thompson, *The Life of William Thomson, Baron Kelvin of Largs*, 2 Vols. (MacMillan, London, 1910), Vol. II, p. 1047.

[34] Kelvin, in *Mathematical and Physical Papers*, Ref. 23, Vol. III, pp. 510–511. In concluding this presidential address to the Institution of Electrical Engineers in 1889, Kelvin regretted that "I must end by saying that the difficulties are so great in the way of forming anything like a comprehensive theory, that we cannot even imagine a finger post pointing to a way that can lead us towards the explanation."

[35] This well-known statement came in Kelvin's address in 1896 during the celebration of his Jubilee as professor of natural philosophy at the University of Glasgow: "One word characterizes the most strenuous of the efforts for the advancement of science that I have made perseveringly during fifty-five years; that word is failure. I know no more of electric and magnetic force, or of the relation between ether, electricity,

and ponderable matter than I knew and tried to teach to my students of natural philosophy fifty years ago in my first session as Professor." (Quoted in Ref. 33, Vol. II, pp. 1072-1073.)

[36]Karl Pearson, *The Grammar of Science* (Black, London, 1892), pp. 214-215.

[37]J. H. Poynting, Brit. Assoc. Adv. Sci. Rep. 620 (1899).

[38]Horace Lamb, Brit. Assoc. Adv. Sci. Rep. 429 (1904).

[39]K. Pearson, Ref. 36, pp. 318-320 and Am. J. Math. **13,** 309-363 (1891).

[40]For discussions of continental thought, see Erwin Hiebert, in *Perspectives in the History of Science and Technology,* edited by Duane H. D. Roller (University of Oklahoma, Norman, OK, 1971), pp. 67-86; and Mary Jo Nye, *Molecular Reality: A Perspective on the Scientific Work of Jean Perrin* (American Elsevier, New York, 1972).

[41]A. W. Williamson, J. Chem. Soc. **22,** 328-365 (1869), as quoted in David M. Knight, *Atoms and Elements: A Study of Theories of Matter in England in the Nineteenth Century* (Hutchinson, London, 1967), p. 114. Also quoted by Alan J. Rocke to help establish the purpose of his article, Hist. Stud. Phys. Sci. **9,** 225 (1978).

[42]J. Maxwell, in *Kinetic Theory,* Ref. 9, Vol. I, p. 164.

[43]D. Knight, Ref. 41, pp. 111-112.

[44]J. Maxwell, in Ref. 31, Vol. II, pp. 376-377 and 482-484. G. G. Stokes, *The Gifford Lectures,* 2 Vols. (Black, London, 1891, 1893), Vol. II, p. 133.

[45]Elizabeth Garber underscores this point in her article, Hist. Stud. Phys. Sci. **9,** 265-297 (1978).

Max Planck and the Beginnings of the Quantum Theory

Martin J. Klein

Communicated by L. Rosenfeld

1. Introduction

On December 14, 1900, Max Planck presented his derivation of the distribution law for black-body radiation to the German Physical Society, and the concept of energy quanta made its first appearance in physics. Considering the enormous consequences which the quantum theory has had, it is astonishing that so little attention has been devoted to detailed study of the reasoning which brought Planck to the first radical step of introducing quanta. There are, of course, many descriptions of the origin of the quantum theory in the literature, but almost all of them are historically inaccurate, uncritical, and quite misleading as to both Planck's own work and the context in which it was done. We do have Planck's retrospective accounts [1] which give a clear and consistent picture of his own view of the development, and there is also an excellent monograph by Rosenfeld [2], too little known, on the early years of the quantum theory, which presents Planck's work in its proper historical setting.

It seems to me that there are still two critical questions, not unrelated, which must be answered, if we are to understand fully the nature of Planck's decisive step and the extent to which it marked a real break with previous thinking. The first is really an historical question: *Was Planck aware of the radiation distribution law which Rayleigh had derived as a necessary consequence of classical physics?* Most authors answer this question in the affirmative and describe Planck's introduction of quanta as his response to the challenge of the "crisis" brought about by the disagreement between classical theory and experimental results and by the internal failure of classical theory as expressed in the "ultra-violet catastrophe". As a matter of fact, there was no such crisis, or perhaps one should say there was no awareness of such a crisis. All of the work on black-body radiation prior to the summer of 1900 was done without benefit of the knowledge of just what classical physics did imply for this problem. It was only in June, 1900 that Lord Rayleigh published a two-page note in which the classical distribution law was first derived, and the very serious implications of Rayleigh's paper were not generally realized for quite some time. Planck makes no reference to Rayleigh's note in his own papers of 1900 and 1901, nor does he refer to Rayleigh in his accounts of the origins of the quantum theory published many years later. It does, however, seem likely that Planck knew of Rayleigh's work, but that he attached no more significance to it than he did to several other papers, published at about the same time, in which more or less *ad hoc* attempts were made to find an equation which would describe the experimental results. The possible reasons for PLANCK's neglect of what now seems to be the critically important contribution made by RAYLEIGH are to be sought in PLANCK's background, in his way of approaching the radiation problem, and also in the manner in which RAYLEIGH had communicated his results.

The second question which I consider to be critical concerns the method which PLANCK actually used in deriving the distribution law: *In what ways did* PLANCK *depart from* BOLTZMANN'S *methods in his statistical calculation of the entropy using energy quanta?* PLANCK himself, in both his original papers and his later accounts, considered that he was using BOLTZMANN'S approach in a rather straightforward way with the discreteness in energy as the only innovation. ROSENFELD [*3*], however, speaks of PLANCK's calculation of the entropy as being "pure heresy" from the classical Boltzmann point of view. PLANCK actually did depart from BOLTZMANN's method in several respects, and it took a number of years for the full implications of his departures to be realized. It has never been pointed out, though, just how much PLANCK was influenced in his derivation by the paper of BOLTZMANN's which was his principal guide in a realm of ideas which had been quite foreign to him before the autumn of 1900.

The discussion of these two questions is the principal theme of the present paper. This discussion requires a brief restatement of the context and background for PLANCK's work, and this is provided in the next section. The final section of this paper deals briefly with another problem which deserves further study: Why did almost five years go by before PLANCK's bold solution of the radiation problem was taken up for further study?

2. Background

In 1897, when PLANCK first turned his attention to the problem of blackbody radiation, he was almost forty years old, and his scientific career had been devoted principally to clarifying the meaning of the second law of thermodynamics and to exploring its consequences. What attracted PLANCK's attention to the radiation problem was the universal character of the distribution law which was required by KIRCHHOFF's theorem. KIRCHHOFF [*4*], and independently BALFOUR STEWART, had shown that the nature of the radiation in thermal equilibrium in an enclosure, whose walls are kept at a fixed temperature, is completely independent of the properties of any material bodies, including the walls, which are in equilibrium with the radiation. The spectral distribution of the radiation then "represents something absolute, and since I had always regarded the search for the absolute as the loftiest goal of all scientific activity, I eagerly set to work" [*5*].

Several properties of the universal function of temperature and frequency which describes this equilibrium spectral distribution had already been established during the preceding two decades. In order to formulate these, it is convenient to introduce the function in question as $\varrho(\nu, T)$, where $\varrho(\nu, T)\, d\nu$ is the energy per unit volume in thermal radiation, at absolute temperature T, which lies in the frequency interval from ν to $\nu + d\nu$. STEFAN [*6*] had found experimentally in 1879 that the total energy density, integrated over all frequencies, is proportional to the fourth power of the temperature; thus the function $\varrho(\nu, T)$ satisfies

PLANCK and quantum theory   461

the equation

$$\int_0^\infty \varrho\,(v,\,T)\,dv = \sigma\,T^4, \tag{1}$$

where σ is a constant. This experimental result of STEFAN's was derived theoretically in 1884 by BOLTZMANN [7], who applied the second law of thermodynamics
to radiation, treating it as a gas whose pressure was the radiation pressure of
MAXWELL's electromagnetic theory. In 1893 WIEN [8] drew another conclusion
from the second law of thermodynamics which imposed a significant limitation
on the energy distribution function $\varrho\,(v,\,T)$. This displacement law of WIEN's
requires that $\varrho\,(v,\,T)$ have the form

$$\varrho\,(v,\,T) = v^3\,f(v/T), \tag{2}$$

where $f(v/T)$ can depend on only the ratio of frequency to temperature★.

In addition to the two laws expressed in Eqs. (1) and (2), there was one other
important result known to PLANCK when he started his work on the radiation
problem. This other result was the distribution law proposed by WIEN [9] in
1896 which gave an explicit form for the function $\varrho\,(v,\,T)$, or the function $f(v/T)$:

$$\varrho\,(v,\,T) = \alpha\,v^3 \exp\,(-\,\beta\,v/T), \tag{3}$$

where α and β are constants. WIEN had given a theoretical argument for the cor-
rectness of Eq. (3), but as RAYLEIGH [10] wrote: "Viewed from the theoretical
side, the result appears to me to be little more than a conjecture." The important thing about WIEN's distribution in the late 1890's was not WIEN's
derivation, but rather the fact that it gave an adequate account of all the experimental results on the energy distribution in black-body radiation which were
then available. It seemed reasonable to suppose that a fundamental theory of
radiation, such as PLANCK proposed to develop, would have to conclude with
an adequate grounding for the Wien distribution law, if the theory were to be in
accord with experiment.

In the first [11] of a series of five papers which PLANCK presented to the
Prussian Academy of Sciences in the years 1897 to 1899, he set forth his program
for a theory of radiation. This program arose naturally from his earlier work
in thermodynamics, since he took as his goal the finding of a basis in electrodynamics for the irreversible approach of radiation to equilibrium. His idea was
that the conservative system consisting of electromagnetic radiation in an enclosure, interacting with a collection of harmonic oscillators, could be shown to
approach an equilibrium state, without the need for any assumptions beyond
the laws of electromagnetism. For PLANCK, such a demonstration would have
completed the understanding of the second law of thermodynamics. PLANCK
thought he saw the basic mechanism for the irreversible behavior of the system
in the way in which an oscillating dipole emits electromagnetic energy as a
spherical wave, changing the character of the radiation incident upon it in an
apparently irreversible manner.

★ It is worth noting that the WIEN displacement law implies the STEFAN-BOLTZ-
MANN law, since $\int_0^\infty \varrho\,(v,\,T)\,dv = \int_0^\infty v^3\,f(v/T)\,dv = T^4 \int_0^\infty x^3\,f(x)\,dx$, and the last integral,
so long as it exists, is just a pure number.

32★

Planck's direct advance along this line of thought was promptly stopped by criticism from Boltzmann [12]. Boltzmann, better than anyone else, was in a position to see the flaw in Planck's reasoning since he had concerned himself for many years with the nature of irreversibility, and he knew from rather bitter experience the subtleties and difficulties which blocked an understanding of this elusive concept. Only months earlier he had had to defend and restate his ideas on the essentially statistical origin of irreversibility against an attack launched by Zermelo, a student of Planck. Boltzmann's answers to Zermelo had apparently still not been grasped by Planck since, in this same paper, he had referred approvingly to Zermelo's work and questioned the success of the kinetic theory's explanation of irreversibility. This situation certainly did not make Boltzmann delay in pointing out to Planck that there was nothing in the equations of electromagnetism which excluded processes inverse to those Planck had considered, so that the laws of electromagnetism did not, by themselves, determine the irreversible approach of radiation to equilibrium, any more than the laws of mechanics, by themselves, determined the irreversible approach of a gas to equilibrium. Additional assumptions were needed, statistical assumptions about the disordered character of the initial state such as Boltzmann had made in the theory of gases. Then, said Boltzmann, one could deduce a theorem for radiation which would be analogous to the second law and would play the role of his H-theorem in the kinetic theory of gases. Planck eventually, though not immediately, accepted Boltzmann's criticism as sound, and formulated an assumption of "natural radiation" which assured irreversibility, much as Boltzmann's assumption of "molecular chaos" had done in kinetic theory.

In Planck's subsequent investigations, he continued to treat the interaction between radiation and oscillating dipoles. The harmonic oscillators were chosen, not because they were thought to be a realistic model for matter, but rather because Kirchhoff's theorem asserted that the equilibrium radiation distribution was independent of the system with which the radiation interacted, and oscillators were the simplest to treat. One important result of this work was the proof of a theorem [13] relating the spectral distribution $\varrho(\nu, T)$ to the average energy $\bar{u}_\nu(T)$ of a harmonic oscillator of frequency ν. This theorem, which was derived by equating the emission and absorption rates of the oscillator at equilibrium, had the form

$$\varrho(\nu, T) = (8\pi\nu^2/c^3)\,\bar{u}_\nu(T), \tag{4}$$

where c is the velocity of light.

It is evident from Eq. (4) that Planck needed only to determine $\bar{u}_\nu(T)$, the average energy of a harmonic oscillator at temperature T, in order to have the explicit form for the distribution law. Remarkably enough, although classical statistical mechanics provided a "well-known" and simple answer for $\bar{u}_\nu(T)$ from the equipartition theorem, Planck made no use of it, nor did he then or later indicate that he realized its existence. (This point will be discussed again in the following section.) Instead Planck took what he later referred to as a "thermodynamic" approach, looking for a suitable relationship between the energy and the entropy of the oscillator, rather than one between the energy and the temperature. This relationship was introduced, in the fifth and final paper [13] of the series under discussion, by means of a definition:

$$S = -\frac{u}{\beta v} \ln \frac{u}{a e v} \tag{5}$$

where S is the entropy of the oscillator, u is its energy [previously called $\bar{u}_v(T)$], β and a are constants and e is the base of the natural logarithms. In the original paper, Planck did not motivate this definition when he introduced it, but from his discussion later in that paper and also from his later reviews of this work, it seems most likely that he was guided by the form of Wien's distribution law, Eq. (3).*

With the entropy of an oscillator defined by Eq. (5), Planck could then determine the entropy of the radiation in equilibrium with it and go on to prove that the total entropy was a monotonically increasing function of time, just the property required of the entropy by the second law. That Planck could also demonstrate that the equilibrium distribution was the Wien law of Eq. (3) is hardly remarkable, since that result followed inevitably from his choice of the entropy expression given by Eq. (5), as already indicated. Planck wrote that he was impressed by the simplicity of the relationship expressed in Eq. (5) and particularly by the fact that $\partial^2 S / \partial u^2$, which entered directly into his calculation, was simply proportional to minus u^{-1}. He was, nevertheless, aware that his choice of a particular expression for entropy as a function of energy determined the resulting distribution law, and he gave arguments which seemed to make that choice uniquely determined by the requirements of consistency with the displacement law and the second law of thermodynamics. His conclusion was expressed in the following sentence [14]. "I believe that it must therefore be concluded that the definition given for the entropy of radiation, and also the Wien distribution law for the energy which goes with it, is a necessary consequence of applying the principle of entropy increase to the electromagnetic theory of radiation, and that the limits of validity of this law, should there be any, therefore coincide with those of the second law of thermodynamics. Further experimental test of this law naturally acquires all the greater fundamental interest for this reason."

These arguments and remarks were made in a paper presented to the Academy on May 18, 1899 and were repeated verbatim in Planck's article [15] in the *Annalen der Physik* which summarized the series of five Academy papers. The *Annalen* paper was received by the editors on November 7, 1899, and it appeared in print early in 1900. By the time Planck corrected the proofs of this paper, the "further experimental tests", which he had called for, were in progress. The results of these tests which had already been published caused him to add a note in proof remarking that experimental deviations from the Wien distribution had been observed. As Planck had already indicated, deviations from Wien's law created a serious problem indeed from the standpoint of his theory, and Planck proceeded to reconsider his arguments in some detail. In a paper

---

* From Eqs. (3) and (4) it would follow that $u = (\alpha c^3/8\pi) \, v \exp(-\beta v/T)$. From this last result, one can solve for $T^{-1} = -\frac{1}{\beta v} \ln \left(\frac{8\pi}{\alpha c^3} \frac{u}{v} \right)$. But since $T^{-1} = \partial S/\partial u$, one can integrate to obtain S as a function of u in just the form of Eq. (5) with a equal to $\alpha c^3/8\pi$.

received by the editors of the *Annalen* on March 22, 1900, he reported his new considerations. The experimental situation was not clear yet since Paschen's latest measurements supported Wien's law, but the work of Lummer & Pringsheim [16], which extended to longer wave lengths, had indicated serious deviations. A new distribution law had actually been proposed by Thiesen [17] (independently of Lummer & Pringsheim's work) which was constructed to fit the data and at the same time to be consistent with the Stefan-Boltzmann law and the displacement law. Planck's review of the assumptions and reasoning in his former work led him to propose new arguments, this time attempting to derive the form he had previously assumed for the energy-entropy relationship. Planck found, however, that in order to satisfy the second law of thermodynamics, *i.e.*, in order to have the entropy increase monotonically in time, it would suffice to have $\partial^2 S/\partial u^2$ be any negative function of the energy u; the specific form $-(g(\nu)/u)$, which was equivalent to Wien's law, was not a thermodynamic necessity. Nevertheless, Planck concluded that Wien's distribution law could still be deduced, if he made a very plausible assumption on the functional dependence of the time derivative of the entropy. Once again he ended his arguments with the Wien distribution, even if they did not have the full weight of thermodynamic reasoning to support them.

By October 1900, however, the experimental picture had changed considerably. The very careful work of Rubens & Kurlbaum with long waves over a wide range of temperatures had shown beyond any doubt that Wien's distribution law was inadequate. These new measurements also indicated clearly that for very long wavelengths the distribution function $\varrho(\nu, T)$ approached a very different form, becoming proportional to the absolute temperature T. Planck had been informed [18] of these results by Rubens & Kurlbaum several days before they were reported to the German Physical Society on October 19, 1900, so that he had the opportunity to reflect on the results and to prepare an extended "remark" for the discussion after Kurlbaum delivered the paper [19]. This "discussion remark" was devoted to *An Improvement of the Wien Distribution* [20], the improvement being a new distribution law, now universally known as the Planck distribution law. The arguments for this new distribution law and the discussion of its immediate results properly belong to the next stage of our discussion.

3. Rayleigh, Planck and Equipartition

Planck's problem was now to determine a distribution law which was consistent both with the positive results of his own work and with the new experimental findings of Rubens & Kurlbaum. Since the quantity $\partial^2 S/\partial u^2$ had figured prominently in his earlier analysis of how the entropy increased in time, it was natural for Planck to center his attention on the form of this function. We have already seen that the negative reciprocal of $\partial^2 S/\partial u^2$ is simply proportional to u when the Wien distribution is valid. The next simplest possibility is to take $\partial^2 S/\partial u^2$, or rather its negative reciprocal, proportional to u^2. It is easy to see that, when this is done, u, and therefore $\varrho(\nu, T)$, will be proportional to T, just as Rubens & Kurlbaum had found to be the case in the long wave length limit. The proper limiting forms for low and high frequencies could then be preserved by taking $-(\partial^2 S/\partial u^2)^{-1}$ proportional to $u(\gamma + u)$, where γ is a (frequency dependent) constant. On these grounds, of simplicity and proper behavior in the limit, PLANCK proposed the distribution law

$$\varrho(\nu, T) = \frac{A \, \nu^3}{\exp(B \, \nu/T) - 1},\qquad(6)$$

where A and B are constants. This is the law which follows from the assumption just mentioned for $\partial^2 S/\partial u^2$, where the frequency dependence is fixed by the displacement law together with Eq. (4).★

The adequacy of PLANCK's proposed distribution law was confirmed immediately. As PLANCK described it later [21], "The very next morning I received a visit from my colleague RUBENS. He came to tell me that after the conclusion of the meeting, he had that very night checked my formula against the results of his measurements and found a satisfactory concordance at every point ... Later measurements, too, confirmed my radiation formula again and again—the finer the methods of measurement used, the more accurate the formula was found to be."

Now one of the key points in determining PLANCK's choice of the distribution formula was that it agreed with the experimental results in being proportional to T in the limit of small ν. PLANCK's formula was not the first in the literature to show this property, and he referred in a footnote to an empirical formula proposed by LUMMER & JAHNKE [22] which had the same property. He did not, however, refer to another paper in which a distribution law proportional to T for long wave lengths was not only proposed but was also related to the basic ideas of statistical mechanics: the paper by Lord RAYLEIGH [10].

RAYLEIGH had published a short note in the June, 1900 issue of the *Philosophical Magazine* under the title *"Remarks upon the Law of Complete Radiation"*. In two pages, he had shown that if the equipartition theorem of statistical mechanics could be applied to "the modes of aetherial vibration", then the distribution law for black-body radiation is uniquely determined to have a form radically different from that of the Wien distribution. RAYLEIGH was well aware of the conditional nature of this conclusion saying, "The question is one to be settled by experiment; but in the meantime I venture to suggest a modification of [the Wien distribution], which appears to me more probable *a priori*. Speculation upon this subject is hampered by the difficulties which attend the Boltzmann-Maxwell doctrine of the partition of energy. According to this doctrine, every mode of vibration should be alike favoured; and although for some reason not yet explained, the doctrine fails in general, it seems possible that it may apply to the graver modes."

RAYLEIGH's method for arriving at the radiation distribution law was essentially different from PLANCK's. His argument concerned itself directly with the radiation and did not need to refer to a material system with which it was in equilibrium. The number of standing waves or allowed modes of electromagnetic

---

★ Taking $\partial^2 S/\partial u^2$ proportional to $-[u(\gamma + u)]^{-1}$, one can integrate to find $\partial S/\partial u$ as a function of u. This quantity, $\partial S/\partial u$, is, however, equal to T^{-1} from the second law so that one easily obtains u in the form $d_1[\exp(d_2/T) - 1]^{-1}$, where d_1 and d_2 are (frequency dependent) constants. Using Eq. (4), it follows that $\varrho(\nu, T) = (8\pi d_1/c^3)\,\nu^2$ $[\exp(d_2/T) - 1]^{-1}$, so that d_1 and d_2 must both be proportional to ν in order to satisfy the displacement law; *i.e.*, Eq. (6) must hold.

vibration of the enclosure, whose frequencies lie in the interval from ν to $\nu+d\nu$, is proportional to $\nu^2 d\nu$, by reasoning which was practically second nature of the author of *The Theory of Sound*. According to "the Boltzmann-Maxwell doctrine", *i.e.*, the equipartition theorem, the average energy of every one of these modes, regardless of its frequency, would be proportional to T at thermal equilibrium, with a universal proportionality constant. It follows at once that the distribution function $\varrho(\nu, T)$ must have the form,

$$\varrho(\nu, T) \propto \nu^2 T. \tag{7}$$

As Rayleigh remarked, this is in accord with the displacement law.

Although Rayleigh did not trouble to point it out explicitly in this note, it must have been quite obvious to him that a distribution law of this form could not possibly hold for all frequencies, since it would lead to an infinite concentration of energy at the high frequencies; the integral of $\varrho(\nu, T)$ over frequency would diverge. This undoubtedly accounts both for Rayleigh's reference to "the graver modes", and also for the conclusion of his paper. After having obtained the result expressed in Eq. (7), Rayleigh added, "If we introduce the exponential factor, the complete expression will be

$$\varrho(\nu, T) \propto \nu^2 T \exp(-\beta \nu/T). \tag{8}$$

Whether [this equation] represents the facts of observation as well as [the Wien distribution] I am not in a position to say. It is to be hoped that the question may soon receive an answer at the hands of the distinguished experimenters who have been occupied with this subject."

This last equation of Rayleigh's, clearly intended as only a guess at how the rigorous result of the classical theory, expressed in Eq. (7), might be modified at higher frequencies, was apparently the only thing in his paper which attracted the notice of those to whom it was addressed. This is evident from the paper [23] which Rubens & Kurlbaum presented to the Prussian Academy on October 25, 1900, less than a week after Planck's new distribution had come to their attention. Rubens & Kurlbaum made a systematic comparison of their results with five different formulas which had been proposed for the distribution law: those of Wien [Eq. (3)], Planck [Eq. (6)], Rayleigh [Eq. (8)] and two others due to Thiesen and to Lummer & Jahnke. They concluded that only Planck's formula and that proposed by Lummer & Jahnke were in agreement with their results, and they gave their preference to Planck's formula on grounds of simplicity. (The Lummer-Jahnke formula contained ν/T to the power 1.3 in the exponent, where the number 1.3 was chosen solely for fitting the results.)

The point to be stressed is that Rubens & Kurlbaum discussed Rayleigh's work in the same tone of voice, so to speak, that they used in dealing with strictly *ad hoc* formulas which had no theoretical foundations. They failed to grasp the fundamental importance of the fact that their results did show that ϱ was proportional to T for low frequencies, in complete agreement with Rayleigh's conclusion that equipartition should apply "to the graver modes". When Rayleigh's paper was reprinted in his *Scientific Papers* two years later, he took the opportunity to remark on the proportionality of ϱ to T for low frequencies.

"This is what I intended to emphasize. Very shortly afterwards the anticipation above expressed was confirmed by the important researches of Rubens & Kurlbaum who operated with exceptionally long waves."

Rubens & Kurlbaum were not the only ones who missed this central point in Rayleigh's paper, which he had probably not underlined sufficiently. I have already mentioned that Planck made no reference to Rayleigh in his October, 1900 communication. Nor did he refer to Rayleigh in his papers introducing the quantum concept which appeared a few months later. Although it might appear from this lack of reference that Planck did not know of Rayleigh's work, as Rosenfeld concludes, I find this a rather unlikely hypothesis. Planck had been devoting virtually all of his efforts to the radiation problem for over three years by this time, and he is not likely to have missed something on this subject written by a leading thinker and published in a major journal. Furthermore, we know that Rubens & Kurlbaum had seen Rayleigh's paper and must have referred to it in preparing their own report during the week after Planck proposed his new equation to them. We also know that Planck kept in close touch with the experimenters' attempts to fit their results with his and other distribution laws. Finally, in his paper [24] of December 14, 1900, Planck referred explicitly to the Rubens & Kurlbaum paper in which Rayleigh's work was quoted. It seems hard to believe, therefore, that Planck was not aware of Rayleigh's article.

It is not hard to understand, though, why Planck might have missed the significance of Rayleigh's reasoning: everything in Planck's background argues against his having been ready to receive Rayleigh's ideas. Rayleigh had formulated his argument in such a way that it would be clear to anyone at home in the writings of Boltzmann and Maxwell. His reference in passing to the difficulties attending the equipartition theorem was intended to suggest the whole bothersome, unsolved problem of the specific heats of gases. Rayleigh had himself written on the equipartition problem at some length earlier in the year. Planck, however, had been thinking along quite another line. He tells us himself that statistical mechanics had not been at all to his liking. In a passage of his *Scientific Autobiography* concerning the energetics controversy, in which he found himself allied with Boltzmann against the dominant school of energeticists headed by Ostwald, he wrote [25], "After all that I have related, in this duel of minds I could play only the part of a second to Boltzmann—a second whose services were evidently not appreciated, not even noticed, by him. For Boltzmann knew very well that my viewpoint was basically different from his. He was especially annoyed by the fact that I was not only indifferent but to a certain extent even hostile to the atomic theory which was the foundation of his entire research. The reason was that at that time I regarded the principle of the increase of entropy as no less immutably valid than the principle of the conservation of energy itself, whereas Boltzmann treated the former merely as a law of probabilities—in other words, as a principle that could admit of exceptions."

Planck's comment on what his attitude toward statistical mechanics was in 1900 is borne out by reference to his writings prior to that date. In all of Planck's work on the meaning and implications of the second law of thermodynamics, he had never used the methods of statistical mechanics, never even referred to

468   MARTIN J. KLEIN:

BOLTZMANN's statistical interpretation of the entropy. PLANCK had begun a series
of papers [26] *On the Principle of Increasing Entropy* in 1887 by saying explicitly
that he wanted to extend the series of conclusions drawn from the second law
"taken by itself, *i.e.*, irrespective of definite conceptions about the nature of
molecular motions". In a lecture [27] in 1891, PLANCK admitted the existence
of kinetic or molecular methods as an alternative to the purely thermodynamic
ones which he found so much more to his taste, but he went on to point at once
to the disappointments of the molecular approach after its initial successes.
"Anyone who studies the works of the two scientists who have probably penetrated most deeply in the analysis of molecular motions, MAXWELL and BOLTZ-
MANN, will not be able to resist the impression that the admirable display of physical ingenuity and mathematical cleverness shown in overcoming these problems
is not in suitable proportion to the fruitfulness of the results achieved." This
same attitude toward the molecular approach is displayed in the preface to the
first edition (1897) of PLANCK's *Treatise on Thermodynamics* [28] in which he
refers to the "insurmountable obstacles" and the "essential difficulties in the
mechanical interpretation of the fundamental principles of thermodynamics".

I think it is fair to conclude that PLANCK's often expressed distrust of the
whole molecular approach of statistical mechanics made it very unlikely that he
would see the point of RAYLEIGH's very condensed discussion, or even that he
would take it very seriously if he had. As we shall see in the next section, it was
probably a very good thing that PLANCK was not constrained in his thinking by
the tight classical web which RAYLEIGH had woven.

4. The Introduction of Quanta

"But even if the absolutely precise validity of the radiation formula is taken
for granted, so long as it had merely the standing of a law disclosed by a lucky
intuition, it could not be expected to possess more than a formal significance.
For this reason, on the very day when I formulated this law, I began to devote
myself to the task of investing it with a true physical meaning. This quest
automatically led me to study the interrelation of entropy and probability—in
other words, to pursue the line of thought inaugurated by BOLTZMANN." "After
a few weeks of the most strenuous work of my life, the darkness lifted and an
unexpected vista began to appear" [29].

These are the words PLANCK used many years later to describe his efforts
from October 19 to December 14, 1900 on which date he presented his results
to the German Physical Society in a paper entitled *On the Theory of the Energy
Distribution Law in the Normal Spectrum* [24]. During those two months PLANCK
had successfully changed the direction of all of his previous thought in recognizing
and adopting BOLTZMANN's insight into the relationship between entropy and
probability. In addition, he had created a concept which was eventually to
change the most basic features of physical theory. These two aspects of what
he had done are almost inextricably woven together in both the paper already
referred to and in the more complete paper published soon afterwards in the
Annalen der Physik [30]. We must, nevertheless, try to separate them here if
we are to appreciate the innovations which PLANCK made.

PLANCK and quantum theory

469

PLANCK's earlier work had shown that only one more key step was necessary for the theory of the radiation spectrum: a sound theoretical determination of the relationship between the energy u and the entropy S of a harmonic oscillator of frequency ν. Once this was known, the average energy of the oscillator could be found, and the distribution function $\varrho(\nu, T)$ would then be fixed with the help of Eq. (4), which relates u and $\varrho(\nu, T)$. (It might be well to re-emphasize the point that harmonic oscillators entered the picture only because PLANCK chose them as the simplest material system which could be in equilibrium with the electromagnetic radiation, availing himself of the freedom given by KIRCHHOFF's theorem.) PLANCK's previous attempts at fixing the relationship between S and u on general thermodynamic grounds supplemented by plausibility arguments had failed, as we have seen. He was quite sure that he knew what this relationship had to be (from the work discussed in the last section); if his conjectured distribution law, Eq. (6), were correct, as it seemed to be, then it implied the equation

$$S = \frac{A'}{B} \left[\left(1 + \frac{u}{A'\nu} \right) \ln \left(1 + \frac{u}{A'\nu} \right) - \frac{u}{A'\nu} \ln \frac{u}{A'\nu} \right], \tag{9}$$

where A and B are the constants of Eq. (6), and $A' = A c^3/8\pi^2$.* In order to establish Eq. (9), new methods were necessary, and as already mentioned, PLANCK found these in BOLTZMANN's work.

According to BOLTZMANN, the entropy of a system in a given state is proportional to the logarithm of the probability of this state. This probability in turn is to be found as the number of complexions, the number of distinct microscopic arrangements, compatible with the given state. PLANCK's task then, once this general approach was accepted, was to find a method for determining W, the number of complexions of his set of oscillators. It is sufficient to consider N oscillators, all of frequency ν, whose total energy U_N is then N times the average energy u of one oscillator. Since the entropy is an additive function, S_N will also be N times S, the entropy of one oscillator:

$$U_N = Nu, \tag{10a}$$

$$S_N = NS. \tag{10b}$$

The entropy of the N oscillator system, S_N, is now set equal to a proportionality constant k times the logarithm of W,

$$S_N = k \ln W, \tag{11}$$

where the additive constant which might appear is set equal to zero. PLANCK then made the same assumption that BOLTZMANN had made, that any one complexion of the system is as likely to occur as any other so that W can be obtained by counting the number of complexions. *But*, in order to carry out this counting procedure, it is essential that the energy to be shared among the N oscillators must not be considered as a continuously varying, infinitely divisible quantity. It must instead be treated as consisting of an integral number of finite equal

---

* From Eqs. (4) and (6), it follows that $u = A'\nu \left[\exp(B\nu/T) - 1 \right]^{-1}$. The entropy equation is derived by solving this last equation for T^{-1}, which is equal to $\partial S/\partial u$, and then integrating.

parts, if meaningful, finite values for W are to be obtained. Planck refers to these as elements of energy ε and writes

$$U_N = P \varepsilon, \qquad (12)$$

where P is a (large) integer representing the total number of elements of energy.

With this assumption made, it is evident that there is a finite number of complexions equal to the number of ways in which the P elements of energy can be divided up among the N oscillators. This number W is given by combinatorial analysis as

$$W = \frac{(N + P - 1)!}{P!(N - 1)!} \cong \frac{(N + P)^{N+P}}{P^P N^N}, \qquad (13)$$

where the second, approximate, form comes from dropping the 1 compared to the large numbers N and P and using Stirling's approximation for the factorials. From Eqs. (11) and (13), the entropy is given by the equation

$$S_N = k \left\{ (N + P) \ln (N + P) - P \ln P - N \ln N \right\}. \qquad (14)$$

When P/N is replaced by u/ε, and S_N/N by S, according to Eqs. (10) and (12), the entropy S of one oscillator in terms of its average energy u has the form

$$S = k \left\{ \left(1 + \frac{u}{\varepsilon}\right) \ln \left(1 + \frac{u}{\varepsilon}\right) - \frac{u}{\varepsilon} \ln \frac{u}{\varepsilon} \right\}. \qquad (15)$$

At this stage in the argument, the size of the energy elements ε is completely arbitrary. In fact, however, S must depend on the frequency of the oscillators as well as on u in a way prescribed by Wien's displacement law, and since k is a universal constant, this frequency dependence must come into ε. The displacement law actually requires that the entropy have the form [*]

$$S = g(u/\nu), \qquad (16)$$

so that the energy element ε must be proportional to the frequency of the oscillator,

$$\varepsilon = h \nu \qquad (17)$$

where h is the second universal natural constant in the theory. The expression for S as a function of u is now fully determined in terms of the constants h and k,

$$S = k \left\{ \left(1 + \frac{u}{h \nu}\right) \ln \left(1 + \frac{u}{h \nu}\right) - \frac{u}{h \nu} \ln \frac{u}{h \nu} \right\}. \qquad (18)$$

This result has exactly the same form as Eq. (9), and the distribution law that goes with it must therefore [**] be

$$\varrho(\nu, T) = \frac{8 \pi \nu^2}{c^3} \frac{h \nu}{\exp (h \nu/kT) - 1}. \qquad (19)$$

---

[*] Eqs. (2) and (4) require that $u = \nu f(\nu/T)$. An equivalent way of writing this result is $T = \nu F(u/\nu)$. Using the fact that $T^{-1} = \partial S/\partial u$, one obtains $\partial S/\partial u = \nu^{-1}[F(u/\nu)]^{-1}$ so that, after integration over u, S is simply a function of u/ν.

[**] Eq. (19) is just Eq. (6) rewritten with the constants A and B determined by comparison of Eqs. (9) and (18). One can also obtain Eq. (19) directly from Eq. (18) by differentiating the latter with respect to u to introduce the temperature and then applying Eq. (4).

The two constants were numerically determined from experimental results: the total energy density or STEFAN's constant σ [from Eq. (1)] determines one combination of h and k, and the ratio of frequency to temperature at the maximum of the distribution determines a second combination of h and k. From the experimental values, PLANCK computed h to be 6.55×10^{-27} erg sec and k to be 1.346×10^{-16} erg/K°. But what significance was to be attached to these constants?

In his December 14, 1900 report to the German Physical Society, PLANCK referred to "other relationships" deduced from his theory "which seem to me to be of considerable importance for other fields of physics and also of chemistry". I think there can be little doubt that this was a reference to the far-reaching importance of the two constants he had calculated, and in particular, to the constant k. At the end of this first paper on the quantum theory, PLANCK pointed out that since k is a universal proportionality factor connecting entropy and $\ln W$, it follows from BOLTZMANN's work on the entropy of a gas that

$$k = R/N_0 \qquad (20)$$

where R is the gas constant which appears in the macroscopic equation of state of an ideal gas, and N_0 is AVOGADRO's number, the number of molecules in a gram mole. Since the value of R was well established, PLANCK's computation of k meant that he had also determined AVOGADRO's number. PLANCK's value for it was 6.175×10^{23} molecules per mole. Other quantities follow directly, such as LOSCHMIDT's constant, the number of molecules per cubic centimeter of gas at standard conditions, and the mean kinetic energy of a molecule. Less obvious, especially in 1900, was the fact that the elementary unit of electric charge e was now also determined as essentially the ratio of the macroscopic FARADAY constant to AVOGADRO's number. PLANCK's value for e was 4.69×10^{-10} esu.

PLANCK thoroughly appreciated the importance of these determinations of the basic natural constants which his theory had made possible. As he said, "All these relationships can lay claim to absolute, not approximate validity, so long as the theory is really correct ... Their test by more direct methods will be a problem (for further research) as important as it is necessary." These words appear at the conclusion of the paper we have been discussing, and when PLANCK rewrote this work a few weeks later for the *Annalen der Physik* [31] he separated these considerations on the natural constants from the principal argument concerning the radiation distribution in order to give them the emphasis they well deserved. In his later writings, PLANCK carefully pointed out on several occasions that although k was understandably referred to as BOLTZMANN's constant, BOLTZMANN had never attached any great significance to it nor had he ever made any estimate of its numerical value.

PLANCK's values of N_0 and e were by far the best estimates of these basic quantities which had yet appeared in the literature. There had, in fact, been no direct determination of N_0; the only estimates of N_0 available were very indirect ones based on oversimplified models from the kinetic theory of gases. (Estimates of N_0 by these methods were first made by BOLTZMANN's senior colleague, LOSCHMIDT, in 1865.) It was not until 1908 that PERRIN began his series of experiments which were to give an essentially direct determination of N_0, but one

472   MARTIN J. KLEIN:

which was less reliable than PLANCK'S. As for e, the natural unit of charge or the charge on the electron, attempts at a direct determination had only just begun in J. J. THOMSON's laboratory in Cambridge, and there was to be no good measurement until MILLIKAN's work almost a decade later [32]. PERRIN's report [33] to the first Solvay Congress in 1911 gives a vivid picture of the status of these fundamental constants at that time and indirectly indicates why PLANCK laid so much emphasis on this aspect of his theory.

5. Boltzmann, Planck, and Quanta

Let us now go back to PLANCK's statistical calculation in which energy quanta were introduced. The calculation has been described in the last section pretty much as PLANCK himself presented it. This was a necessary preliminary to a discussion of the question raised in the introduction; how did PLANCK depart from BOLTZMANN's method? I think that this question can best be handled by comparing PLANCK's analysis with BOLTZMANN's own treatment of a closely related problem in the memoir [34] to which PLANCK refers repeatedly. This is BOLTZMANN's great memoir of 1877, *"On the Relation between the Second Law of Thermodynamics and the Theory of Probability"*, in which the statistical interpretation of entropy is set forth at length, separated from the difficulties of the kinetic treatment of the approach to equilibrium. The very first problem which BOLTZMANN treats in this paper, as a simple introduction to his concepts and methods, bears a remarkable resemblance to PLANCK's own problem [35].

BOLTZMANN considers a simple model of a gas consisting of N molecules, in which the energy of each individual molecule can take on only certain discrete values which form an arithmetic progression $0, \varepsilon, 2\varepsilon, \dots, M\varepsilon$. His comments on this model are revealing. "This fiction does not, to be sure, correspond to any realizable mechanical problem, but it is indeed a problem which is much easier to handle mathematically and which goes over directly into the problem to be solved, if one lets the appropriate quantities become infinite. If this method of treating the problem seems at first sight to be very abstract, it nevertheless is generally the quickest way of getting to one's goal in such problems, and if one considers that everything infinite in nature never has meaning except as a limiting process, one cannot understand the infinite manifold of possible energies for each molecule in any way other than as the limiting case which arises when each molecule can take on more and more possible velocities."

BOLTZMANN then turns his attention to the possible states of this gas model when it is assigned a total energy of $P\varepsilon$, where P is a large integer. Any such state is characterized by the set of integers $w_0, w_1, \dots, w_M$ which give the number of molecules having energy $0, \varepsilon, \dots, M\varepsilon$. The w's are subject to the two constraints,

$$\sum_{r=1}^{M} w_r = N \tag{21}$$

and

$$\sum_{r=1}^{M} r\, w_r = P, \tag{22}$$

which express the fixed number of particles and the fixed total energy. Each of the states, characterized by a set of numbers $\{w_r\}$, can be achieved in many ways, which BOLTZMANN refers to as complexions, depending on *which* of the

PLANCK and quantum theory   473

molecules are found with energies $0, \varepsilon, \ldots$. The number of complexions for a given distribution $\{w_r\}$ is readily recognized to be given by the expression

$$W_B = \frac{N!}{w_0! \, w_1! \ldots}. \tag{23}$$

BOLTZMANN makes the basic assumption that any particular complexion (in which, for example, molecule number one has energy 17ε, molecule number two has energy 3ε, *etc.*) is as likely to occur as any other particular complexion. (He makes this assumption plausible by a comparison to the game of lotto, or bingo.) It follows, then, that the probability of occurrence of a state characterized by the set $\{w_r\}$ is equal to W_B for this state divided by $\sum W_B$, where the sum is over all sets $\{w_r\}$ compatible with Eqs. (21) and (22).

Now it is essential in BOLTZMANN's procedure that he asks for that state, *i.e.*, that set of numbers $\{w_r\}$, for which W_B is a maximum, since it is that most probable state which he will identify with the state of thermodynamic equilibrium. It is not necessary to repeat here the calculation in which W_B is maximized and the $\{w_r\}$ in the equilibrium state are shown to obey the "Boltzmann distribution", *i.e.*, w_r in the equilibrium state is proportional to $\exp(-\beta' r \varepsilon)$, where the constant β' is shown to be proportional to T^{-1}. It is important, however, to recall that at an appropriate stage in the calculation BOLTZMANN takes the limit in which ε goes to zero and M goes to infinity in such a way that the molecules can really take on *all* values for their energy. For BOLTZMANN the ε is an artifice which makes the calculation possible (and makes the continuum intelligible!).

We can now compare BOLTZMANN's procedure with PLANCK's way of handling his problem. (The formal similarity in the problems has been deliberately stressed by using a common notation.) The first evident difference between the two procedures is in the meaning attached to the quantity W. In BOLTZMANN's discussion, the quantity W_B, in Eq. (23), is the number of complexions, detailed assignments of the energies of the individual molecules, compatible with a given distribution. It is proportional to the probability of one state $\{w_r\}$ compared to another. PLANCK, on the other hand, never introduces quantities analogous to the $\{w_r\}$. For him the W of Eq. (13) is a probability *by definition:* he has no "model" for understanding this probability in any sense analogous to BOLTZMANN's. Thus when PLANCK sets $k \ln W$ equal to the entropy, he says [30], "In my opinion, this stipulation basically amounts to a definition of the probability W; for we have absolutely no point of departure, in the assumptions which underlie the electromagnetic theory of radiation, for talking about such a probability with a definite meaning."

PLANCK, then, does not find the number of complexions which belong to the most probable state at all. Instead he takes for his W a quantity which is the total number of complexions for *all* sets $\{w_r\}$ which satisfy the constraints of Eqs. (21) and (22). In other words, PLANCK's W is equal to the quantity BOLTZMANN called $\sum W_B$, and the equivalent of Eq. (13) appears in BOLTZMANN's memoir when he calculates $\sum W_B$, the normalization factor for his probabilities.

It is natural, though rather pointless, to ask why PLANCK deviated from BOLTZMANN's procedure at this particular stage. Had he carried on with BOLTZMANN's method, he would have arrived at exactly the same result for the average

474   Martin J. Klein:

energy of an oscillator, namely $\varepsilon/\{\exp(\varepsilon/kT) - 1\}$. Rosenfeld [36] suggests that Planck actually started with Eq. (9) for the entropy of an oscillator required by his conjectured distribution law, and went from that to the corresponding form for S_N, the entropy of N oscillators. If S_N were to be given by an expression of the form $k \ln W$, the form of W was then determined to be something like $(N+P)^{N+P}/N^N P^P$. This last result could then be recognized as a legitimate approximation to $(N+P-1)!/(N-1)! \, P!$, a standard formula of the theory of combinations, which, as we have just seen, actually appeared in the Boltzmann paper to which Planck was referring. This rather plausible conjecture is well confirmed by what Planck himself says in his *Naturwissenschaften* article [1(b)] written in 1943.

There is one other aspect of Planck's combinatorial procedure which deserves a comment here. Some years after Planck's work, when Einstein had driven the theory a long step further by showing that radiation itself could behave as if it consisted of energy quanta, physicists tried to reinterpret Planck's reasoning along this same line. This attempt to consider Planck's energy elements as, in some sense, particles of energy seemed a plausible one, but it was quite inconsistent with the combinatorial treatment which had to be given these "particles" if one were to obtain the Planck distribution law. This was pointed out by Ehrenfest in 1911 [37] and again, in more detail, in 1914 [38]. (The latter paper is especially noteworthy as it contains the simple and graphic derivation of the basic combinatorial formula, Eq. (13), which is now universally given.) What Ehrenfest showed, in effect, was that "particles" which have to be counted according to Eq. (13) are not independent particles in any ordinary sense. They are, in fact, particles which obey the Bose-Einstein statistics, but that concept could not be clarified until many years later [39].

This first, combinatorial, deviation from Boltzmann is less striking than the one we shall now consider. As we have already seen, Boltzmann too used "energy elements" ε in order to carry out his combinatorial procedure, but Boltzmann was always ready to take the limit $\varepsilon \to 0$, once the discreteness was no longer necessary to the analysis. It is obviously of the very essence of Planck's work that ε could not be allowed to vanish, if the proper distribution law were to be reached. Planck apparently did not even consider the possibility of taking this limit. This is undoubtedly related to Planck's apparent unawareness of the equipartition theorem and all it implied, which we have already seen.

This aspect of Planck's work seems to have been recognized first in 1905, and it came out very clearly in the course of an exchange between Jeans and Lord Rayleigh in the columns of *Nature*. In the May 18, 1905 issue, Rayleigh [40] repeated his calculation of five years before (see Sect. 3 above), but this time he took care to include all of the proportionality constants which he had not bothered with earlier. The distribution law he obtained was

$$\varrho(v, T) = (8\pi v^2/c^3)\,(kT), \tag{24}$$

and, as he pointed out, this is exactly the same as the form that Planck's law, Eq. (19), takes in the limit of low frequencies (*i.e.*, when $hv/kT \ll 1$), a limit which Planck had not explicitly considered. It followed that the quantity k, and therefore Avogadro's number N_0, could in principle be determined from the experi-

PLANCK and quantum theory

475

mental value of $\varrho(\nu, T)$ at low frequencies using Eq. (24) without any reference to the Planck distribution itself. RAYLEIGH went on to say, "A critical comparison of the two processes (*i.e.* his own and PLANCK's) would be of interest, but not having succeeded in following PLANCK's reasoning, I am unable to undertake it. As applying to all wave lengths, his formula would have the greater value if satisfactorily established. On the other hand, the reasoning which leads to [Eq. (24)] is very simple, and this formula appears to me to be a necessary consequence of the law of equipartition as laid down by BOLTZMANN and MAXWELL. My difficulty is to understand how another process, also based upon BOLTZMANN's ideas, can lead to a different result."

Actually RAYLEIGH had made an error of a factor of eight in his calculation which was soon pointed out by JEANS. (Eq. (24) is, however, correct, and it is the limiting form of PLANCK's law.) RAYLEIGH [*41*] readily admitted his error but returned to the same point: "But while the precise agreement of results in the case of very long waves is satisfactory so far as it goes, it does not satisfy the wish expressed in my former letter for a comparison of processes. In the application to waves that are not long, there must be some limitation on the principle of equi-partition."

RAYLEIGH's repeated request for a critical discussion was finally met, at least in part, by JEANS [*42*] in *Nature* for July 27, 1905. (JEANS was then, and for a number of years thereafter, doing his utmost to account for black-body radiation on strictly classical grounds, preserving the truth of the equi-partition theorem in general, and treating deviations from it as due to the absence of true thermodynamic equilibrium.) JEANS undertook a severe criticism of PLANCK's arguments on two principal points. These points are just the two we have been discussing, where PLANCK broke with BOLTZMANN. Thus JEANS attacked PLANCK's use of W as a "probability", pointing out that no population was given from which probabilities could be calculated, and that one could not introduce such a population with an *a priori* probability law consistent with PLANCK's arguments. JEANS' second point was that PLANCK had no right to refrain from taking the limit in which ε is zero. If this were done, PLANCK's expression for the average energy of an oscillator would reduce to kT in accord with the equipartition theorem. JEANS recognized that PLANCK had fixed ε as equal to $h\nu$ by the use of WIEN's displacement law, but he argued that nothing in the displacement law determined the value of h, "whereas statistical mechanics gives us the further information that the true value of h is $h=0$". If by "true" one means in agreement with the equipartition theorem, then JEANS was correct. JEANS erred only in supposing that "the methods of both are in effect the methods of statistical mechanics and of the theorem of equipartition of energy". That was JEANS' method, but it was certainly not PLANCK's.

The most revealing exposure of the deep chasm which PLANCK had opened between his own ideas and BOLTZMANN's was made by EHRENFEST [*43*]. He pointed out that PLANCK's work really re-opened the whole question of the statistical foundation of the second law of thermodynamics, since PLANCK's energy elements amounted to a radical change in the *a priori* weight function introduced into phase space. BOLTZMANN's statistical mechanics had been built on the assumption that regions of equal volume in phase space were to be assigned

476   MARTIN J. KLEIN:

equal *a priori* weights, but if the energy were to be a discrete variable, some new basic assumption would be needed to rebuild the foundations.

6. Conclusion

A revolutionary idea is not always recognized as such, not even by its propounder. PLANCK'S concept of energy quanta went practically unrecognized in the literature of physics for over four years. His radiation formula was accepted as describing the experimental facts in a simple and adequate way, but the theory which he had proposed as a basis for this formula drew no attention until 1905. Although this virtual ignoring of what later came to be recognized as a major advance in physics may seem odd, there are several possible explanations for it. The most obvious is that the theory of radiation was not the center of interest in physics in 1900. I need only mention that X-rays were discovered in 1895, radioactivity in 1896, the electron in 1897, radium in 1898, *etc.*, to remind the reader of the series of exciting discoveries which were being made and which were drawing the attention of a substantial fraction of the best minds in physics.

In addition to the competition of these unparalleled advances, PLANCK'S work suffered from another serious hazard. At just this time, when not only the atom but even its constituent parts were taking the center of the experimental stage, a substantial and influential group of theorists on the continent had set themselves against the whole atomic theory [*44*]. With varying degrees of emphasis, such men as OSTWALD, MACH and DUHEM were denying the significance of the whole program of the kinetic theory of gases and the general atomic, mechanical view of physics which BOLTZMANN had championed. BOLTZMANN carried on a vigorous polemic, particularly against OSTWALD, in the latter 1890's, feeling himself obliged to justify "The Indispensability of Atomism in the Natural Sciences". (PLANCK had joined BOLTZMANN in the attack on the "energetics" school in 1896, despite his own lack of sympathy with BOLTZMANN'S work at that time, pointing out that the "energeticists" misconceived the proper meaning of the second law of thermodynamics.)

The seriousness of the attacks made by OSTWALD and others against the kinetic molecular theory can be measured by the tone of the preface which BOLTZMANN wrote in 1898 for the second volume of his *Gastheorie* [*45*]. "When the first part of this book was printed I had almost completely finished a manuscript of the present second and last part in which the more difficult parts of the subject were not treated. Just at that time [1896] the attacks against the kinetic theory multiplied. I am now convinced that these attacks are based completely on misconceptions and that the role of the kinetic theory in science is far from being played out ... In my opinion it would be a loss to science if the kinetic theory were to fall into temporary oblivion because of the present, dominantly hostile mood, as, for example, the wave theory did because of NEWTON'S authority. I am conscious of how powerless the individual is against the currents of the times. But in order to contribute whatever is within my powers so that when the kinetic theory is taken up once again, not too much will have to be rediscovered, I have also treated the most difficult parts of the kinetic theory in the present volume, those which are most liable to misunderstandings, and I have tried to give as easily intelligible an exposition of these as possible, at least in outline."

PLANCK and quantum theory   477

Now PLANCK had quite deliberately associated his thinking with BOLTZMANN's in his papers on quanta at just the time when BOLTZMANN himself was under the dark cloud cast by the school of "energetics". (This gives a touch of poignancy to PLANCK's remark, "As an offset against much disappointment, I derived much satisfaction from the fact that LUDWIG BOLTZMANN, in a letter acknowledging my paper, gave me to understand that he was interested in, and fundamentally in agreement with, my ideas." [46]) There is little doubt that the intentionally Boltzmann-like tone of PLANCK's papers on quanta contributed to the delay in their recognition, at least on the continent.

In England, as we have already seen, RAYLEIGH and JEANS did some probing into the significance of PLANCK's work in 1905. It is certainly fair to say, however, that neither RAYLEIGH nor JEANS was sympathetic to the idea of energy quanta nor was either of them interested in developing the idea any further. Such a revolutionary idea could take root and grow only in a mind keen enough to see its implications and bold enough to develop them immediately in a variety of directions. Such a mind was ready, fortunately for science, and in the work of ALBERT EINSTEIN the full significance of PLANCK's concept began to show itself.

In the first [47] of the three great papers which he wrote in the spring of 1905, EINSTEIN pointed out the distribution law which was required by classical physics, the "Rayleigh-Jeans law" (Eq. (24) above) and stressed both its inconsistency with experiment and its internally paradoxical nature, due to the infinite radiation energy which it implies. This was apparently done quite independently of RAYLEIGH's 1900 paper to which EINSTEIN does not refer and which there is no reason to believe he knew. EINSTEIN also pointed out, as RAYLEIGH did simultaneously in the *Nature* letter discussed in the last section, that PLANCK's determination of AVOGADRO's number is really independent of his radiation formula, and that the same result could be found from the classical formula using the experimental results on black-body radiation. All these things, as well as the deep gap between PLANCK's quanta and classical physics were, as EINSTEIN was to write many years later [48], "quite clear to me shortly after the appearance of PLANCK's fundamental work". In addition to these points, which RAYLEIGH saw too, EINSTEIN showed that the non-classical part of PLANCK's radiation formula, the Wien limit, so to speak, had a remarkable consequence. It implied, as EINSTEIN demonstrated by a characteristically simple argument, that radiation itself behaved as if it consisted of energy quanta whose magnitude was given by $h\nu$. This went well beyond any conclusion which PLANCK himself had drawn, since PLANCK had quantized only the energy of the material oscillators and not the radiation. EINSTEIN went on to show in a few pages how the photoelectric effect, STOKES' rule for fluorescence, and the photoionization of gases could all be very simply understood by treating light as composed of energy quanta $h\nu$. None of these phenomena had been explicable on the basis of the electromagnetic wave theory of light.

EINSTEIN was well aware that all of this marked the beginning of a new era in physics, and he indicated that awareness by referring to his work in the title of his paper as offering "a heuristic viewpoint". He saw that thoroughgoing changes in the foundations were needed, but even "without having a substitute for classical mechanics, I could nevertheless see to what kind of consequences this law of temperature radiation leads".

478   Martin J. Klein:

The effect of Planck's quantum theory on physics in this period is perhaps best expressed in Einstein's words [48]: "It was as if the ground had been pulled out from under one, with no firm foundation to be seen anywhere upon which one could have built."

Acknowledgement. This work was supported in part by the National Carbon Company, a division of Union Carbide Corporation.

References

[1] There are three such retrospective accounts: (a) Planck's Nobel Prize Lecture: Die Entstehung und bisherige Entwicklung der Quantentheorie (Leipzig 1920); English translation in Planck's: A Survey of Physics (New York n.d.); (b) Naturwissenschaften **31**, 153—159 (1943); (c) Wissenschaftliche Selbstbiographie (Leipzig, 1948); English translation in Planck's Scientific Autobiography (New York, 1949). — All three are reprinted in Max Planck: Physikalische Abhandlungen und Vorträge, Bd. III. Braunschweig 1958. This collection will be referred to throughout as Phys. Abh.

[2] Rosenfeld, L.: La première phase de l'évolution de la Théorie des Quanta. Osiris **2**, 149—196 (1936).

[3] Rosenfeld, L.: Max Planck et la définition statistique de l'entropie, in Max-Planck-Festschrift 1958 (Berlin, 1958), pp. 203—211.

[4] See E. T. Whittaker, A History of the Theories of Aether and Electricity (London, 1951), vol. 1, p. 371.

[5] Planck, M.: Ref. 1 (c) English transl. pp. 34—35.

[6] Stefan, J.: Wien. Ber. **79**, 391 (1879).

[7] Boltzmann, L.: Wied. Annalen **22**, 291 (1884). See also M. Planck, The Theory of Heat Radiation (Reprinted New York, 1959), Chapter 2.

[8] Wien, W.: Wied. Annalen **52**, 132 (1894).

[9] Wien, W.: Wied. Annalen **58**, 662 (1896).

[10] Lord Rayleigh: Phil. Mag. **49**, 539 (1900). Reprinted in his Scientific Papers (Cambridge, 1903), vol. IV, pp. 483—485.

[11] Planck, M.: S.-B. Preuss. Akad. Wiss. (**1897**), 57; — Phys. Abh. Bd. I, S. 493.

[12] Boltzmann, L.: S.-B. Preuss. Akad. Wiss. (**1897**), 660, 1016.

[13] Planck, M.: S.-B. Preuss. Akad. Wiss. (**1899**), 440; — Phys. Abh. Bd. I, S. 560. Especially S. 575—581. — The theorem in the form quoted in the text appears in Bd. I, S. 724.

[14] Planck, M.: Phys. Abh. Bd. I, S. 597.

[15] Planck, M.: Ann. Phys. **1**, 69 (1900); — Phys. Abh., Bd. I, S. 614.

[16] Lummer, O., & E. Pringsheim: Verh. dtsch. Physik. Ges. **1**, 215 (1899).

[17] Thiesen, M.: Verh. dtsch. Physik. Ges. **2**, 37 (1900).

[18] See reference 1 (b). Planck, M.: Phys. Abh., Bd. III, S. 262.

[19] See Planck, M.: Phys. Abh., Bd. III, S. 404.

[20] Planck, M.: Verh. dtsch. Physik. Ges. **2**, 202 (1900); — Phys. Abh., Bd. I, S. 687.

[21] Planck, M.: Reference 1 (c). English translation p. 40. Phys. Abh., Bd. III, S. 394.

[22] Lummer, O., & E. Jahnke: Ann. Phys. **3**, 288 (1900).

[23] Rubens, H., & F. Kurlbaum: S.-B. Preuss. Akad. Wiss. (**1900**), 929.

[24] Planck, M.: Verh. dtsch. Physik. Ges. **2**, 237 (1900); — Phys. Abh., Bd. I, S.698.

[25] Planck, M.: Reference 1 (c). English translation p. 32. Phys. Abh., Bd. III, S. 387.

[26] Planck, M.: Wied. Ann. **30**, 562 (1887); — Phys. Abh., Bd. I, S. 196.

[27] Planck, M.: Z. phys. Chem. **8**, 647 (1891); — Phys. Abh., Bd. I, S. 372.

[28] Planck, M.: Treatise on Thermodynamics (Reprint of English translation of Seventh German Edition, New York, 1945) p. vii.

PLANCK and quantum theory   479

[29] PLANCK, M.: Reference 1 (a). English translation p. 166; — Phys. Abh., Bd. III, S. 125.

[30] PLANCK, M.: Ann. Phys. 4, 553 (1901); — Phys. Abh., Bd. I, S. 717.

[31] PLANCK, M.: Ann. Phys. 4, 564 (1901); — Phys. Abh., Bd. I, S. 728.

[32] See MILLIKAN, R.A.: The Electron (Chicago 1917).

[33] PERRIN, J.: Les Preuves de la Réalité Moléculaire, in La Théorie du Rayonne-ment et les Quanta, edit. by P. LANGEVIN & M. DE BROGLIE, pp. 153—250 (Paris 1912).

[34] BOLTZMANN, L.: Wien. Ber. 76, 373 (1877). Reprinted in his Wissenschaftliche Abhandlungen, Bd. II, S. 164 (Leipzig 1909).

[35] See the discussion in R. DUGAS, La Théorie Physique au sens de Boltzmann (Neuchâtel, 1959), especially Ch. 6 (Seconde Partie) and Ch. 4 (Troisième Partie).

[36] ,ROSENFELD, L.: Reference [2], p. 167.

[37] EHRENFEST, P.: Ann. Phys. 36, 91 (1911).

[38] EHRENFEST, P., & H. KAMERLINGH ONNES: Proc. Amsterdam Acad. 17, 870 (1914). The two preceding papers are reprinted in P. EHRENFEST, Collected Scientific Papers, ed. by M. J. KLEIN (Amsterdam, 1959), on pages 185 and 353 respectively.

[39] See KLEIN, M. J.: Proc. Amsterdam Acad. B 62, 41 (1959).

[40] Lord RAYLEIGH: Nature 72, 54 (1905).

[41] Lord RAYLEIGH, Nature 72, 243 (1905).

[42] JEANS, J.: Nature 72, 293 (1905).

[43] EHRENFEST, P.: Phys. Z. 15, 657 (1914). See also his article in Naturwissenschaf-ten, 11, 543 (1923). Both are reprinted in his Collected Scientific Papers, pages 347 and 463.

[44] See DUGAS: Reference 35, Chap. 6—8 (Première Partie) and Chap. 5 (Troisième Partie).

[45] BOLTZMANN, L.: Vorlesungen über Gastheorie, Bd. II, S. V. (Leipzig 1898).

[46] PLANCK, M.: Reference 1 (a). English tranlation p. 167. Phys. Abh., Bd. III, S. 126.

[47] EINSTEIN, A.: Ann. Phys. 17, 132 (1905).

[48] EINSTEIN, A.: Philosopher-Scientist, edit. by P. A. SCHILPP, p. 45 (New York 1949).

Department of Physics
Case Institute of Technology
Cleveland, Ohio

(Received September 1, 1961)

THOMAS S. KUHN

Revisiting Planck

IT IS NOW over five years since the publication of *Black-body theory and the quantum discontinuity, 1894–1912,*[1] a book which provides the most fully realized illustration of the concept of history of science basic to my historical publications. The same conception underlies my more philosophical writing—is, indeed, what ties these apparently disparate aspects of my work together. But what I see as the best and, technical difficulty aside, the most representative of my historical works has been widely received, even among those who praise it, as a misfit, a problem child, among my publications. Under those circumstances the invitation to prepare for *HSPS* a summary of the book's central thesis and of the main arguments relevant to its evaluation has been particularly welcome. With the editors' encouragement, I shall try also to indicate what, beyond the historical facts of the matter, seems to me to be at stake. Need for a restatement of this sort results only partly from the technicality which has kept many commentators, friendly as well as hostile, from quite recognizing what the book is about. In some circles those difficulties have been compounded by misperceptions of the relationship between it and my earlier *Structure of scientific revolutions.*

Taken as a whole, the book is a narrative account of Planck's invention of the black-body theory known by his name and of that theory's development during the years when it and a closely-related theory of specific heats were the two exemplary applications of a still-to-be-developed quantum theory. Though the account is probably more thorough and detailed than previous treatments of its subject, its objective was of another sort: a fundamental reinterpretation of Planck's thought and of the stages in its gradual transformation. Most of the

*Department of Linguistics and Philosophy, Massachusetts Institute of Technology, Cambridge, MA 02139. The final form of this paper has greatly benefited from suggestions directed to an earlier draft by Peter Galison, John Heilbron, Paul Horwich, Andrew Pickering, and Norton Wise.

1. Oxford and New York, 1978.

232   KUHN

detail in the book serves the reinterpretation. Only in the first chapter of Part I and the last chapter of Part II were extra topics included to make the narrative more nearly comprehensive and balanced.

In returning to the subject here, I have concentrated on the reinterpretation of Planck, abandoning both balance and the narrative mode. The paper that results consists of four parts. Section 1 outlines the principal elements of the new interpretation: the long-forgotten way in which Planck and many of his readers initially understood his black-body theory. Section 2 summarizes the three main lines of argument favoring that interpretation. Section 3 concludes my return to Planck by analyzing those passages in his early black-body papers that have persistently been read as the introduction of quantization. In the final section, an addendum, I ruminate on the historiographic/philosophical position that these arguments both presuppose and reinforce. The treatment is skeletal throughout; fleshed out versions of all but Section 4 will be found in the book, together with documentation and ac.-knowledgements. Parenthetical page references, below, indicate relevant passages in the book.

1. PLANCK'S FUNDAMENTAL INNOVATION

Planck, it has ordinarily been said, introduced at the end of 1900 the concept of a linear electrical oscillator with energy restricted to integral multiples of the energy quantum $h\nu$, ν being the oscillator frequency and h the universal constant later known by Planck's name. He had discovered that restricting energy levels to a discontinuous spectrum was essential to the derivation of the black-body radiation law he had introduced shortly before. Though Planck doubtless hoped that the discontinuity would prove eliminable and may even have conceived it as primarily a mathematical artifice, he as yet saw no route to his black-body law that did not require the hypothesis of discontinuity, of energy quantization. That view of Planck's development has been standard for years, and I am among those who have propagated it and based research upon it. But I am now quite certain it is wrong and that Planck's discovery should instead be described in the following manner.

In order to derive his black-body law Planck had to subdivide the energy continuum into cells or elements of size ϵ. Boltzmann had introduced such divisions in 1877 when presenting a probabilistic derivation of the entropy and velocity distribution of a gas, and Planck was following Boltzmann closely. But there was, Planck discovered, a crucial difference between the requirements he and Boltzmann had to place on these elements. For Boltzmann the precise size of ϵ made no difference. It did have to be large enough so that the cells would contain many molecules and also (though this requirement Boltzmann seems not to have recognized) small enough to permit the substitution of integrals for sums. But any value satisfying those constraints would do (pp. 59f.). Planck's derivations, in contrast, required that the cell size be proportional to oscillator frequency. With h as the constant of proportionality, Planck's cells were fixed at the small finite size $h\nu$.

Only by fixing the cell size in this way could Planck derive his distribution law. The restriction puzzled him. But it was for him a restriction on cell size, not on resonator energy, and it did not therefore bring to mind anything like quantization. With respect to energy, Planck's oscillators (he called them "resonators") were like Boltzmann's molecules. They moved freely through and between energy cells as required by Newton's and, in Planck's case, by Maxwell's laws (pp. 104f., 130–134). What differentiated Planck's derivation from Boltzmann's was not that the former required the violation of these or other laws of classical physics but rather the entry of the constant h, which Planck referred to as the "quantum of action" and which he would later describe as giving a physical structure to phase space (pp. 129, 250f.).

In fact, of course, Planck's law cannot be derived in this way. Planck failed to notice that his argument was valid only for frequencies such that $h\nu \ll kT$. But that failure was standard at the time. Boltzmann, whom Planck was following, had overlooked the equivalent approximation in his own derivation. Until Einstein pointed it out at Salzburg in 1909, neither he nor anyone else appears to have been aware that Planck's mathematics had slipped (pp. 185f.). And until that slip was discovered, Planck's view of what he had done was both coherent and independent of discontinuity.

2. THREE SORTS OF EVIDENCE

The preceding interpretation of Planck's early theory is clearly non-standard. Also, it could be wrong. No single piece of available evidence demands it, and evidence incompatible with it could yet be discovered, a letter, for example, or an unpublished manuscript. As things now stand, however, evidence for the reinterpretation seems to me overwhelming. The relevant arguments take three main forms.

First, the reinterpretation makes the development of Planck's black-body research both more nearly continuous and also a deeper, more elegant piece of physics than it appears in the standard version. That development began in 1894 when Planck launched an attempt to apply to electromagnetic radiation a formulation of the second law of thermodynamics that he had originally introduced in his doctoral thesis. His main aim in doing so was to show that reversible equations (in this case those of wave theory) could be used to explain irreversible

234   KUHN

processes, and his approach was very close, though Planck may not at the start have known it, to one that Boltzmann had developed for gases in 1872.

Boltzmann's approach was not yet the probabilistic one mentioned in the previous section but an earlier one that required tracing molecular trajectories and averaging over the possible molecular collisions in a gas. More specifically, Boltzmann showed, or claimed to show, that if $f(q_i, p_i, t) \, dq_i \, dp_i$ was the fraction of gas molecules in the phase-space cell $dq_i \, dp_i$ at time t, then the integral over phase space of $f \log f$ was a function, $H(t)$, which could only decrease with time until it reached a minimum value that it thereafter retained. When the minimum was reached the gas was in equilibrium and the distribution function had acquired the Maxwellian form. That result constituted Boltzmann's H-theorem, which demonstrated, subject to a certain controversial restriction, the irreversible approach of a gas to equilibrium. Boltzmann carried it one crucial step further. Using the standard kinetic theory definitions of heat and temperature, he showed that S, the entropy of the gas in equilibrium, was just the negative of H. Decreases in H therefore corresponded precisely to increases in S (pp. 39–42). A second route to the distribution law was thus available. It could be obtained by applying the standard thermodynamic relation, $\partial S / \partial E = 1/T$.

Planck attempted to develop a similar theorem for radiation in a cavity that also contained damped resonators tuned to all radiation frequencies. He sought, that is, a function S of the field and resonator parameters that could only increase with time. By 1899 he had found one, though it, too, was subject to a restriction like Boltzmann's. Next, he attempted to complete the Boltzmann program, which he by then knew well, and at once encountered, as he would again in 1900, a key difference between the radiation and the gas-theoretical cases (pp. 73–86).

What Planck now sought was a function that specified the distribution of radiant energy over frequency at each temperature. But temperature does not appear in either Maxwell's equations or those governing a damped resonator, and Planck had no model, like that of kinetic theory, which permitted him to introduce it. He could not therefore directly derive a distribution function, nor could he, for the same reason, derive the entropy. But he could and did take an indirect route to those goals, one that he acknowledged from the start was open to question. He assumed that his function S, just because it increased monotonically to a maximum, was *the* thermodynamic entropy, and he then obtained a distribution function with the aid of the relationship $\partial S / \partial E = 1/T$. He had sought, he said, for other functions that, like entropy, could only increase monotonically with time, and he had found none. His function, he believed, was unique, and it must therefore coincide with the entropy of the radiation system (pp. 86–91).

The distribution law at which Planck arrived in this way was the so-called Wien law, which until late in 1899 corresponded well with available experiments. During the following twelve months, however, as measurements were extended further into the infra-red, the experimental accuracy of that law became more and more questionable, and other laws were proposed. The resulting uncertainties extended, of course, to Planck's derivation, and he immediately sought a more direct way of establishing an entropy function, his own or another, to insert in his treatment of irreversibility.

A first such attempt was announced in March and led again to the Wien law (pp. 92–97). Then, as evidence against that law continued to accumulate, Planck produced, first, a particularly promising candidate for the distribution law, and, then, in December 1900 and January 1901, a probabilistic derivation of that law. Those two derivation papers are the ones in which he is supposed to have introduced the concept of a discrete energy spectrum for resonators, thus of quantization and discontinuity. But the papers themselves make no explicit mention of such concepts, and his next relevant paper is not easily reconciled with the assumption that he nevertheless had them in mind.

That paper, which appeared late in 1901, was described in its title as a "Supplement" to the one in which, at the start of 1900, he had presented his proof of irreversibility and his demonstration that the Wien law would follow if his candidate for entropy function were unique. After a brief introduction, both the paragraphs and formulas of the "Supplement" were numbered to continue where those of the earlier paper had stopped. What he showed in those paragraphs was that his new entropy function, like the older one he had thought unique, could only increase monotonically to a maximum with time. The role of his new probabilistic argument was, as he saw it, simply to fill a gap in the theory he had completed in 1899. It demonstrated that the new function was *the* thermodynamic entropy, thus replacing the always questionable uniqueness assumption.

There followed a four-and-one-half year interval in which Planck published nothing on radiation theory. Then, in 1906, for the first edition of his *Lectures on the theory of thermal radiation,* he redeveloped precisely the treatment he had provided in 1900. The first three chapters of the *Lectures* derived the properties of cavity radiation and damped oscillators from the equations of "Electrodynamics and thermodynamics"; the fourth used probability theory to derive an entropy function; and the fifth, titled "Irreversible radiation processes," combined these results to show that the entropy function Planck had derived could only increase to a maximum with time. A bibliography

236   KUHN

of Planck's writings on thermal radiation concluded the volume (pp. 116–120). Neither in it nor in Planck's earlier papers is there any mention of discontinuity, any talk of a restriction on resonator energy, any formula like $U = nh\nu$, with U the energy of a single resonator and n an integer. Not until 1908 do formulations of that sort occur in Planck's known work, published or unpublished. What has nevertheless suggested that he conceived the energy spectrum as discrete is the manner in which he computes the probability of an energy distribution, the probability which he then converts to entropy by taking its logarithm. That evidence is striking, but it is also based on a persistent misunderstanding of Planck's derivation, a misunderstanding that will be the subject of the next section.

Glimpses of the possibility of the narrative just sketched are what initially persuaded me that Planck's first theory did not involve quantization or discontinuity. Told in this way, the story makes better historical sense than the long-standard version. First, as almost always happens in historical reinterpretation, the new narrative is more nearly continuous than its predecessor. The stages that prepare the way to fundamental innovation are seen to be more numerous than they seemed before, but also individually smaller, more fully prepared, more obviously within the reach of an exceptionally capable person. Consider the successive scholarly retellings of the story of Newton, Galileo, Darwin, or Freud.

Second, this reinterpretation eliminates a number of the apparent textual anomalies and inconsistencies that have led to talk of Planck's conservatism, of his confusion, and of his good luck in finding within Boltzmann's work the probabilistic formula he needed while failing entirely to see how properly to derive that formula from his model. Planck's so-called "second theory," a reformulation first announced in 1911, was not for him a retreat, as is often said, but rather the first theory from his pen to make use of discontinuity at all. His apparently unproblematic use of the equations of classical physics (in particular to derive the famous formula, $u_\nu = (8\pi\nu^2/c^3)\, U_\nu$, relating energy density in the field to resonator energy) within a theory that set these equations aside, was not an example of inconsistency, for such a theory did not then exist. And, as I shall argue in the next section, Planck understood both probability theory and its role in his derivation very well. In short, the Planck who appears in the reinterpretation is a better physicist—less a sleepwalker, deeper and more coherent—than the Planck of the standard story. Though the reinterpretation changes the nature of his contributions to the quantum theory, it does not at all diminish his role in that theory's creation.

A second sort of evidence for the reinterpretation is likely in the long run to prove decisive. Whatever Planck's theory may have been in 1900 and 1901, the theory presented in his 1906 *Lectures* is the continuum theory I have just attributed to him. My most learned and authoritative critic concedes this point. No other commentator of whom I am aware has challenged it. Either Planck held the same view when he introduced his theory at the end of 1900, or his view had changed in the six intervening years.

The last alternative is extremely unlikely. The evidence that Planck believed in a discrete energy spectrum in 1900 and 1901 is his method of calculating the probability of a particular distribution. He used the identical method in 1906, and described it in virtually identical words. The main presently relevant differences between the earlier and later treatments are, as will appear in Section 4, below, the inclusion in the latter of an occasional clarifying phrase. If Planck had changed his mind, found a continuum version of his previously discontinuous theory, would he not have altered his derivation, or at least his phraseology, accordingly?

There is another difficulty as well. By the turn of the century Planck's publication pattern was well established. Typically, he published reports on research programs in progress. Then, when he thought the program finished, he would sometimes present a systematic summary of the whole in an article or in a book. Between 1901 and 1906, however, he published nothing at all on black-body theory. What he did then publish in the way of new research dealt primarily with dispersion theory. An important footnoote in his *Lectures* suggests that he took up that topic in search of clues to an explanation of the universal constant h, but nothing in his papers suggests that he found any, and he makes no reference to this work in his lectures (p. 133). It is unlikely that Planck would have brought out a new version of his theory in book form without either publishing reports on the research that had led to it or mentioning that the version was new, not to be found in the papers listed at the end of his book. His behavior was very different when, in and after 1911, he did explicitly change his mind. Before describing his second theory in a much-revised second edition of his *Lectures,* Planck had, in the two preceding years, published four different papers on two different versions of his new viewpoint (pp. 236f.).

Turn, finally, to a third line of evidence, Planck's behavior at the time when, in my view, he was at last persuaded that his derivation demanded discontinuity. The point itself was not original with Planck: Einstein and Ehrenfest had insisted upon it in 1906 (pp. 166–169; 182–185). But Planck himself was not convinced for another two years. What then changed his mind appears to have been interaction with Lorentz who, in April of 1908, gave a major address on the black-body problem to an international assembly of mathematicians in Rome

238   KUHN

(pp. 189–196). It is in a letter to Lorentz about that lecture, a letter written in October 1908, that Planck's first known statement of the discreteness of the energy spectrum and the need for discontinuity occurs. The excitation of resonators, he writes,

> does not correspond to the simple known law of the pendulum; rather there exists a certain threshold: the resonator does not respond at all to very small excitations; if it responds to larger ones, it does so only in such a way that its energy is an integral multiple of the energy element $h\nu$, so that the instantaneous value of the energy is always represented by such an integral multiple.
> In sum, I might therefore say, I make two assumptions:
> 1) the energy of the resonator at a given instant is $gh\nu$ (g a whole number or 0);
> 2) the energy emitted and absorbed by a resonator during an interval containing many billion oscillations (and thus also the average energy of a resonator) is the same as it would be if the usual pendulum equation applied (p. 198).

These are the sorts of statements that one expects, but cannot find, anywhere in Planck's previous writings.

Statements of the same sort appear regularly in Planck's later correspondence and in his published writings after late 1909. Then, early in 1911, he presented a first version of his second theory, which restricted the discontinuity he had just acknowledged to the emission process (pp. 199f., 235f.). These conceptual changes were accompanied by two significant alterations in Planck's technical vocabulary. The phrase "energy element," which had referred to a mathematical subdivision of the energy continuum, was replaced by "energy quantum," the term for a physically separable and indivisible atom of energy. Only twice before had Planck used the latter phrase in his extant writings, both times in correspondence with people who used it themselves. From 1909, however, he used it regularly. Simultaneously, Planck banished the term "resonator," which had previously been his standard word for the hypothetical entities which, by absorption and reemission, redistributed energy in the field. Writing to Lorentz at the end of 1909, he said: "Of course you are right to say that such a resonator [one that responds discontinuously to stimulation] no longer deserves its name, and that has moved me to strip it of its title of honor and call it by the more general name 'oscillator'." Thereafter, Planck used the term oscillator exclusively (pp. 200f.).

Changes like these, which began only in late 1908, are a third sort of evidence for the proposed reinterpretation of Planck's early theory. There are others, for example his autobiographical narratives (pp. 131), but the three sketched here are primary.

3. THE COUNTER-EVIDENCE: PLANCK'S DERIVATION

What, then, is the evidence that favors the longstanding interpretation of Planck's first derivations of his black-body law? There is some, and it is strong enough so that two of Planck's most proficient early readers, Lorentz and Ehrenfest, described that derivation as positing discontinuity (p. 138). But I am nevertheless persuaded that that longstanding description rests on a misunderstanding of Planck's derivation, the first presentations of which were extremely obscure. And Lorentz, at least, corrected his description after the publication of Planck's *Lectures*. In 1908, for example, he wrote that "according to Planck's theory [elsewhere Lorentz calls it "Planck's first treatment of this subject"] resonators receive or give up energy to the ether in an entirely continuous manner (without there being any talk of a finite energy quantum)" (p. 139). Most other early readers, apparently including both Larmor and Einstein, had read Planck that way from the beginning (pp. 135–40, 182–186).

Look now at Planck's derivation, which is often said to be no derivation at all. Working backward from his distribution law Planck derived a corresponding entropy function. Looking for a probability W, of which the logarithm would be proportional to that entropy, he discovered an appropriate combinatorial formula, probably in an 1877 paper of Boltzmann's that he is known to have read. According to the standard account, however, he was unable to justify his use of that formula, and his remarks on the reasons for introducing it were no more than hand-waving. I have offered a new account of the derivation, one that makes it good physics, and that account seems to have been generally accepted. But what has not ordinarily been recognized about it—partly because I had not, at the time of writing, fully grasped the point myself—is the way in which the reinterpretation of Planck's derivation eliminates even the appearance of discontinuity from his early papers.

To understand either Planck's derivation or the standard way of misreading it, one must first understand Boltzmann's probabilistic derivation of the distribution law for gases. Boltzmann examines the ways of distributing total energy E over a gas consisting of N molecules. For this purpose, he divides the energy continuum into P cells of size ϵ, so that $P\epsilon = E$. Then he distributes the N molecules at random over these cells, asking at the end how many molecules have been placed in the first cell, how many in the second, how many in the k'th, and so on up to the P'th. The answer to that question is specified by a set of integers, $w_1, w_2, \cdots w_k, \cdots w_P$, and this set specifies a particular distribution of molecules along the energy continuum. Not just any set of w_k's will do, however. To meet the physical

240   KUHN

conditions of the problem, the w_k's must satisfy the so-called constraints on number of molecules and on total energy: they must, that is, satisfy the equations $\sum_k w_k = N$ and $\sum_k k w_k = P$. If, for example, $w_1 = N$, then all the molecules lie in the first cell, and all the other cells must be empty. Correspondingly, if a single molecule lies in the P'th cell, then it possesses all available energy, and all other molecules must lie in the first cell. There are, of course, a great many distributions that do satisfy these constraints and thus represent possible conditions of the gas. Each of them can, furthermore, since the molecules are indistinguishable among themselves, be achieved in numerous ways. Boltzmann asks in how many ways each possible distribution can be achieved, and records the now standard answer, $N!/w_1! w_2! \cdots w_P!$

That number is proportional to the probability of the corresponding distribution; the equilibrium distribution is the one with greatest probability; and the logarithm of its probability is proportional to the entropy of the gas. When the corresponding computations are carried through, they disclose a central feature of the formula for molecular entropy, one that separates Boltzmann's problem from Planck's. The cell size ϵ turns out to affect only the size of a disposable additive constant, so that ϵ may be eliminated from the entropy formula. (James Jeans and many more recent commentators are mistaken in saying that Boltzmann eliminates ϵ by letting it go to zero; indeed, for finite N, his derivation does not permit his taking the limit, since his cells must remain large enough to contain many molecules.) In an aside, Boltzmann notes also that the total number of ways of achieving all possible distributions of N molecules over P cells is just $(N + P - 1)!/(N - 1)!P!$, the very formula that Planck's derivation will require.

Planck's derivation has quite properly been taken to be modeled on Boltzmann's. But their problems are not the same, and there are two very different ways in which Planck's modeling could have been achieved. The traditional interpretation is, I believe, based on the wrong one: it takes Planck to have been asking for the most probable way of distributing some given energy E over N resonators all at the same frequency. His distribution law can be derived in that way. Lorentz did so in 1910; Planck adopted that derivation in 1913; and it has been standard since, a fact that has facilitated the traditional reading. But if that first interpretation is the one Planck had in mind between 1900 and 1906, then, though his method implies quantization, his derivation makes no sense. The second way of relating Planck's problem to Boltzmann's does make sense of the derivation, but it simultaneously eliminates the appearance of quantization.

Boltzmann had considered all the molecules of his gas. Planck considered all the resonators in his cavity, N at frequency ν, N' at frequency ν', N'' at frequency ν'', and so on. To the first set he attributed energy E, to the second E', to the third E'', and so on, again. The sum of all these energies was fixed at E_T; a distribution was a particular division of E_T among the resonators at each frequency; and it was the division of energy among particular frequencies that Planck varied to maximize the entropy of the system as a whole. This interpretation of Planck's derivation fits the texts closely, and it also models the problem Planck needed to solve: the distribution of energy, not over the resonators at a single frequency, but over sets of resonators at different frequencies. Unfortunately, the problem thus formulated is quite cumbersome, and Planck adopted a drastic shortcut to reach a solution. That shortcut, referred to in his first papers but only explained in his *Lectures*, is what made his first derivation papers so hard to understand.

Look now at what is at stake in the choice between these two ways of understanding Planck's derivation. According to the traditional view Planck should have begun by distributing P units of energy, ϵ or $h\nu$, at random over N resonators; next he should have computed the number of ways of achieving some particular distribution; and, finally, he should have determined the most probable distribution, computed the number of ways in which it could be obtained, and set the logarithm of that number proportional to entropy. Planck does take the first step, illustrating what a distribution means by supposing that the first resonator receives 7 elements, the second 38, the third 11, the fourth 0, the fifth 9, and so on, until all P are distributed. Those passages are the ones that provide evidence for quantization. But Planck's next step is to ask, not for the number of ways of obtaining some particular distribution, the number he ought then have maximized. Rather he asks for the total number of ways of obtaining all distributions, records the formula $(N+P-1)!/(N-1)!P!$ as the answer, and then, without further comment, sets its logarithm proportional to entropy. That is why so many of his later readers have felt that, having previously determined the combinatorial form his distribution law required, Planck substituted hand-waving for a derivation.

Now suppose that Planck, rather than distributing energy over resonators at a single frequency, were instead distributing it to the sets of resonators at different frequencies, E to those at frequency ν, E' to those at frequency ν', and so on. His problem would then be to discover in how many ways that could be done for a particular division, E, E', E'', $\cdots$, and then to discover which particular division could be achieved in the largest number of ways. The straightforward, but cumbersome, way to do that would be to begin as Boltzmann had begun. At each frequency, ν, ν', etc., divide the continuum of available energy, E, E', etc., into P, P', etc., units of size ϵ, ϵ', etc. Next, at each frequency, distribute the N, N', etc., resonators over the P, P', etc., cells and find out in how many ways the corresponding

242   KUHN

distributions could be achieved. It is only after this point that Planck's derivation begins to diverge from Boltzmann's, for Planck must next compute, not the most probable, but the total number of ways in which the N, N', etc., resonators can be distributed over each of the P, P', etc., cells. The product of all those numbers is the total number of ways of achieving the distribution that attributes energy E to the resonators at frequency ν, E' to those at ν', E'' to those at ν'', etc. And it is this product that must be maximized by varying E, E', E'', etc., subject to the constraint on total energy (pp. 103–110). The logarithm of the resulting number is, again, proportional to entropy, but in the radiation case, unlike the molecular, the dependence of entropy on cell size is not restricted to an additive constant. Cell size, thus, plays an essential role in determining the entropy of radiation, and ϵ cannot be dropped from the entropy formula.

If Planck had traveled that route he would have gotten the answer he did get, and no one would have suggested that he was restricting resonator energy to an integral number of elements or that he did not understand what he was doing. But Planck instead, after an extremely condensed sketch of the full method, pointed out that it would be "obviously very roundabout" (p. 106), and he announced that he was therefore resorting to a drastic shortcut. His problem did not require knowledge of the number of ways of achieving any particular distribution of N resonators over P energy cells, but only of the total number of ways in which all possible distributions of a given amount of energy could be achieved. That number could have been computed directly, by summing over the number of ways of achieving each possible Boltzmann distribution. But the computation required was lengthy, and its outcome was known to be the same as the number of ways in which P energy elements could be distributed over N resonators. Planck chose the latter, briefer technique of computation for the early accounts of his derivation.

Choice of that shortcut considerably obscured the conceptual structure of Planck's derivation, and the obscurity was compounded by the way he described his distribution procedure. He failed, that is, to point out that his distribution of energy over resonators was only a way of computing the *number* he needed, that it did not correspond to the distribution technique (resonators over energy) that produced the individual distributions over which his full derivation would have summed. But neither his choice of technique nor his imperfect description of it made his derivation unsound or restricted it to the case of a discrete energy spectrum. What Planck was computing was the total number of ways of distributing resonators of a given frequency to cells of size $h\nu$ along the energy continuum.

All of that can be seen in Planck's original papers of 1900 and 1901, but only with difficulty. The same interpretation is, however, inescapable if one turns to the first edition of Planck's *Lectures*. There, as in his papers, Planck writes of distributing P whole units, ϵ, of energy to N resonators, 4 to one of them, 3 to another, and so on. But he also provides a full description of the "roundabout" method, spells out the nature of his shortcut, and clarifies the phrase, "the number of resonators with energy of a given magnitude," by adding "(better: which lie in a given 'energy region')." Elsewhere, after dividing the phase plane of a one-dimensional harmonic oscillator into annular rings of area $\Delta U = h\nu$, Planck asks for "the probability that the energy of a resonator lies between U and $U + \Delta U$" (pp. 128–30). Here there can be no talk of quantization or discontinuity.

Two assorted remarks bring this explication of Planck's derivation to a close. First, there is one other putative difficulty with Planck's derivation, his apparent failure to maximize his count of the number of ways of realizing a particular distribution. That difficulty, too, proves to be only apparent. The function of maximization is to discover the equilibrium state, and there are other ways to do that, one of which Planck utilizes (pp. 107–108). Though he is straining the limits of turn-of-the-century physics, Planck is not groping in the dark. Second, as previously noted, Planck's result is incompatible with the continuity of resonator energy. He ought to have arrived at the Rayleigh-Jeans law. There is a mistake in his derivation, the one that Einstein pointed out in 1910. Boltzmann's method of counting ways of realizing a distribution is valid only if the distribution function is effectively constant within each cell. In Planck's case this requires that $h\nu \ll kT$, an approximation that does not generally hold. The difficulty remained invisible to all but Einstein, however, and even he did not recognize it until about 1910.

4. HISTORIOGRAPHIC/PHILOSOPHICAL ADDENDUM

I shall carry these argument sketches no further. Fuller versions are to be found in my book, together with some additional sorts of evidence. But *Black-body theory* has also raised issues of another sort. The concept of historical reconstruction that underlies it has from the start been fundamental to both my historical and my philosophical work. It is by no means original: I owe it primarily to Alexandre Koyré; its ultimate sources lie in neo-Kantian philosophy. But it is not everywhere accepted or even understood, and some significant responses to my reinterpretation of Planck seem less a reaction to its thesis or arguments than to the concepts of history and method that gave rise to them.

244   KUHN

Twenty-five years ago, two pages from the end of *Structure of scientific revolutions,* I wrote: "If we can learn to substitute evolution-from-what-we-do-know for evolution-toward-what-we-wish-to-know, a number of vexing problems may vanish in the process." The intent of that remark was philosophic: to understand scientific progress we need not suppose that science moves closer and closer to the truth; the same phenomena follow from the assumption that science, at any time, simply evolves from its current position under the pressure of currently available argument and observation; if one adopts that more minimal explanation, then significant epistemological difficulties disappear. The same sentence has, however, an historiographic reading, which is for me the source of the philosophic one. Not accidentally, that reading is most fully expressed two pages from the start of the book, where I described the "historiographic revolution" whose implications I aimed to make explicit.

"Historians of science," I suggested in that place, "have begun to...ask, for example, not about the relation of Galileo's views to those of modern science, but rather about the relationship between his views and those of his group, i.e., his teachers, contemporaries, and immediate successors in the sciences. Furthermore, they insist upon studying the opinions of that group and other similar ones from the viewpoint—usually very different from that of modern science—that gives those opinions the maximum internal coherence and the closest possible fit to nature." The corresponding historiographic reading of the sentence from the end of the book is, then: if we can learn to refrain from attributing to past scientists a grasp of views that had not emerged when they wrote—from equating discoveries with their canonical formulations—vexing problems may vanish in the process. On this reading, the problems that vanish are mostly textual anomalies found in passages that do not quite make sense when read in ways that became standard only after they were written.

This point is often heard as a mere plea for scholarly accuracy: historians should be careful not to attribute to their subjects beliefs that conflict with things they said or wrote. But I have in mind both something more and something different. Creative scientists can be, and typically are, responsible for the emergence of beliefs that they did not hold themselves, at least not during the period when their discoveries were made. If one is to learn how those discoveries came about, how new knowledge emerged, then one must find out how the discoverers themselves thought about what they were doing. Often it turns out that not just their beliefs but their very modes of thought were different from the ones to which their discoveries gave rise, and the latter difference is central to *Structure of scientific revolutions.*

The wrenching experience of entering into an older mode of thought is the source of my references to gestalt switches and revolutions; difficulties in translating the discoverer's language into our own are what led me to write also of incommensurability; and paradigms were the concrete examples needed—since definition in words was impossible—to acquire the language of the older mode. I do my best, for urgent reasons, not to think in these terms when I do history, and I avoid the corresponding vocabulary when presenting my results. It is too easy to constrain historical evidence within a predetermined mold. If history (or ultimately philosophy) is to be learned from the texts that are my main sources, then I must minimize the role of prior conviction in my approach to them.[2] Often I do not know for some time after my historical work is completed the respects in which it does and does not fit *Structure*.

Nevertheless, when I do look back, I have generally been well satisfied by the extent to which my narrative fit the developmental schema that *Structure* provides. *Black-body theory* is no exception. The start of the revolution that produced the old quantum theory is moved from the end of 1900 to 1906; most of the book deals with the period before the revolution occurred. The preceding crisis, to the extent that there was one, resulted from the difficulties in reconciling Planck's derivation with the tenets of classical physics. Planck's change of vocabulary—from "resonator" to "oscillator" and from "element" to "quantum"—is the central symptom of incommensurability. It signals the changed meaning of the quantity $h\nu$ from a mental subdivision of the energy continuum to a physically separable atom of energy. That my critics continue to apply the term "energy quantum" to the pre-1906 papers and lectures in which Planck consistently used "energy element" reveals something of the difficulty of reversing the gestalt switch that took place during that year and those that followed. Among the numerous paradigms (in the sense of concrete examples) to be found in the book, Boltzmann's probabilistic derivation of the entropy of a gas is of particular interest, for it illustrates the problem to which the concept of paradigm was a response. The derivation was not reduced to rules but instead served as a model to be applied by means of analogy. As a result, when its application was transferred from gases to radiation, Planck and Lorentz could invent different analogies with which to effect the change.

These illustrations of the substantive applicability of *Structure* can be extended, but for this paper it is the book's historiographic applications

2. On this subject see also my essay review, "The halt and the blind: Philosophy and history of science," *British journal for the philosophy of science, 31* (1980), 181–192.

246   KUHN

that are relevant. My way of using concepts like revolution and gestalt switch was drawn from and continues appropriately to represent what historians must often go through to recapture the thought of a past generation of scientists. Concerned to reconstruct past ideas, historians must approach the generation that held them as the anthropologist approaches an alien culture. They must, that is, be prepared at the start to find that the natives speak a different language and map experience into different categories from those that they themselves bring from home. And they must take as their objective the discovery of those categories and the assimilation of the corresponding language. "Whig history" has been the term reserved for failure in that enterprise, but its nature is better evoked by the term "ethnocentric."

What has preceded is prefatory, a quick overview of the close parallelism between an historiographic and an epistemological position. Perhaps the parallelism is obvious, but it has corollaries which are not. Entry into a discoverer's culture often proves acutely uncomfortable, especially for scientists, and sophisticated resistance to such entry ordinarily begins within the discoverer's own retrospects and continues in perpetuity. That resistance—with its manifestations, causes, and consequences—is the central concern of this closing section.

Systematic distortions of memory, both the discoverer's memory and the memory of many of his contemporaries, are a first manifestation of resistance. Another, regularly found among members of later generations, is the attribution of real or supposed anomalies in the discoverer's behavior to "confusion." I shall take up these techniques in turn, beginning with two examples of a discoverer's systematic misconstruction of the memory of his own discovery. Both are from my own experience, and both illustrate the suppression of concepts that played an essential role in the discovery and their replacement by the projection onto the past of concepts that emerged only in the discovery's aftermath.[3]

The first experience involves Otto Stern, the physicist who conceived, and in 1922 carried out with Walther Guerlac, the first direct demonstration of space-quantization. While the required apparatus was still under construction, Stern published a paper demonstrating that a beam of silver atoms passing through an inhomogeneous magnetic field would only be broadened if classical theory held but would be split in two if the atoms were spatially quantized. That splitting is what the experiment showed, and it followed, as Stern had shown, directly from current theory. Four years later, however, relevant theory changed

3. On this subject see also Section XI, "The invisibility of revolutions," in *The structure of scientific revolutions.*

drastically. To understand why the beam split in two then required a knowledge of electron spin, a conception not even dreamt of when the experiment was made. Furthermore, the later theory predicted that, in the absence of spin, the beam of silver atoms should be split, not into two parts, but into three. At the time of my meeting with Stern I did not yet know the theoretical paper, and I therefore asked why he had expected the beam to be split in two. He responded sharply that, before the discovery of spin, he and Gerlach obviously could not have known what to expect. Their experiment was done simply to see what would happen. In that respect, he insisted, the experiment was a fishing expedition.

My second example is more complex. The famous paper that announced the Bohr model of the hydrogen atom was submitted from Copenhagen on 6 March 1913 and published the following July, the first installment of a three-part series. I first read it during the fall of 1962 in preparation for interviews with its author. Not surprisingly, the paper includes a full description of the quantized Bohr model for the hydrogen atom as it would be taught in an elementary physics course today. But it also includes a number of phrases incompatible with that model. In particular, Bohr sometimes wrote as though the hydrogen spectrum were emitted by an electron falling into the ground state from outside the atom and strumming all the stationary states that it passed along the way.[4]

These anomalous remarks, together with Bohr's repeated assertion that he had not known the Balmer formula until February 1913, suggested an unexpected hypothesis, subsequently fully confirmed by the discovery of an unpublished manuscript. Many months before he attempted an explanation of spectra, Bohr had developed a quantized version of the Rutherford atom for chemical applications of the sort made familiar by J. J. Thomson. That model, which I was quite sure had had only a ground state, provided the basis for the second and third installments of the 1913 series. The first, which developed the Bohr model for hydrogen and derived the Balmer formula from it, was a last-minute insertion.

My first few interviews with Bohr dealt with the background for his atomic model, and I asked what sorts of connections he had made between the Rutherford atom and the quantum during the period before his attention was directed to the Balmer formula. He replied that he could not have had developed ideas on the subject before turning to spectra, and his assistant later reported to me that, after I had

4. These passages and much else are discussed in J. L. Heilbron and T. S. Kuhn, "The genesis of the Bohr atom," *HSPS, I* (1969), 211–290.

248   KUHN

left the room, Bohr shook his head and said of our exchange, "Stupid question. Stupid question."

All that occurred at our first interview. For the next one, I included a similar question in a list submitted to Bohr in advance, and it was received in much the same way as the original. One last attempt to retrieve memories of an early quantized Rutherford atom occurred late in the third interview. This time, however, when Bohr said again that there could have been no concrete model without the Balmer formula in hand, I for the first time showed him the passages in his famous paper that had led me to enquire. He looked them over and then muttered half to himself, "Perhaps it was a mistake to put the paper into print so fast. Perhaps I should have waited until I had it right." Then, he went over to his personal collection of reprints, took from it a paper he had presented to the Danish Academy of Sciences six months after the publication of his original paper, and handed it to me with the words, "It's alright there, isn't it?" About the earlier model not a word was ever forthcoming.[5]

These stories typify the autobiographical reports of the participants in a discovery and often of their contemporaries as well. Not always but quite usually, scientists will strenuously resist recognizing that their discoveries were the product of beliefs and theories incompatible with those to which the discoveries themselves gave rise. Similar resistance is encountered among later generations, but memory and its distortion are no longer involved.

Later accounts of a discovery typically redescribe it, but again in the conceptual vocabulary of the period in which they are prepared. The result is the linearized or cumulative histories familiar from science textbooks and from the introductory chapters of specialized monographs. Those accounts, however, almost never withstand detailed comparison with documents from the period of the discovery. And when discrepancies are pointed out, one or another variant of a second standard technique for resisting the past is often deployed. Faced with apparent anomalies in the work of the discoverer, scientists and at least an occasional historian protect their version of the discovery by invoking the discoverer's "confusion" during the early stages of its emergence. It is only because he was confused, they explain, that his words

5. The words in quotation marks are paraphrases from memory. A confidential memorandum I prepared for the file later in the year cannot now be found. Search for it has, however, turned up a second memorandum, one I had entirely forgotten. Written in haste the day *before* the third interview, it recounts a report from Bohr's assistant that Bohr now said "he should not have published so soon. Parts II and III," he went on, "should have been entirely suppressed and Part I should have been withheld until recast in the form of the lecture delivered in Copenhagen in December."

fail to fit their story.

These appeals to confusion are damaging, but not because discoverers are never confused. Typically, they are, and Bohr's discovery of the Bohr atom is a clear example. When he wrote the paper announcing his discovery, he had two incompatible models in mind, and he occasionally confused them, mixed the two up. No reading of his first reports on his invention will eliminate the resulting contradictions, and those contradictions, which testify to his confusion, provide essential clues to the reconstruction of his route to the discovery. The standard appeal to confusion dismisses those clues, rejects them as challenges to historical reconstruction, and permits the attribution of confusion to stand as the end of the story. That is the first part of the damage.

For the second, more serious part, compare the case of Planck. Again there are anomalies in the early papers; again they provide clues to an unsuspected state of mind; and, again, dismissing them discards evidence essential for historical reconstruction. Thus far the damage is the same. But in Planck's case, unlike Bohr's, the anomalies do not take the form of internal contradictions, and they therefore provide no reason to suppose that Planck himself felt or had reason to feel confused. If it is nevertheless appropriate to apply the term to him, that is by virtue of a second standard use of the word "confused," one independent of the state of mind of the person to whom it is applied.

Consider, for example, the case of a student who, having read a textbook derivation of the black-body distribution law, then wrote it up in a way like the one found in Planck's early papers. That student would be confused, not in the sense of being pulled about by conflicting elements in his thought, but in the sense of having seen only dimly or confusedly the structure of the derivation that had been set before him by the text. That, I believe, is the sense of "confused" in the minds of the people, mostly scientists, who complain, for example, that I try too hard to make the thought of a Planck or a Boltzmann logical and coherent. Why, I am repeatedly asked, can I not simply acknowledge that they were confused?

That way of talking about a discoverer makes no sense. Taken literally, it suggests that the discovery, of which its author is said to have had only a confused view, had already been made, was somehow already there, in the discoverer's mind. Occasionally that implication is explicit. The discoverer, I am then told, was relying on intuition; his view of his discovery was still so clouded that he could only grope his way to it; that is why he described what he had in mind in such odd and inconsistent ways, appeared so much a sleepwalker as he proceeded towards his discovery.

Doubtless, few of those who explain anomaly by resort to confusion would go quite so far, but all must encounter the identical difficulty.

250   KUHN

What licensed our calling the student confused was our knowledge of the concepts he brought to the text and of the proper way to fit the two together. If only he had clearly seen that much himself, he would not, any more than we, have described the derivation as he did. When, in the absence of internal contradiction, we apply the label "confused" to Planck, we are again using ourselves as measure. We assume that Planck brought to his problem the same concepts as we do, and we explain his anomalous behavior as we would explain similar behavior of our own. But the concepts we bring to the black-body problem are themselves products of the discovery Planck had not yet made. To claim for them a role in the emergence of his discovery is again to make him a sleepwalker or else clairvoyant. That is an incoherent notion of discovery—one that makes discovery dependent on prior grasp of what is to be discovered. No other result of the resort to confusion is so damaging.

So much for manifestations of resistance. What can possibly be their cause? Why should so many scientists and also an occasional historian fight so hard, often without quite realizing they are doing so, to recast past developments in the language of modern concepts? Why should attempts to reconstruct a conceptual past in its own terms so often be regarded as subversive? Answers to those questions are to be sought in two directions. The first is independent of any special characteristics of science, and I shall here merely identify it. The second involves the sciences uniquely, and about it I shall say a bit more.

I have already suggested that the past of science should be approached as an alien culture, one that the historian strives first to enter and then to make accessible to others. Entry into another culture, scientific or not, is regularly resisted, however, and the standard form of resistance is to carry one's own culture with one and assume that the world conforms. Remember the not-so-mythical English tourist who thinks that English-spoken-loudly will suffice for life in France. Motives for such resistance are familiar to anyone who has experienced the anxieties of foreign travel. Cultures shape life forms and with them the world in which the participants in the culture live. Entry into another culture does not simply expand one's previous form of life, open new possibilities within it. Rather, it opens new possibilities at the expense of old ones, exposing the foundations of a previous life form as contingent and threatening the integrity of the life one had lived before. Ultimately the experience can be liberating, but it is always threatening.

That is the general case, and it needs much exploration. The transition between cultures always threatens something, but what is at stake is not always clear. Where the cultures are scientific, however, one object of the threat stands out. In the socialization or professionalization of scientists the concept of the unit discovery plays a profound and generally functional role. The knowledge they acquire in school and university is transmitted to them in such units, both experimental and theoretical; unit discoveries are the bricks from which, in a familiar image, the edifice of science is piecemeal built. When the student later enters the profession, it is with the understanding that success is measured by the size and number of bricks the individual is able to put in place. That is why, since at least the seventeenth century, attempts to establish priority in discovery have played so large a part in scientific development. The concept of the unit discovery is constitutive of the scientific life as we know it.

Among historians, however, it is by now virtually a truism that that concept will not do. Discoveries are extended processes, seldom attributable to a particular moment in time and sometimes not even to a single individual. Usually there is little doubt who was responsible for them—who brought things to a point from which there could be no retreat. Planck himself is a clear case in point: given his distribution law and his technique for deriving it, the recognition of discontinuity was bound soon to occur. But, as we have seen, it then typically turns out that the person responsible for a discovery did not believe in it himself, at least not if belief in the discovery is belief in its subsequent canonical formulation. Even when the discoverer's own contribution was substantially complete, he still thought of what he had done in a way significantly different from the way in which that contribution came later to be described.

The priority controversies for which science is known are again especially revealing. The description of the work an individual did or the date at which he did it is rarely a/issue. Matters of that sort are usually easily settled. Instead, what participants debate is whether one or another piece of work can properly be said to constitute *the* discovery. Most historians would insist that the question has no answer, that the debate itself rests on a misconception. But scientists can continue the argument in perpetuity, insisting that one party or the other must be *the* discoverer. Implicitly or explicitly, what is at stake for them is the concept of the unit discovery, a concept that will not withstand application to actual practice, but on which much of the reward system of science as well as important elements of the scientist's conception of self are nevertheless based.[6] Doubtless, other things

6. The following fragment, atypical only in having reached the published record, is illustrative. Early in 1979 I presented to an Einstein centennial symposium a brief account of Einstein's role in the genesis of the concept of quantization. Eugene Wigner, who was not convinced, commented on my paper as follows: "In spite of all my admiration of Professor Kuhn, I would like to contradict him a little bit. I think Planck did wonderful work, and even the discovery of his equation is wonderful. I do not believe he believed

252   KUHN

besides the unit discovery are threatened by the resurrection of older modes of scientific thought. But I shall stop with this single example which, at the very least, illustrates the form that explanations of resistance must take.

I conclude with a few brief remarks on a question I do not quite understand. "Suppose you were right," I have been asked, "suppose that the first suggestion of a restricted energy spectrum did come, not from Planck, but from Einstein and Ehrenfest; what difference could it possibly make, and to whom?" For me there are two sorts of answer. I wrote the book as an historian of ideas, and my primary object was just to get the facts straight. Different specialties differ in the facts they must get straight, and no one is obliged to engage in this one. But a person who assigns no importance to such questions ought not even make the attempt.

Most historians would, I think, be content to stand on that answer alone, and the ground it provides is bedrock. For me, however, something else is also involved. I believe that history can provide data to the philosopher of science and the epistemologist, that it can influence views about the nature of knowledge and about the procedures to be deployed in its pursuit. Part of the appeal of the standard account of Planck's discovery is, I think, the closeness with which it matches a still cherished view of the nature of science and its development. Though I appreciate both the charms and the functions of that view, understanding requires that it be recognized as myth.

---

in the details of any derivation of it, and this was natural since the physics of that time was full of contradictions....So that I feel the quantum was discovered, at least from what I read, by Planck and even though we admire Einstein, this is not for what we admire him most." (Cf. Harry Woolf, ed., *Some strangeness in the proportion* (Reading, MA, 1980), p. 194.) A few weeks later the following paragraph appeared in *Science news* (30 Mar 1979, p. 112): "Indeed, at the centennial symposium the historian of science Thomas S. Kuhn proposed to credit Einstein with the very concept of the quantum of radiation, an idea usually attributed to Max Planck. This provoked a rebuke from Eugene Wigner (from Wigner a not-so-gentle rebuke) to the effect that people ought to be allowed to keep the credit they have justly earned. Not everything in physics need be annexed to Einstein."

Note: When my article was written, the book it discusses (*Black Body Theory and the Quantum Discontinuity, 1894-1912*) existed only in the original hardcover edition issued by the Oxford University Press in 1978. It has since been, in the spring of 1987, reissued as a paperback by the University of Chicago Press.

*LINDA WESSELS**

SCHRÖDINGER'S ROUTE TO WAVE MECHANICS

THE European physics community barely had time to comprehend the significance of matrix mechanics in late 1925 and early 1926 when wave mechanics burst onto the scene. This new theory was presented in a series of five papers which appeared at the rate of almost one a month, each completed and posted for publication before the ideas of the next were fully formed.[1] Their author, Erwin Schrödinger, had followed the fortunes of atomic theory and had contributed several papers on the subject;[2] but he could not be counted among those who had undertaken its sustained development. He has claimed that his acute awareness of the unsystematic and *ad hoc* character of its foundations kept him from becoming deeply involved in that development.[3] How, then, did he come to make such a major contribution to atomic theory, indeed to set the mold by which our modern theories of microphysics are fashioned?

Schrödinger has left few traces of how his ideas evolved as he worked towards wave mechanics. Unlike those physicists who developed matrix mechanics, he worked virtually alone, and apparently corresponded only infrequently and then quite briefly on his progress. He did confer weekly with Hermann Weyl on the mathematical problems he encountered, but no written record of these meetings remains.[4] In his papers on wave mechanics

*Department of History and Philosophy of Science, Indiana University, Goodbody Hall 130, Bloomington, IN 47401, U.S.A.

[1]E. Schrödinger, 'Quantisierung als Eigenwertproblem, Erste Mitteilung', *Annln. Phys.* **79** (1926) 361–376, sent for publication January 26, 1926; 'Quantisierung, Zweite Mitteilung', *ibid.* **79**, 489–527, sent February 22; 'Über das Verhältnis der Heisenberg-Born-Jordanschen Quantenmechanik zu der meinen', *ibid.* **79**, 734–756, sent March 17; 'Quantisierung, Dritte Mitteilung', *ibid.* **80**, 437–490, sent May 6; 'Quantisierung, Vierte Mitteilung', *ibid.* **81**, 109–139, sent June 19. These papers (plus others) were published together in Schrödinger's *Abhandlungen zur Wellenmechanik* (Leipzig: J. A. Borth, 1927). The second edition of *Abhandlungen* was translated as *Collected Papers on Wave Mechanics*, J. F. Shearer and W. M. Dean, trans. (London: Blackie, 1928). Passages from Schrödinger's papers on wave mechanics will be cited below by reference to pages in *Abhandlungen* and *Collected Works*. Schrödinger mentioned the sequential writing of his papers in 'Vorwort' to *Abhandlungen*, p. iii, *Collected Works*, p. v.

[2]See W. T. Scott, *Erwin Schrödinger: An Introduction to His Writings* (Amherst: University of Massachusetts Press, 1967) for a discussion of Schrödinger's scientific papers and a bibliography of Schrödinger's works. P. Hanle provides a more thorough historical analysis of parts of Schrödinger's early work in his dissertation, *Erwin Schrödinger's Statistical Mechanics, 1912–1925* (Unpublised Ph. D. Dissertation Yale University, 1975).

[3]E. Schrödinger, 'Antrittsrede des Hrn. Schrödinger', *Preuss. Akad. Wiss. Sitzb.* **C–CI** (1929), partially translated in J. Murphy's 'Biographical Introduction' to E. Schrödinger, *Science, Theory and Man* (New York: Dover, 1957), xiii–xviii.

[4]Interview with Annemarie Schrödinger, conducted by T. S. Kuhn, April 5, 1963, Archive for

Stud. Hist. Phil. Sci., Vol. 10 (1977), No. 4, pp. 311–340.
Pergamon Press, Ltd. Printed in Great Britain.

Schrödinger acknowledged his debt to Louis de Broglie, who introduced and first applied the notion of matter waves,[5] and several times he reported that a favorable reference to de Broglie's work in a paper by Albert Einstein had been crucial to his development of wave mechanics.[6] He did not explain precisely how the work of de Broglie and the footnote by Einstein influenced his discovery, however.

Recent scholarship has contributed a great deal to our understanding of portions of Schrödinger's route to wave mechanics. In his study on Einstein and the wave-particle duality, Martin Klein has examined Schrödinger's work on gas theory in 1925, and identified in one of Schrödinger's papers the concept of matter wave that later appeared in his first papers on wave mechanics.[7] V. V. Raman and Paul Forman have demonstrated a striking similarity between a discovery Schrödinger made in 1922 concerning Bohr's atomic theory and de Broglie's treatment several years later of matter waves in the atom; they have argued that this similarity made Schrödinger more receptive to de Broglie's work than most physicists of that period.[8] Paul Hanle has recently uncovered additional evidence to support this argument.[9] Raman and Forman have also presented evidence that prior to his discussion of matter waves in connection with gas theory Schrödinger applied the notion of matter waves to atomic physics in a way suggested by this similarity. The results of these studies are crucial to the account of Schrödinger's route offered here. Analysis of the papers discussed by Klein will reveal a sustained effort by Schrödinger to work out an approach to the quantization of gases that he had framed in response to certain foundational problems. The concept of matter wave was adopted as a basis for carrying through that approach. The events described by Raman and Forman mark an important stage in the development of Schrödinger's ideas, and when seen as part of the whole story will highlight a significant change in Schrödinger's understanding of how to formulate a wave theory of matter. Traditional accounts have assumed that the derivation of the wave equation in Schrödinger's second paper on wave mechanics

the History of Quantum Physics (hereafter AHQP). See *Sources for History of Quantum Physics: An Inventory and Report,* prepared by T. S. Kuhn, J. Heilbron, P. Forman and L. Allen (Philadelphia: Am. Phil. Sci., 1967) for a partial catalogue of this collection and a list of its locations.

[5]Letter from Schrödinger to A. Einstein, April 23, 1926, in *Letters on Wave Mechanics,* K. Przibram (ed.), M. Klein, trans. (New York: Philosophical Library, 1967), p. 26; letter from Schrödinger to W. Wien, June 18, 1926, AHQP; E. Schrödinger, 'Antrittsrede'; 'Quantisierung, Erste Mitteilung', *Abhandlungen,* p. 12, *Collected Works,* p. 9; 'Quantisierung, Zweite Mitteilung', *ibid.* pp. 27 and 20; 'Der stetige Übergang von der Mikro- zur Makromechanik', *ibid.* pp. 56 and 41; 'Über der Verhältnis', *ibid.* pp. 63 and 46.

[6]Letter from Schrödinger to A. Einstein, April 23, 1926; *Abhandlungen,* pp. 56 and 63, *Collected Papers,* pp. 41 and 46.

[7]M. Klein, 'Einstein and the Wave-Particle Duality', *Nat. Philosopher* 3 (1964), 3–49.

[8]V. V. Raman and P. Forman, 'Why Was it Schrödinger Who Developed de Broglie's Ideas?' *Hist. Stud. Phys. Sci.* 1 (1969), 291–303.

[9]P. Hanle, 'Erwin Schrödinger's Reaction to Louis de Broglie on Quantum Theory', *Isis,* **68** (1977), 606–609.

mirrors the process by which he discovered that equation.[10] This assumption is consistent with the results of Klein, Raman and Forman and Hanle. It will be argued that certain parts of the derivation cannot reflect Schrödinger's earlier ideas, however; the status of the remaining parts is not so clearly determined.

The story of Schrödinger's route reveals that underlying every step along the way was either a question about the physical significance of certain formal expressions, or a desire to capture theoretically an intuitive model of physical processes. Throughout his life Schrödinger's work was shaped by his interest in foundational and conceptual matters, and his concern over physical significance and physical models. Much of his early work probed the issues of atomism: in several papers on fluctuation phenomena he examined observable consequences of an atomic structure for matter; in a paper of 1914 an analysis of elasticity was used to support the atomic hypothesis; and two notebooks from 1918 and 1919 show Schrödinger exploring the implications of the atomic view through the work of Marian Smoluchowski.[11] In a philosophical treatise written in 1925, Schrödinger attacked Mach's view that such 'metaphysical speculation' must give way to concern for facts and economy: 'to have only this goal in mind would not suffice to keep the work of research going forward in any field whatsoever'.[12] The influence of Schrödinger's attitude is apparant in his reaction in 1924 to the theory of Neils Bohr, H. A. Kramers and John C. Slater, in which the virtual radiation field of distant atoms determined the probabilities of transition from one electron orbit to another.[13] Schrödinger was enthusiastic about the assumption of irreducibly statistical processes, but objected to the authors' reluctance to give a coherent physical picture for the theory. He wrote to Bohr:

> I cannot completely go along with you when you keep calling [this radiation] 'virtual'. . . For what is the 'real' radiation if it is not that which 'causes' transitions, i.e., which creates the transition probabilities? Another sort of radiation will surely not be assumed in addition. Indeed, from a purely philosophical standpoint, one might even venture to wonder which of the two electronic systems has a greater reality — the 'real one' which describes the stationary states or the 'virtual one' that supplies the virtual radiation and scatters impinging virtual radiation.[14]

[10]For example, M. Jammer, *The Conceptual Development of Quantum Mechanics* (New York: McGraw-Hill,1966), pp. 326 – 380; A. Hermann, 'Schrödinger — Eine Biographie', *Die Wellenmechanik, Dokumente der Naturwissenschaft,* Band 3, A. Hermann (ed.) (Stuttgart: Ernst Battenberg, 1963), p. 180; J. Gerber, 'Geschichte der Wellenmechanik', *Archs. Hist. exact Sci.* **5** (1968 – 1969), 347 – 416; E. Cassirer, *Determinism and Indeterminism in Modern Physics,* O. T. Benfy, trans. (New Haven: Yale University Press, 1956), pp. 45 and 50.

[11]See W. T. Scott, *Introduction,* on the published works; P. Hanle, *Schrödinger's Statistical Mechanics,* also discusses the notebooks on Smoluchowski.

[12]E. Schrödinger, 'Seek for the Road', *My View of the World* (Cambridge: Cambridge University Press, 1964), p. 4.

[13]N. Bohr, H. A. Kramers and J. C. Slater, 'Über die Quantentheorie der Strahlung', *Z. Phys.* **24,** (1924), 69 – 87; also published as 'The Quantum Theory of Radiation', *Phil. Mag.* **47** (1924), 785 – 802. For an historical perspective on the theory see M. Jammer, *Conceptual Development,*

In his published article on the Bohr – Kramers – Slater theory Schrödinger did not talk of 'virtual' radiation at all, but clothed the theory in his own physical interpretation.[15]

Schrödinger has traced his concern with foundations and physical significance to the influence of Ludwig Boltzmann,[16] whose work was also the source of Schrödinger's lifelong interest in statistical mechanics. Though Boltzmann committed suicide the summer before Schrödinger entered the University of Vienna, the Boltzmann statistical tradition was still strong while he was a student.[17] Schrödinger apparently immersed himself in Boltzmann's writings, and he later described Boltzmann's statistical approach as 'my first love in science; no other has ever had such an impact on me, nor will ever again'.[18] Between 1914 and 1926 Schrödinger wrote eight articles on the specific heats of solids and gases; two of these were review articles, published in 1917 and 1919, that established his international reputation. In five less influential papers of this period Schrödinger used statistical methods to study magnetic properties of matter.[19] In early 1924, a little over one year after he arrived at the University of Zurich, Schrödinger published his first work on the quantum statistics of ideal gases.[20]

I. Toward a Foundation for Einstein's Gas Theory

In 1925 Schrödinger's concern over physical significance combined with his new interest in the quantum statistics of gases to yield a series of three papers on quantum gas theory.[21] These papers focused specifically on the

pp. 181 – 187; M. Klein, 'The First Phase of the Bohr – Einstein Dialogue', *Hist. Stud. Phys. Sci.* **2** (1970) 1 – 40; R. H. Stuewer, *The Compton Effect* (New York: Neal Watson, 1975), pp. 291 – 302.

[14]Letter from Schrödinger to N. Bohr, May 24, 1924, AHQP, Bohr Scientific Correspondence, reel 16; 'Ich kann nicht ganz mitgehen, wenn Sie [diese Strahlung] stets als "virtuell" betiteln. . . Welches ist denn die "reelle" Strahlung, wenn nicht die, welche die Übergänge "bewirkt", d.h. Übergangswahrscheinlichkeiten schafft? Auch wird eine andere Art ja überhaupt nicht angenommen. Ja, vom rein philosophischen Standpunkt, könnte man sogar wagen zu zweifeln, welchem von den beiden Elektronensystemen die grössere Realität zukommt — dem "reeallen", das die stationären Bahnen beschreibt, oder dem "virtuellen", das die virtuelle Strahlung abgibt und auftreffende virtuelle Strahlung zerstreut'.

[15]'Bohrs neue Strahlungshypothese und der Energiesatz', *Naturwissenschaften* **12** (1924), 720 – 724.

[16]Letter from Schrödinger to A. S. Eddington, March 22, 1940, AHQP, reel 41.

[17]P. Hanle, *Schrödinger's Statistical Mechanics,* examines in detail the influence of the Boltzmann statistical tradition on Schrödinger.

[18]'Antrittsrede', p. C: 'Sein Ideenkreis spielt für mich die Rolle der wissenschaftlichen Jugendgeliebten, kein anderer hat mich wieder so gepackt, keiner wird es wohl jemals tun'.

[19]W. T. Scott, *Introduction;* P. Hanle, *Schrödinger's Statistical Mechanics.*

[20]'Gasentartung und mittlere Weglängen', *Phys. Z.* **24** (1924), 41 – 45.

[21]These papers and 'Gasentartung' are the subject of P. Hanle's 'The Coming of Age of Erwin Schrödinger: His Quantum Statistics of Ideal Gases', *Archs. Hist. exact Sci.* **17** (1977), 165 – 192. In the account that follows I skip over many aspects of Schrödinger's papers that are not relevant to tracing his route to wave mechanics. They receive full and revealing treatment in Hanle's work,

foundations, statistical and physical, of a new theory of ideal gases, introduced by Einstein September 1924 and February 1925.[22] Einstein had applied to gas molecules the 'new statistics' recently devised by S. N. Bose to obtain Planck's radiation law on the assumption of light 'particles' or quanta. The basic difference between the new statistics and traditional Boltzmann – Gibbs statistics lay in their methods of counting distributions in phase space. According to an approach introduced by Max Planck in 1906[23] and used extensively since then, a quantized statistical treatment is given by dividing the phase space of a system into cells of volume h^3, where h is Planck's constant; the number of equally probable distributions of N particles in a phase space volume V is determined by counting the number of ways the N particles could be distributed over the V/h^3 cells in V. Traditionally, this meant counting the number of ways N things could be assigned to V/h^3 cells (where some cells may be empty), so that a distribution was characterized by the number of particles in each available cell. In the Bose – Einstein approach, however, a distribution was characterized by the number of cells containing each possible number (0, 1, . . . , N) of particles; thus the number of ways of distributing N particles was determined by counting the number of ways V/h^3 cells could be sorted into N categories (where some categories may be empty). For many physicists, the traditional counting procedure seemed 'natural', while the procedure introduced by Bose and Einstein looked *ad hoc*. Paul Ehrenfest's strong objections were conveyed to Einstein privately.[24] Planck's public response was to counter Einstein's results with a description of gases founded on traditional statistics.[25]

In his first gas theory paper of 1925, 'Remarks on the Statistical Definition of Entropy for an Ideal Gas',[26] Schrödinger examined Planck's definition of entropy. He argued that if developed consistently, the crucial insight underlying Planck's definition would lead to Einstein's count of equiprobable states rather than Planck's. In 1921, Planck had derived an expression for entropy by dividing by $N!$ the thermodynamic probability obtained from a Boltzmann – Gibbs treatment of a gas of N particles.[27] To justify this

where he shows the relation of Schrödinger's papers to other work on gas theory and its relevance to later developments in that area.

[22]'Quantentheorie des einatomigen idealen Gases', *Preuss. Akad. Wiss. Sitzb.* (1924) 261 – 267, and (1925), 3 – 14.

[23]M. Planck, *Vorlesungen über die Theorie der Wärmestrahlung* (Leipzig: J. A. Barth, 1906) p. 54.

[24]M. Klein, 'Einstein and the Wave-Particle Duality', pp. 33 – 34.

[25]M. Planck, 'Zur Frage der Quantelung einatomiger Gase', *Preuss. Akad. Wiss. Sitzb.* (1925), 49 – 57.

[26]'Bemerkungen über die statistiche Entropiedefinition beim idealen Gas', *Preuss. Akad. Wiss. Sitzb.* (1925), 434 – 441.

[27]M. Planck, 'Absolute Entropie und Chemische Konstanten', *Annln Phys.* **66** (1921), 364 – 372. Planck again defended this division by N! in 'Zur Quantenstatistik des Bohrschen Atommodells', *Annln Phys.* **75** (1925) 673 – 684, partially in reaction to Schrödinger's 'Gasentartung'.

316   *Studies in History and Philosophy of Science*

apparently arbitrary correction, Planck argued that in the derivation of thermodynamic probability it is assumed that a permutation of essentially indistinguishable molecules gives rise to different distributions of the gas molecules. Therefore, the resulting expression for the number of genuinely distinct distributions is too large by the factor $N!$. In his response to Einstein, Planck again relied on the division by $N!$, and again supported this division by an appeal to the indistinguishability of molecules.[28]

Schrödinger argued that if the classical enumeration failed to recognize the identity of those states arising from a permutation of indistinguishable molecules, as Planck claimed, then the right number of independent states could not be obtained by simply dividing the traditional enumeration by a correction factor. 'In order for two molecules to be able to exchange their roles, they must surely have roles that are actually different; otherwise we would just not have counted such states as different in the earlier enumeration, and therefore need not and cannot now "correct away" a multiplicity that never even existed.'[29] According to Schrödinger, the usual way of counting states should be replaced by one that individuates only genuinely distinguishable states. The latter, he claimed, amounts to counting 'all distributions of the total energy'[30] of the gas, which, he argued, corresponds to Einstein's procedure.

Schrödinger did not infer from this that Planck was mistaken in focusing on the indistinguishability of permutations, since the description of gases generated by the new statistics was plausible on both theoretical and experimental grounds. But neither did he conclude that traditional statistical methods should be replaced by those of Bose and Einstein. Like Planck and Ehrenfest, he saw in Bose — Einstein statistics 'a radical departure from Boltzmann — Gibbs statistics'[31] that stood in need of explanation. In the penultimate section of his paper, Schrödinger examined an approach to quantum gas theory that promised to yield Einstein's results but on the basis of only traditional statistical methods.

Schrödinger proposed that 'definite quantum states with completely fixed energy levels are to be assigned not to the individual gas molecules, but to the body of the gas as a whole'.[32] Planck had used a version of this approach in his response to Einstein; it was a natural way of motivating the 'correction' for the indistinguishability of molecular permutations. But, Schrödinger argued,

[28]Note 26 above.
[29]'Statistische Entropiedefinition', p. 437: 'Damit zwei Moleküle ihre Rollen tauschen können, müssen sie doch wirklich verschiedene Rollen haben; andernfalls haben wir solche Zustände bei der früheren Abzählung ja gar nicht als verschieden gezählt, daher brauchen und dürfen wir eine Vielfachheit, die für uns gar niemals bestanden hat, auch jetzt nicht "wegkorrigieren"!'
[30]*Ibid.*, p. 437: 'aller Verteilungen bei der Gesamtenergie'.
[31]*Ibid.*, p. 440: 'ein radikales Abgehen von der Boltzmann — Gibbschen Art von Statistik'.
[32]*Ibid.*, p. 438: 'dass nicht den einzelnen Molekülen, vielmehr dem Gaskörper als Ganzem bestimmte Quantenzustände mit ganz bestimmten Energieniveaus zuzuschreiben sind'.

Planck's version rested on two questionable assumptions: (1) the quantum conditions prescribe only the energy levels for molecules, and (2) the individuality of molecules can be assumed. Schrödinger had already argued that strict regard for indistinguishability made the second untenable. The first rested on the traditional assumption that the energies of the gas molecules are quantized, which, said Schrödinger, 'seems to me by no means necessary in the new foundational viewpoint'.[33] Even if quantized molecular levels were assumed, they would not be determined by restricting the molecular energies directly, as Planck had done. Rather:

> the energy levels . . . of the gas molecules must now, of course, be derived from the energy level distribution of the body of gas as a whole, exactly the opposite of how it was previously done.[34]

According to Schrödinger, a correct development of the holistic approach would yield a single gas state corresponding to each set of distributions generated on the molecular approach by permutations of indistinguishable molecules. This is exactly the correspondence Schrödinger had earlier suggested would yield a counting of states equivalent to Einstein's. Thus Schrödinger might have offered the holistic approach as a means of deriving Einstein's results from a proper regard for the distinguishability of gas states. He did begin his presentation of the approach by observing that 'one will be led almost necessarily to Einstein's form of statistics when a statistical foundation is given for Planck's "division by $N!$".' But then, commenting only that 'one can nevertheless also take a quite different way', he raised the following possibility. Perhaps the problem with earlier traditional treatments was the usual assumption that:

> quantum theoretically one ascribes to each individual molecule a series of distinct states and energy values, . . . and that each possible quantum state of the whole gas is obtained by assigning each individual molecule, independently from the others, to one of these molecular quantum states; on the other hand, one can also take the view that such a thoroughgoing independence of gas molecules from one another . . . does not exist quantum theoretically.[35]

Schrödinger did not bother to say what was obvious to him and his readers —

[33]*Ibid.*, p. 439: 'scheint mir bei der veränderten Grundauffassung in keiner Weise zwangsläufig'.

[34]*Ibid.*, p. 439: 'müssen jetzt, gerade umgekehrt wie früher, die Energiestufen . . . der Gasmoleküle aus der Energiestufenverteilung des Gaskörpers abgeleitet werden'.

[35]*Ibid.*, p. 438: 'jedem einzelnen Molekül eine Reihe ausgezeichneter Zustände und Energiewerte quantentheoretisch vorgeschrieben sind . . . und dass jeder mögliche Quantenzustand des Gaskörpers einfach dadurch erhalten wird, dass man jedes einzelne Molekül, unabhängig von den übrigen, auf einen von diesen Molekülquantenzuständen versetzt; kann man andrerseits auch der Ansicht sein, dass eine so weitgehende Unabhängigkeit der Gasmoleküle voneinander . . . quantentheoretisch nicht besteht'.

318   *Studies in History and Philosophy of Science*

that the proposed new assumption was the natural starting point for developing a physical understanding of Einstein's results. In their papers of 1925 both Planck and Einstein had noted that certain expressions of Einstein's theory implied the presence, in Einstein's words, of 'a mutual interaction of the molecules whose nature is at present completely mysterious'[36] — mysterious because it did not involve any exchange of energy between the 'interacting' particles.

There was also a natural way to proceed from this new view: discover the nature of the molecular interaction and then derive Einstein's results directly, without recourse to the Bose – Einstein statistics. Schrödinger could not take this approach.

> For the moment there is no possibility of understanding the remarkable sort of molecular interaction that would lead to the elimination of the permutation number from the statistical calculus. The gas molecules must be something quite different, as Planck and Einstein have emphasized, and must interact in a way quite different from what we have imagined in the past.[37]

Immediately after proposing the new view he turned instead to the discussion of the holistic approach outlined above. Thus quantization of the gas as a whole was presented as a technique for deriving Einstein's results by taking into account, albeit indirectly, the molecular interaction apparently underlying those results.

Schrödinger offered additional arguments for the new approach: the resulting entropy function would be additive and would behave properly at low temperatures, two characteristics previously achieved with traditional statistical methods only through Planck's division by $N!$. But, he finally admitted, he could see no way to actually carry out the holistic quantization. He could think of no physically plausible assumptions to replace the two unacceptable ones on which Planck's version had depended. 'I take [these] difficulties to be so great', he concluded, 'that it makes it impossible to carry through in a nonarbitrary way the beautiful idea of beginning not with the quantization of the single molecules, but with that of the whole gas'.[38] In the last sentence of 'Statistical Definition of Entropy' Schrödinger could only suggest a return to an idea he had considered in a previous paper, that of

[36]A. Einstein, 'Quantentheorie des einatomigen idealen Gases', 1925 communication, p. 6: 'eine gegenseitige Beeinflussung der Moleküle von vorläufig ganz rätselhafter Art'.
[37]'Statische Entropiedefinition', pp. 440–441: 'Es fehlt vorläufig jede Möglichkeit, die merkwürdige Art von Wechselwirkung zwischen den Molekülen zu verstehen, welche zur Ausschaltung der Permutationszahl aus dem statistischen Kalkül führen soll. Die Gasmoleküle müssten dann, wie von Planck und Einstein auch betont wird, ganz etwas anderes sein und ganz anders aufeinander einwirken, als wir uns das bisher vorgestellt hatten'.
[38]*Ibid.*, p. 440: 'Ich halte [diese] Schwierigkeiten für so gross, dass sie eine willkürfreie Durchführung des schönen Gedankens: nicht von der Quantelung der Einzelmoleküle sondern von der des ganzen Gases auszugehen — unmöglich machen'.

Schrödinger's Route to Wave Mechanics   319

quantizing the mean free path of the gas molecules.[39] He would adopt the holistic approach in both of his next papers on quantum gas theory, however, and the successful execution of this approach would mark a first step towards wave mechanics.

Schrödinger's second gas theory paper of 1925, 'The Energy Levels of Ideal Monatomic Gases', was completed several months after the first.[40] The quantum restriction for total gas energy used there had been suggested by Einstein:[41] Planck's assumption that $N!$ molecular distributions correspond to one gas state was recast as the requirement that each gas level corresponded to a phase space cell of volume $N!h^{3N}$. Straightforward calculations gave energy, pressure and entropy in terms of the gas volume, temperature and number of molecules. An analysis of the behavior predicted for gases at low temperatures suggested that the lowest level should be assigned zero energy. The resulting 'gas laws' approached the classical laws at higher temperatures and predicted, at least qualitatively, the usual quantum phenomena at low temperatures.

Nevertheless, in the final section, 'Physical Significance', Schrödinger argued that these laws could not describe the thermodynamic behavior of real gases. His calculations had identified the energy of a gas level with the translational energy of non-interacting point particles. But, Schrödinger pointed out, at the low temperatures and high pressures where quantum phenomena occur, the 'cohesive forces' and molecular volume cannot be ignored. 'At best, therefore, one can attribute to the laws found in this way the significance of an extrapolation by analogy of the classical [ideal] gas laws to the low temperature region'.[42] The value of his results lay not in their descriptive power, he argued, but in the opportunity they provided for theoretical analysis: by comparing them with quantum theories based on more realistic molecular models, one could distinguish between the respective contributions of 'the van der Waal correction and the quantum influence' to low temperature phenomena.

Schrödinger did not actually offer such a comparison, however. His earlier quantization of mean free path had taken into account both molecular size and interaction (collisions). But, he noted, those results were obtained by

[39]Note 20 above.

[40]'Die Energiestufen der idealen einatomigen Gasmodelle', *Preuss. Akad. Wiss. Sitzb.* (1926), 23 – 26. This paper was given to Einstein to communicate to the Academy of Science in Berlin, which he did on January 7, 1926. The paper must have been completed much earlier than that, however, since it predates Schrödinger's third 1925 gas theory paper (note 54 below), which was received by the editor of *Phys. Z.* on December 15, 1925. A reference to 'Energiestufen' in a November 16 letter from Schrödinger to Landé (note 52 below) puts an even earlier limit on when the paper was finished.

[41]*Ibid.*, p. 23: 'Hr. Prof. Einstein hatte nun die Güte, mir brieflich folgenden Vorschlag mitzuteilen . . .'

[42]*Ibid.*, p. 35: 'Den so gefundenen Gesetzen kann man daher bestenfalls die Bedeutung einer sinngemässen Extrapolation der klassischen Gasgesetze in das Gebiet tiefster Temperatur beimessen'.

quantizing individual molecules, while the new ones rested on quantization of the whole gas. 'A comparison with the theory . . . developed here is not possible without something further'.[43] Apparently Schrödinger did not have in mind a comparison of the phenomenological laws resulting from the two approaches, since that would have been possible. A comparison of their implications for molecular behavior, however, would have required a further development of the holistic method. In 'Statistical Definition of Entropy' Schrödinger had argued that on this approach the properties of molecules would be inferred from the properties of the gas as a whole, but the quantization of a gas by partitioning phase space left the connection between gas levels and molecular properties unspecified. Schrödinger had not yet discovered a holistic strategy that could guide the inference to molecular behavior.

II. First Applications of de Broglie's Matter Waves

Schrödinger's third gas theory paper of 1925 rested on Louis de Broglie's notion of 'matter waves'. While many European physicists were vaguely aware in 1925 of de Broglie's theory of light and matter, few had studied it in detail. According to Schrödinger it was a passage in Einstein's paper of February 1925 that called his attention to the significance of de Broglie's work. According to evidence recently discovered by Hanle, however, Schrödinger did not even obtain a copy of de Broglie's thesis until late October of 1925, more than three months after he had studied Einstein's paper in some detail.[44] Thus the initial impact of this reference to de Broglie could not have been great. Additional impetus may have come in the fall of 1925 from Peter Debye, then at the Zürich Institute of Technology, who suggested that Schrödinger present an exposition of de Broglie's work for the combined university – institute of technology physics seminar.[45]

Writing to Einstein on November 3, 1925, Schrödinger indicated two aspects of de Broglie's work that attracted his attention: the clarification it provided for that section of Einstein's paper containing the reference to de Broglie, and the similarity between de Broglie's explanation of the quantization of atomic orbits and a property of Bohr orbits that Schrödinger himself had discovered several years earlier. During the next month or so Schrödinger pursued the separate but parallel paths suggested by these two aspects of de Broglie's work,

[43]*Ibid.*, p. 36: 'ist ein Vergleich mit der hier durchgeführten Theorie . . . nicht ohne weiteres möglich'.
[44]P. Hanle, 'Schrödinger's Reaction'.
[45]'Peter J. W. Debye — An Interview', with E. E. Salpeter, D. R. Corson and S. H. Bauer, *Science* **145** (1964), 554 – 559; and F. Bloch, 'Reminiscences of Heisenberg and the Early Days of Quantum Mechanics', *Physics Today* **29** (1976), 23 – 27.

seeking to apply the matter wave concept to quantum gas theory on the one hand, and to atomic theory on the other.

Schrödinger began with atomic theory. In 1922 he had discovered 'A Remarkable Property of the Quantum Paths of a Single Electron'[46] which emerged when he examined some implications of Herman Weyl's work for the electron orbits of old quantum theory. In Weyl's extension of general relativity the square of the length of a vector could change as the vector underwent a parallel displacement.[47] Schrödinger found that when such a vector was transported by parallel displacement around an electron orbit, the only change in its square was an integral multiple increase in a certain exponential factor. The formal expression of this property bore a striking resemblance to de Broglie's connection between the quantum electronic orbits and his particle-wave electrons. According to de Broglie, a particle of rest mass m_o and velocity v was associated with an inherent periodic phenomenon of frequency

$$\frac{1}{h} \, m_o c^2 \, \frac{1}{\sqrt{1-\beta^2}}$$

that was always in phase with an accompanying wave with frequency

$$\frac{1}{h} \, m_o c^2 \, \sqrt{1-\beta^2}$$

where c is the speed of light, and $\beta = \dfrac{v}{c}$;

the path of the particle lay along a ray of its accompanying phase wave. He had shown that the atomic orbits of single electrons in the old quantum theory were just those electron paths for which the phase waves were in resonance, that is, the circumference of the orbit was equal to an integral number of wave lengths of the phase wave.[48] The similarity between the mathematical expression of this connection and that of Schrödinger's 'remarkable property' of 1922 was so striking that Fritz London remarked on it to Schrödinger in 1926.[49] More recently Raman and Forman identified it as a source of

[46]'Über eine bemerkenswerte Eigenschaft der Quantenbahnen eines einzelnen Elektrons', *Z. Phys.* **12** (1922), 13 – 23.

[47]H. Weyl, *Raum, Zeit, Materie* (4th edn., Berlin, 1921).

[48]L. de Broglie, 'Recherches sur la Théorie des Quanta', *Annls Phys.* **3** (1925), 22 – 128. See also F. Kubli, 'Louis de Broglie und die Entdeckung der Materiewellen', *Archs Hist. exact Sci.* **7** (1970 – 1971), 26 – 68, and E. MacKinnon, 'De Broglie's thesis: A critical retrospective', *Am. J. Phys.* **44** (1976), 1047 – 1055.

[49]Letter from F. London to Schrödinger, *ca.* December 10, 1926, in V. V. Raman and P. For-

Schrödinger's interest in phase waves even before they were aware of his letter to Einstein.[50]

In his letter to Einstein Schrödinger remarked that de Broglie's version of the mathematical expression was much more revealing than his own because it emerged in the framework of a comprehensive theory. He was quick to explore this new insight. Twelve days later Schrödinger wrote a long letter to his friend Alfred Landé about Landé's recent attempt to trace the statistical interdependence implied by Einstein – Bose statistics to a superposition phenomenon:[51]

> It pleases me greatly to hear that your paper is intended to be a 'return to wave theory'. I also am very much inclined to do so. Recently I have been deeply involved with Louis de Broglie's ingenious thesis. It is extraordinarily stimulating, but nonetheless some of it is very hard to swallow. I have vainly attempted to make myself a picture of the phase wave of an electron in an elliptical orbit. The 'rays' are almost certainly neighboring Kepler ellipses of equal energy. That, however, gives horrible 'caustics' or the like as the wave front. At the same time, the length of the wave ought to be equal to one Zeeman or Stark cycle![52]

Schrödinger had tried to determine the structure of the phase wave for an atomic electron of given energy by constructing a wave that was refracted enought to make its rays coincide with the Bohr – Sommerfeld orbits of that energy, and that had a wave length equal to integral fractions of the distances around those orbits. Apparently he found that under such extreme refraction the wave fronts were not well behaved.

In the face of these difficulties Schrödinger turned from this attempt at a wave theory of the atom to the other line of research suggested by his reading of de Broglie's thesis.[53] On December 15, about a month after his letter to

man, 'Why Was it Schrödinger?', p. 304.

[50]See V. V. Raman and P. Forman, 'Why Was it Schrödinger?' for a more detailed exposition of Schrödinger's paper of 1922, and of the similarity between Schrödinger's discovery and de Broglie's atomic model.

[51]A. Landé, 'Lichtquanten und Ḳohärenz', *Z. Phys.* **33** (1925) 571 – 579.

[52]Letter from Schrödinger to A. Landé, November 16, 1925, in The Niels Bohr Library, Center for History and Philosophy of Physics of the American Institute of Physics, New York. The quoted portion is translated in V. V. Raman and P. Forman, 'Why Was it Schrödinger', p. 313.

[53]Schrödinger's November 16 letter to Landé indicates that he did not begin thinking about how to use phase waves in gas theory until after his construction of atomic phase waves. The central idea of Landé's paper was to explain the interaction implied by Bose – Einstein statistics by assuming a superposition phenomenon for both light quanta and gas molecules. That part of his paper concerning gas theory addressed the very problem Schrödinger had been working on most of the summer, and the solution offered by Landé involved classical wave considerations in an essential way. If when he wrote the November 16 letter Schrödinger had even begun trying to apply de Broglie's idea to gas theory, we would expect him to describe this attempt, or at least to remark that his thoughts on gas theory were moving along lines similar to Landé's. Instead, Schrödinger made only a passing reference to his own ideas on gas theory, mentioning 'Energy Levels of Gases' as his latest word on the subject. When he did note a similarity between his and Landé's views, it was not in connection with gas theory at all, but concerned his unsuccessful attempt to apply de Broglie's wave notion to atomic theory.

Landé, Schrödinger completed the third quantum gas theory paper of 1925, 'On Einstein's Gas Theory'.[54] He once again adopted the holistic approach, but now in a way quite different from that of 'Energy Levels of Gases'. He abandoned Einstein's recent suggestion that gas energy be quantized by partitioning phase space, and took his cue from Einstein's earlier reference to de Broglie's notion of phase waves. Einstein noted that the fluctuations predicted by his theory had the form of interference fluctuations in radiation. 'I believe that this is more than a mere analogy', he commented; 'de Broglie has shown in a very important work how a (scalar) wave field can be coordinated with a material particle or a system of material particles'.[55] Instead of treating the gas as a collection of individual particles, Schrödinger applied methods appropriate to wave phenomena: he quantized the gas by adapting a procedure that previously had been applied only to cavity radiation.

In 1910 Debye had derived Planck's formula for cavity radiation by attributing to the radiation a set of characteristic frequencies, v_1, v_2, . . ., v_s, . . ., and imposing two restrictions: (1) the energy in the sth mode of oscillation is either 0, hv_s, $2hv_s$, . . ., nhv_s; and (2) the sum of the energies of all modes is constant and equal to the total energy of the radiation, E, i.e. $\Sigma_s n_s hv_s = E$ (where n_s is the integer determining the energy of the sth mode).[56] Bose's application of the new statistics to a particle model of radiation contrasted sharply with Debye's application of classical statistics to a wave-oscillator model. Schrödinger saw in this difference a clue for fashioning an alternative to Einstein's use of the new statistics on a particle model of gases.

> The following simple idea is a guide to this [new approach]: Einstein's theory of a gas is obtained by applying to the gas molecules that form of statistics that leads to the Planck radiation law when it is applied to 'atoms of light' [as Bose had done]. However, one can also obtain the Planck radiation law by using 'natural' statistics, if one applies them to so-called 'aether oscillators', that is, to the degrees of freedom of the radiation. The light atoms then appear only as the energy levels of the aether oscillators. . . .One must therefore simply form a picture of the gas like the picture of cavity radiation that does not correspond to the extreme light-quantum representation: the natural statistics. . . will then lead to Einstein's gas theory.[57]

[54]'Zur Einsteinschen Gastheorie', *Phys. Z.* **27** (1926), 95 – 101.

[55]A. Einstein, 'Quantentheorie des einatomigen idealen Gases', 1925 communication, p. 9: 'ich glaube, dass es sich dabei um mehr als um eine blosse Analogie handelt. Wie einem materiellen Teilchen bzw. einem System von materiellen Teilchen ein (skalares) Wellenfeld zugeordnet werden kann, hat Hr. L. de Broglie in einer sehr beachtenswerten Schrift dargetan'.

[56]P. Debye, 'Der Wahrscheinlichkeitsbegriff in der Theorie der Strahlung', *Annln Phys.* **33** (1910), 1427 – 1434.

[57]'Zur Einsteinschen Gastheorie', p. 95: 'Dazu führt folgender einfacher Gedanke: die Einsteinsche Gastheorie wird erhalten, indem man auf die Gasmoleküle die Form der Statistik anwendet, die, auf die "Lichtatome" angewendet, zum Planckschen Strahlungsgesetz führt. Aber man kann das Plancksche Strahlungsgesetz auch durch "natürliche" Statistik gewinnen, indem

Studies in History and Philosophy of Science

Schrödinger began with the assumption that the gas as a whole was capable of taking on only certain characteristic oscillations. The first restriction imposed by Debye on radiation modes was applied unchanged to the gas modes. Instead of adopting Debye's second restriction, however, Schrödinger required that $\Sigma_s n_2 = N$, where N is the number of molecules in the gas sample. The rationale for this restriction is found in the way Schrödinger related the quantum states of the gas as a whole and the states of the constituent molecules. He assumed that each vibration mode of the gas corresponded to a different molecular energy level, the first mode corresponded to the lowest molecular level, the second mode to the next molecular level, etc. The energies of the molecular levels were set equal to the energy quanta, $h\nu_1$, $h\nu_2$, . . . , of the corresponding gas modes. Then, according to Schrödinger, attributing an energy $n_s h\nu_s$ to the sth mode of vibration amounted to assigning n_s molecules to the molecular level of energy $h\nu_s$. Schrödinger's requirement that $\Sigma_s n_s = N$ just amounted to ensuring that the total number of molecules remained constant. The difference between this restriction on gas modes and Debye's second restriction on radiation modes exactly mirrored the difference between a restriction on the total number of gas particles imposed by Einstein in his new statistical treatment of gases and a restriction on the total energy of 'light particles' imposed by Bose in his treatment of radiation.

In specifying the relation between gas modes of oscillation and molecular energy levels Einstein's solved the problem that in 'Energy Levels of Gases' had prevented a comparison with earlier theories. Once the two restrictions had been used to identify the permitted gas states and thermodynamic properties had been determined by the usual statistical methods, Schrödinger could express his results in terms of molecular energy levels. In this form they could be compared with Einstein's, and were found to agree. So far the results were expressed only as functions of unspecified molecular energy levels, however. The agreement would be complete only if they assigned the same energies to these levels.

It was for this final step, the evaluation of permissible molecular energies, that Schrödinger turned to the details of de Broglie's theory:

We will calculate [the molecular energy levels], following L. de Broglie, from the idea that a moving molecule, with velocity $v = \beta c$ and rest mass m, is nothing more than a 'signal', one might say 'the wave crest', of a wave system whose frequency lies in the neighborhood of

man sie auf die sog. "Ätherresonatoren", d.i. auf die Freiheitsgrade der Strahlung anwendet. Die Lichtatome treten dann nur als die Energiestufen der Ätherresonatoren auf. . . . Man muss also einfach das Bild des Gases nach demjenigen Bilde der Hohlraumstrahlung formen, das noch nicht der extremen Lichtquantenvorstellung entspricht; dann wird die natürliche Statistik. . . zur Einsteinschen Gastheorie führen'.

Schrödinger's Route to Wave Mechanics   325

$$v = \frac{mc^2}{h\sqrt{1-\beta^2}}$$

and for whose phase velocity *u* a dispersion law holds that is given by the above equation in conjunction with

$$u = \frac{c}{\beta} = \frac{c^2}{v}$$

(where *v* plays the role of the signal velocity, as one can easily calculate and de Broglie has shown).[58]

Classical wave theory was used to calculate the wave lengths of the characteristic oscillations of the gas given the boundary conditions determined by the gas volume. De Broglie's theory associates a frequency and wave with each molecular level, which Schrödinger assumed would also be the frequency and velocity of the corresponding gas mode. He could then use de Broglie's equations to compute the characteristic frequencies of the gas from the allowed wave lengths. Since the energy quantum for each gas mode, *hv*, was equal to the energy of the corresponding molecular level, the spectrum of molecular energy levels was determined. It coincided exactly with the spectrum given by Einstein.

In 'Statistical Definition of Entropy' the problem of discovering the 'quite different' nature of molecules and molecular interactions implied by Einstein's theory had been set aside because it was too difficult. The holistic approach was suggested, which would bypass the problem by focussing on the gross manifestations of molecular behavior — the properties of the gas as a whole. If this approach led to Einstein's theory it would place that theory on a less problematic foundation, since it acknowledged the physical implications of the theory and yet required only traditional statistics. When Schrödinger finally developed a successful holistic strategy, however, it served as more than just a way to avoid the problems of understanding the nature of molecules. It suggested a solution for that problem:

[58]*Ibid.*, p. 97: 'Wir berechnen es in engem Anschluss an L. de Broglie aus der Vorstellung, dass mit der Geschwindigkeit *v* = *βc* bewegtes Molekül von der Ruhmasse m nichts weiter ist als ein ''Signal'', man könnte sagen ''der Schaumkamm'', eines Wellensystems, dessen Frequenz *v* in der Nachbarschaft von

$$v = \frac{mc^2}{h\sqrt{1-\beta^2}}$$

liegt und für dessen Phasengeschwindigkeit u ein Dispersionsgesetz gilt, das durch vorstehende Gleichung, in Verbindung mit

$$u = \frac{c}{\beta} = \frac{c^2}{v}$$

gegeben wird (*v* spielt dann die Rolle der Signalgeschwindigkeit, wie man leicht nachrechnet und de Broglie gezeigt hat).

This [way of deriving Einstein's theory] means nothing other than taking seriously the de Broglie – Einstein undulatory theory of moving particles, according to which the particles are nothing more than a kind of 'wave crest' on a background of waves.[59]

de Broglie had shown that the relation between his wave and particle velocities (pages 324 and 325 above) was equivalent to the usual relation between the mean velocity of a group of waves and the velocity of energy propagation, or 'signal velocity', of the group. It was well known that an appropriately constructed set of two-dimensional waves could be superimposed to give a negligible resultant amplitude everywhere except in a very small region, and that the resulting 'wave packet' would travel with the signal velocity of the waves. In a section called 'On the Possibility of Representing Molecules or Light Quanta by the Interference of Plane Waves',[60] Schrödinger raised the natural question: could particles be reduced to packets of phase waves? He identified two problems. First, while it was clear that a thin pencil or 'signal' could be constructed from a group of parallel plane waves, 'one is perhaps in doubt for a moment whether and how it is possible to restrict the signal to a region that is small in all three directions'.[61]. Schrödinger suggested using a technique originally designed to construct a cone from the rays of plane waves:

> According to Debye and [Max] v[on] Laue, one achieves this by beginning with a plane wave and allowing not only the frequency to vary, but also the wave normal — the latter over a small spatial angle $d\psi$, and integrating together a continuum of infinitesimal wave functions within this region of frequency and wave normal.[62]

He did not show how this method would yield a packet, but immediately pointed out that a further problem would remain unsolved. Such a construction

> does not, according to the classical wave laws, ensure that the 'model of light quanta' so produced . . . also remains together for a period of time. Rather, it is dispersed into larger and larger regions . . .[63]

[59]*Ibid.*, p. 95: 'Das heisst nichts anderes, als Ernst machen mit der de Broglie – Einsteinschen Undulationstheorie der bewegten Korpuskel, nach welcher dieselbe nichts weiter als eine Art ''Schaumkamm'' auf einer den Weltgrund bildenden Wellenstrahlung ist'.

[60]'Über die Möglichkeit, Moleküle oder Lichtquanten durch Interferenz ebener Wellen darzustellen', *Ibid.*, pp. 100 – 101.

[61]*Ibid.*, p. 100: 'ist man vielleicht einen Augenblick im Zweifel, ob und wie es möglich ist, das Signal auf einen in allen drei Richtungen kleinen Raumteil zubeschränken'.

[62]*Ibid.*, p. 101: 'Man erreicht dies nun nach Debye und v. Laue dadurch, dass man in der einen ebenen Wellenfunction, von welcher man ausging, nicht nur die Frequenz, sondern auch die Wellennormale über einen kleinen Bereich, einen kleinen Raumwinkel $d\omega$ variieren lässt und ein Kontinuum infinitisemaler Wellenfunctionen innerhalb dieses Frequenz und Wellennormalenbereiches zusammenintegriert'.

[63]*Ibid.*, p. 101: 'ist nach den klassischen Wellengesetzen natürlich nicht zu erreichen, dass das so erzeugte ''Modell eines Lichquants'' . . .auch dauernd beisammen bleibt. Vielmehr zerstreut es sich . . . auf immer grössere Räume'.

He had at least sketched a solution to the first problem; for the second he offered little more than an expression of hope: 'If one could avoid this latter result through a quantum theoretical modification of the classical wave laws, then it appears that a way is opened up to be free of the dilemma of light quanta.'[64]

Schrödinger referred to the new wave representation as 'the de Broglie – Einstein wave theory of moving particles', but his use of the wave concept was not the same as de Broglie's. For de Broglie particle and wave were separate entities. They moved together, each affecting the behavior of the other, and giving rise to phenomena that in the case of matter had traditionally been associated with a particle, in the case of radiation with a wave. For Schrödinger, on the other hand, a 'particle' was only a special part of a wave (or group of waves). Whether Schrödinger recognized this difference is not clear. He may have been misled by the fact that de Broglie seemed to attach great importance to the equality of particle and phase wave signal velocities. What significance the equality did have in de Broglie's dualistic wave-particle model is not clear.[65] Schrödinger saw in it the basis for a pure wave theory of matter. To develop such a theory he turned back to the problem of atomic structure.

III. The Wave Equation

In his first attempt to apply de Broglie's notion of matter waves to atomic theory, Schrödinger had constructed the phase wave of the orbital electron. As he reported to Landé in mid-November, the results were not promising. In his second attempt at a wave theory of the atom, however, Schrödinger did not try to reproduce the old electron orbits or even to construct appropriately contorted waves. Instead he adopted a more traditional method; he tried to derive wave functions describing allowed vibration modes by imposing boundary conditions on the solutions of a wave equation. With this change Schrödinger's model of the atomic waves shifted away from de Broglie's picture of well defined rays along Bohr – Sommerfeld orbits. Schrödinger later characterized the difference between de Broglie's atomic phase waves and the matter waves inherent in his new approach as the difference between travelling and standing waves.[66] Little information is available on what inspired this change in approach. Felix Bloch has recalled that in the Zürich colloquium

[64]*Ibid.*, p. 101: 'Wenn man durch eine quantentheoretische Modifikation der klassischen Wellengesetze dies letzte Folgerung vermeiden könnte, so schiene ein Weg zur Befreiung von dem Lichtquantendilemma angebahnt'.

[65]See E. MacKinnon, 'De Broglie's Thesis'.

[66]*Abhandlungen*, pp. 12 – 13, *Collected Works*, p. 9.

where Schrödinger reported on de Broglie's work, Debye suggested a search for the wave equation governing phase waves.[67] The new approach may also have suggested to Schrödinger by his application of de Broglie's ideas to gas theory, where he had determined characteristic standing waves by imposing boundary conditions in the classical manner.

From the beginning Schrödinger sought a wave equation that was consistent with the special theory of relativity.[68] This was natural. The previous decade of research on fine structure had shown that a full explanation of spectra required a relativistic atomic theory. Furthermore, de Broglie's theory had been developed in the framework of relativistic assumptions. To settle for a non-relativistic theory would have been to discard the generality that de Broglie's approach seemed to promise. By late December 1925 Schrödinger had found the relativistic equation now referred to as the Klein–Gordon equation.[69] Two days after Christmas he wrote to Wilhelm Wien, who as editor of *Annalen der Physik* had asked Schrödinger to write an article on Einstein's new approach to gas theory. Schrödinger replied that he already had three papers being published on that subject, and that now he was deeply involved in a new project:

> At the moment I am struggling with a new atomic theory. If only I knew more mathematics! I am very optimistic about this thing, and expect that, if only I can . . . solve it, it will be very beautiful. I think that I can specify a vibrating system that has as eigenfrequencies the hydrogen term frequencies — and in a relatively natural way, not through *ad hoc* assumptions. But you don't actually get these term

[67]F. Bloch, 'Reminiscences'.

[68]Evidence of this relativistic attempt is found in certain passages of Schrödinger's original four part work on (non-relativistic) wave mechanics. In the first part Schrödinger followed the presentation of his successful application of the non-relativistic equation to the hydrogen electron with the remark: 'if one works out the relativistic Kepler problem exactly according to the procedure above, it leads strangely to *half-integral partial* quanta'. [*Abhandlungen*, p. 12; *Collected Works*, p. 9.] Later in the same paper he attempted to motivate the relation $E = hv$ by considerations involving an approximation which, he claimed, 'is completely avoided if one develops the relativistic theory . . . Unfortunately, however, the unobjectionable development of this theory still meets with certain difficulties already mentioned above'. [*Ibid.*, pp. 14–15 and 10.] In a footnote in the second part of his series on wave mechanics, Schrödinger again mentioned that the construction of a relativistic equation 'presents numerous difficulties'. [*Ibid.*, pp. 42 and 30.] When in the fourth part Schrödinger finally presented his relativistic equation, he referred back to his comments in the first part, and again emphasized that 'there is still *missing* the *supplement* needed for the determination of the numerically correct splitting of the hydrogen lines. . .' [*Ibid.*, pp. 163 and 110.] For further evidence of Schrödinger's relativistic attempt see the items referred to in notes 70 and 72–74 below.

[69]By the time Schrödinger published his relativistic equation in 'Quantisierung, Vierte Mitteilung' [*Abhandlungen*, p. 163, *Collected Works*, p. 119.], others had taken his non-relativistic equation as their starting point, and attempted to find a relativistic generalization. Oskar Klein and Walter Gordon independently arrived at the *same* equation Schrödinger had found earlier, which is why it is now called the Klein–Gordon equation. [O. Klein, 'Quantentheorie und fünfdimensionale Relativitätstheorie', *Z. Phys.* 37 (1926), pp. 895–906; W. Gordon, 'Der Comptoneffekt nach der Schrödingerschen Theorie', *Z. Phys.* 40 (1926), pp. 117–133.]

Schrödinger's Route to Wave Mechanics   329

frequencies themselves, i.e., not $-R/n^2$, but $mc^2/h - R/n^2$ (m is electron mass) . . . if, say

$$\nu_n = \frac{mc^2}{h} - \frac{R}{n^2}, \quad \nu_m = \frac{mc^2}{h} - \frac{R}{m^2}$$

then

$$\nu_n - \nu_m = R\left(\frac{1}{m^2} - \frac{1}{n^2}\right)$$

. . . I hope that I can soon report in a little more detailed and understandable way about the matter. At present I must learn a little mathematics in order to completely solve the vibration problem — a linear differential equation, similar to Bessel's, but less well known.[70]

Schrödinger's enthusiasm was short-lived, however. When he finally found solutions of the relativistic equation for the hydrogen atom, they yielded half-integral quantum numbers that gave a completely incorrect fine structure spectrum.[71] Though understandably discouraged, he nevertheless wrote up his results. David Dennison, an American who studied in Europe in the late 1920s, has recalled a meeting with Schrödinger in 1927. Schrödinger spoke about the failure of his original relativistic version of wave mechanics, and mentioned that he had written a report of this work which, according to Dennison, he had never attempted to publish.[72] A letter written by Schrödinger in 1956 to his

---

[70]Letter from Schrödinger to W. Wien, December 27, 1925, AHQP: 'Im Augenblick plagt mich eine neue Atomtheorie. Wenn ich nur mehr Mathematik kennte! Ich bin bei dieser Sache sehr optimistisch und hoffe, wenn ich es nur . . . bewältigen kann, so wird es sehr schön. Ich glaube, ich kann ein schwingendes System angeben — u. auf verhaltnismässig natürlichen Wege, nicht durch *ad hoc* Annahmen — das die Wasserstoffterm frequenzen zu Eigenfrequenzen hat. Aber nicht eigentlich diese selbst also nicht $\frac{R}{n^2}$ sondern $\frac{mc^2}{h} - \frac{R}{n^2}$ (m = Elektronenmass). . . . wenn etwa

$$\nu_n = \frac{mc^2}{h} - \frac{R}{n^2}, \quad \nu_m = \frac{mc^2}{h} - \frac{R}{n^2}$$

then

$$\nu_n - \nu_m = R(\frac{1}{m^2} - \frac{1}{n^2})$$

. . .Ich hoffe, ich kann bald ein wenig ausfürlich und verständlicher über die Sache berichten. Verläufig muss ich noch Mathematik lernen, um das Schwingungsproblem ganz zu übersehen — eine lineare Differenzialgleichung, der Bessel'schen änlich, aber weniger bekannt'.

[71]See L. I. Schiff, *Quantum Mechanics* (3rd edn., New York: Addison-Wesley, 1968), pp. 467–471.

[72]Interview with D. Dennison, conducted by T. S. Kuhn, January 30, 1964, AHQP.

330   *Studies in History and Philosophy of Science*

former student, Wolfgang Yourgrau, refers to the paper mentioned by Dennison, and indicates that it was in fact submitted for publication:

> . . . the Schrödinger theory, relativistically framed (without spin), gives a *formal* expression of the fine structure formula of Sommerfeld, but is *incorrect* owing to the appearance of half integers instead of integers. My paper in which this is shown has . . . never been published; it was withdrawn by me and replaced by the non-relativistic treatment.[73]

Soon after the relativistic equation proved inadequate, Schrödinger discovered that the non-relativistic version of the equation could explain the gross spectrum of the hydrogen atom.[74] Less than a month elapsed between this discovery and January 26, when Schrödinger sent his first paper on wave mechanics to Wien for publication. This paper and the one that followed it reveal some of the ways that Schrödinger's ideas developed as he found the wave equation and explored its implications.

IV. Unveiling of Wave Mechanics

One of the most striking characteristics of Schrödinger's first paper is the almost complete lack of any reference to the wave picture that motivated his discovery. Schrödinger started with a form of the classical Hamilton equation for a particle of mass m and energy E: $(\Delta\psi(q_i))^2 - \frac{2m}{K^2}(E-V)\psi(q_i)^2 = 0$, where the q_i are the position coordinates, V the potential acting on the particle, and

[73]W. Yourgrau and S. Mandelstam, *Variational Principles in Dynamics and Quantum Theory* (2nd edn., London: Pitman, 1969), p. 114.

[74]There has been some confusion in the literature over the length of the period separating Schrödinger's work on the relativistic and non-relativistic equations. In a report of his conversation with Schrödinger in Stockholm, when they jointly received the 1933 Nobel Prize for physics, Dirac recalled a discussion of the original relativistic attempt. He reports that when the failure of the relativistic equation was discovered, 'Schrödinger was very disappointed and concluded that his method was not good and abandoned it. Only after some months did he return to it . . .' [P.A.M. Dirac, 'Professor Erwin Schrödinger — Obituary', *Nature* **189** (1961), 355–356.] M. Jammer, in *Conceptual Development*, p. 258, has taken the last quoted phrase in Dirac's report as the basis for his claim that several months elapsed between Schrödinger's relativistic and non-relativistic attempts, and presumably it also stands behind a similar claim made by A. Hermann in 'Erwin Schrödinger — Eine Biographie'. It might seem that this position is supported by Annemarie Schrödinger, who told Kuhn in a 1693 interview (note 4 above) that her husband had first discovered wave mechanics while they were vacationing in Arosa in the summer of 1925. Kuhn's further questioning revealed, however, that Frau Schrödinger remembered clearly only the fact that the discovery took place in Arosa, but could not really remember exactly when this had been. She recalled that it had been Hermann who had told her it must have beeen in the summer. J. Gerber, in 'Geschichte der Wellenmechanik', has also followed Hermann in this dating of Schrödinger's relativistic attempt. A. Hermann has recently changed his mind on this point, primarily on the basis of the December 27 letter from Schrödinger to W. Wien (note 70 above), which describes a relativistic equation and was written in Arosa. See Hermann, 'Erwin Schrödinger', *Lexikon Geschichte der Physik A – Z* (Cologne Aulis Verl. Deubner, 1972), pp. 343–345, and 'Schrödinger, Erwin', *Dictionary of Scientific Biography,* Vol. XII (New York Scribners, 1975), pp. 217–223.

$\psi(q_i)$ is related to Hamilton's characteristic function $S(q_i)$ by $S = K \ln \psi$, where K is an unspecified constant. In classical particle mechanics one would have solved this equation for ψ, placed boundary conditions on these solutions, and then derived relations among ψ, the q_i and the energy. Schrödinger deviated from the classical treatment at this point:

> Now we do *not* look for a solution of [the above] equation, but . . . seek a function ψ which is everywhere real, single-valued, finite and continuously twice differentiable, and which makes the integral of [the expression on the left side of the equation] taken over the whole coordinate space an *extremum. We replace the quantum requirements by this problem.*[75]

Common variational techniques yielded the following equations:

$$\nabla^2\psi + \frac{2m}{K^2}(E\text{-}V)\psi = 0 \text{ and } \int df\partial\psi\,\frac{\delta\psi}{\delta n} = 0,$$

where $\delta\psi$ was the variation of ψ, df the differential an element of the closed surface at infinity, and δn the differential of a line element perpendicular to this surface. Setting K equal to $(\frac{h}{2\pi})^2$ in the first equation, Schrödinger obtained the non-relativistic wave equation. The second placed boundary conditions on ψ. Schrödinger applied the equations to hydrogen by taking as V the Coulomb potential on the single electron of the hydrogen atom, and solving for ψ. The eigenvalues E agreed with the energy spectrum of hydrogen determined by the Bohr theory.

Schrödinger urged acceptance of his wave equation solely on the basis of its success with hydrogen and the fact that these results were obtained without the artificially imposed quantum conditions of the old quantum theory. It was natural for Schrödinger to use these arguments. The successful application to the test case of the hydrogen atom was a necessary first step in the defense of any new atomic theory. Wolfgang Pauli had demonstrated that matrix mechanics could handle this case almost four months after Heisenberg's original formulation of the theory.[76] It was only after the theory had been given its more general and rigorous matrix form by Born, Heisenberg and Jordan that Pauli's results became well known and won for the new quantum mechanics a wide acceptance.[77] Schrödinger's solution of the hydrogen case was very strong

---

[75] *Abhandlungen*, p. 2: *Collected Works*, pp. 1 – 2; Schrödinger's italics.

[76] W. Pauli, 'Über das Wasserstoffspektrum vom Standpunkt der neuen Quantenmechanik', Z. *Phys.* **36** (1926), 336 – 363 (received January 17, 1926); translated in *Sources of Quantum Mechanics*, B. L. van der Waerden (ed.) (New York, 1968), pp. 387 – 415.

[77] B. L. van der Waerden, *Sources*, p. 58.

evidence in favor of his theory. The elimination of the quantum conditions was also something to be emphasized, since it would appeal to many who, like Schrödinger, saw in the *ad hoc* character of these conditions a major problem in old quantum theory.

What is surprising is Schrödinger's failure to provide an intuitive motivation for the wave equation. It was not presented as the product of a search for a theory of matter waves. It was not derived as a *wave* equation at all, but as the result of placing an unmotivated formal requirement on Hamilton's equation for *particles*. While in the last section of his paper Schrödinger did raise the possibility of linking his theory with a wave model and even referred to de Broglie's notion of phase waves, his discussion of this possibility began with the remark: 'It is, of course, strongly suggested that we should try to connect the function with some *vibration process* in the atom . . .'[78] As a result he left the impression that the idea of giving a wave interpretation to his theory was inspired by the fact that the equation for ψ had the form of a wave equation. The wave picture that had been crucial to his discovery of his theory now appeared as a byproduct of it.

Why did Schrödinger introduce his new theory in this way? The idea that the wave equation be generated from a variation principle came from Debye. Erwin Fues, an assistant of Schrödinger's, attended the colloquium in which Schrödinger first presented his new theory to the Zürich physics group. 'I remember clearly', wrote Fues in 1952, 'that the suggestion came from him [Debye] to give the foundations of the theory the form of a variation principle'.[79] According to Fues, Debye argued that the most important theories in physics can all be formulated in this way.

The variational basis could have been presented in addition to an intuitive one, however. There are two reasons Schrödinger might have had for not founding his theory on a wave picture in his first paper. First, Schrödinger was introducing the theory to an audience with a wide variety of prejudices for and against classical models. Many physicists were convinced that the key to solving the problems of quantum physics was to surrender all hope for a theory based on physical pictures. It was good strategy to avoid the touchy question of physical pictures in the first presentation. Schrödinger could still argue on other grounds that his theory had strong intuitive appeal. After raising the possibility of a 'vibration process in the atom' he explained:

> I originally intended to found the new quantum conditions in this more intuitive manner, but finally gave them the above neutral mathematical form, because it brings more clearly to light what is really essential. The essential thing seems to me to

[78] *Abhandlungen*, p. 12; *Collected Works*, p. 9.
[79] Letter from Fues to T. S. Kuhn, July 2, 1963, AHQP, reel 66: 'Ich erinnere mich deutlich, dass von ihm [Debye] die Anregung ausging, der Begründung der Theorie die Form eines Variationsprinzips zu geben. . .'

be, that the postulation of 'whole numbers' no longer enters into the quantum rules mysteriously, but that we have traced the matter a step further back, and found the 'integralness' to have its origin in the finiteness and single-valuedness of a certain space function.[80]

The issue of interpretation could be addressed later, after he had obtained convincing agreement with experiment.

A second, perhaps even stronger reason for not emphasizing the wave picture might be found in the problems facing a wave interpretation of the new theory. The stability of the wave packets had not been established yet. Schrödinger was also aware of an awkward incongruity between the wave model that had inspired his search for a wave equation and the theory that resulted. In the case of a single classical particle ψ could be interpreted as a wave function describing a matter wave. For a system of n classical particles, however, ψ was a function of $3n$ spatial coordinates and therefore described a wave in a $3n$-dimensional space that could not be identified with ordinary physical space. To give his theory a wave interpretation Schrödinger would either have to show how the ψ in $3n$-dimensional space determined n waves in 3-dimensional space, or reformulate the theory so that it would yield directly the required n wave functions. (Eventually each of these escape routes were to be explored, but neither would prove successful.[81]) In his first paper Schrödinger was careful to define ψ as the solution to equations formulated in a 'coordinate space'; he was already quite aware of the problem.

In his second paper on wave mechanics, completed a month later, Schrödinger offered a more intuitive presentation of wave mechanics. Its basic outline is familiar to students of quantum theory. The starting point was a formal analogy, first discovered by William Rowen Hamilton in the 19th century, between the fundamental extremal laws for geometrical optics and classical particle mechanics. According to Fermat's Principle the path between points A and B of a light ray in a dispersive medium with index of refraction n lies along that line for which $\int_A^B n\,ds$ is an extremum; Maupertuis' Principle stipulates that the path between points A and B of a particle of mass m and

[80]*Abhandlungen*, p. 12; *Collected Works*, p. 9.
[81]The obvious solution would be to rewrite the equations of wave mechanics so that even for a system of several 'particles', only three-dimensional wave functions would be determined. C. Eckart has reported that at one time he attempted this and remarked that it was something that initially 'everybody' was trying to do. [Interview with C. Eckart, conducted by J. Heilbron, May 31, 1962, AHQP.] In Schrödinger's electrodynamic interpretation of wave mechanics, introduced in 'Über das Verhältnis' and developed in 'Vierte Mitteilung', the square of a 'many-particle' wave function could be used to determine the electric charge continuum surrounding the nucleus. By late 1926, Schrödinger had begun to realize that there were major difficulties with this interpretation, however. [L. Wessels, *Schrödinger's Interpretations of Wave Mechanics* (Unpublished Ph.D. Dissertation, Indiana University, 1975), Chapt. 4.] He occasionally noted that a wave interpretation could be given to the three-dimensional functions determined by the method of second quantization, but never worked out this possibility in detail.

velocity v lies along the line for which $\int_B^A mv\, ds$ is an extremum. If the particle momentum varies with spatial location in the same way that the optical index of refraction does, then the path determined for the particle coincides with the ray determined for the light wave.

Schrödinger used the analogy first to argue the need for a wave mechanics. He pointed out that in a medium with an extremely non-uniform index of refraction a light ray is no longer well defined; geometrical optics fails when the curvature of the ray became large. When this happened in classical physics one turned to a pure wave theory, a theory in which rays need not be assumed. Schrödinger noted that according to recent analyses of Bohr's theory and Heisenberg's own interpretation of matrix mechanics, as the curvature of the path of an electron approached the large values required for an orbit around an atomic nucleus, classical mechanics failed and the notion of the 'path' of the electron was no longer well defined. 'Perhaps', Schrödinger suggested,

> this failure is fully analogous with the failure of geometrical optics. . . Perhaps our classical mechanics is the *complete* analogue of geometrical optics, and as such is wrong, i.e., not in complete agreement with reality. It fails whenever the radii of curvature and dimensions of the path are no longer large compared with a certain wave length. . . Then it becomes a question of searching for an 'undulatory mechanics' — and the obvious way to do that is by a wave theoretical extension of Hamilton's idea.[82]

He then used Hamilton's analogy to determine the velocity and frequency of the electron wave. He took Hamilton's characteristic function for the electron as the locus of the wave fronts. Hamilton's equation determined the velocity of these fronts. Because Hamilton's analogy held in the geometrical optics limit of the matter wave theory, the frequency of the wave had to be directly proportional to the energy of the corresponding particle, $E = Kv$. The analogy also provided a way to estimate K. Schrödinger determined the ray curvature of the constructed wave at which a geometrical optics approximation would fail, and set this equal to the path curvature required of a classical electron circling the atomic nucleus. Using experimentally determined values of the atomic radius to evaluate the path curvature, Schrödinger showed that K must be on the order of Planck's constant. Finally a form for the wave equation was chosen on the basis of its simplicity and these wave parameters were inserted.

The result was the equation that Schrödinger had generated in his first paper from an extremal principle. This time, however, it emerged as the product of a new physical picture of matter, and further comments showed that it was the same picture that Schrödinger had suggested several months earlier in

---

[82]*Abhandlungen*, p. 25; *Collected Works*, p. 18.

'Einstein's Gas Theory'. For he again proposed that the classical material particle be replaced by wave packets. He again pointed out that the velocity of de Broglie's phase wave was related in the same way to both the velocity of its accompanying particle and the signal velocity of a group of similar phase waves. Schrödinger was careful to construct his waves in a multi-dimensional coordinate space, which meant that a system of classical particles would be represented by an 'image point' in this space rather than by a set of points in physical space. But the relation still held in a more general form — the velocity of the image point in coordinate space was equal to the signal velocity of waves associated with that point by Hamilton's analogy. 'This fact', wrote Schrödinger,

> can be used to produce a more intimate connection between wave propagation and the motion of the image point that was held up to now. One can try to build up a wave group which will have relatively small dimensions in every direction. Such a wave group will then presumably obey the same laws of motion as a single image point of the mechanical system. It could give, so to speak, a *substitute* [einen Ersatz] for the image point, so long as we can look on it as being approximately confined to a point, i.e. so long as we can neglect any spreading out in comparison with the dimensions of the path of the system.[83]

Schrödinger's discussion of the difficulties associated with this idea and of their possible solutions echoed that of 'Einstein's Gas Theory'. He first addressed the problem of constructing the wave packet in more than two dimensions. Generalizing the procedure of Debye and von Laue, Schrödinger showed how to construct a cone of rays by the appropriate superposition of the multi-dimensional waves determined by Hamilton's analogy. A packet of these waves would follow along this cone with the same velocity as the corresponding image point, and it would be centered around the place where the waves were in phase. Schrödinger then argued that the cone could be constructed so that this point of phase agreement moved 'according to the same laws as the image point of the mechanical system', and therefore according to the laws of classical mechanics.[84] The second problem posed in the gas theory paper, that of ensuring the stability of the packet, remained unsolved, however. To demonstrate such stability, it would have to be shown

---

[83]*Ibid.*, pp. 27 and 20.

[84]One might ask how this result fits with Schrödinger's assumption that the situations where classical mechanics fails are also exactly those situations where rays are not well formed. Packets that might emerge in such situations could not be constrained to follow rays. Schrödinger also assumed, however, that in those circumstances where classical mechanics failed, that is in the states previously represented by the lower Bohr orbits, no packets would even be generated. Only in those states represented by high quantum numbers would the packets appear; there also the wave rays would arise, and classical mechanics would begin to take hold. See *Abhandlungen*, pp. 34 – 36, *Collected Works*, p. 26, and the letter from Schrödinger to H. A. Lorentz, written June 6, 1926, in K. Przibram, *Letters*, p. 63.

that the packet constructed at a point of phase agreement retained a relatively small size as it moved along the cone of rays. Schrödinger only remarked that this would be 'a very difficult task'. The question of wave packet stability would continue to plague Schrödinger in his attempts to give a wave interpretation for his new theory.[85]

Most traditional accounts of the origins of wave mechanics have assumed that the intuitively-based presentation in Schrödinger's second paper does reflect some of the considerations that guided him in his discovery of the theory.[86] It is true that in subsequent discussions of the theory Schrödinger never offered anything other than Hamilton's analogy as a reason to search for a wave theory, nor did he suggest other ways to construct such a theory.[87] But the whole presentation cannot be accepted as an account of the way Schrödinger developed his wave equation. Erwin Fues, who, as mentioned, worked under Schrödinger in Zürich, has pointed to parts of the second paper on wave mechanics that could only have been worked out after Schrödinger had written the first. Corresponding with Thomas Kuhn in 1963, Fues noted that in these two papers Schrödinger provided different justifications for the relation $E = h\nu$:

> Schrödinger . . . asks why the frequency in his wave equation only appears in the first power, whereas normally it occurs in the usual wave equations of physics in the second power. [In the first paper] he answered the question by an approximation of the relativistic expressions . . . *But later,* [*in the second*], he again freed himself from this artifice, and as a result of this went back to the de Broglie relations.[88]

A closer examination shows that Schrödinger's justification for $E = h\nu$ in the second paper was more complicated than Fues has suggested, but it also shows that there was a significant difference between the two ways this issue was handled.

In the last section of his first paper, Schrödinger asked why the energy in his equation was proportional to the frequency of the proposed oscillation rather than to the square of the frequency, as in classical theory. His answer relied on

[85]See L. Wessels, *E. Schrödinger's Interpretations,* pp. 200 – 212 and 267 – 268.

[86]See note 10 above.

[87]*Abhandlungen,* p. v: *Collected Works,* p. ix; E. Schrödinger, *Four Lectures on Wave Mechanics* (London: Blackie, 1928), pp. 1 – 9; E. Schrödinger, 'Die Grundgedanken der Wellenmechanik', *Was ist ein Naturgesetz? Beiträge zum naturwissenschaftlichen Weltbild* (Munich and Vienna: Oldenbourg, 1962), Schrödinger's 1933 Nobel Prize addresses, translated as 'The Fundamental Idea of Wave Mechanics', *Science, Theory and Man.*

[88]Letter from E. Fues to T. S. Kuhn, October 31, 1963, AHQP, reel 66: 'Schrödinger . . . fragt, warum die Frequenz in seiner Wellengleichung nur in der ersten Potenz vorkommt, während sie doch in den üblichen Wellengleichungen der Physik meist quadratisch auftritt.

Zunächst beantwortet er die Frage durch eine Näherung aus dem relativistischen Hausrat. . .

Aber später, in der zweiten Mitteilung. . . befreit er sich wieder von dieser Künstelei und kommt in Anschluss daran wieder auf die de Broglieschen Beziehungen zurück'.

Schrödinger's Route to Wave Mechanics 337

the original relativistic version of wave mechanics. So far '*no natural zero level has been laid down for the parameter E*', he explained. The energy associated with a particular electron is not really E, but 'E increased by a certain constant' C, which 'is intimately connected with the rest energy of the electron'. The larger value, $C + E$, could be set proportional to the square of the frequency in the classical manner: $C + E = Kv^2$ (where K was an unspecified constant). Since C was much larger than E, the following approximation held: .

$$v = \frac{1}{\sqrt{K}} \left(\sqrt{C} + \frac{1}{2\sqrt{C}} + ... \right).$$

Ignoring all terms where C appeared with exponent less than $-\frac{1}{2}$: $2\sqrt{CK}v = E + 2C$. Schrödinger argued that the additional $2C$ could never be detected because the frequency of radiation emitted by an atom was proportional to the difference between the energies of two levels.[89] In the second paper, however, the proportionality of energy and frequency emerged naturally out of the wave front construction that was guided by Hamilton's analogy, and that analogy also gave an estimate of the constant of proportionality. 'Thus', Schrödinger said, 'we get the frequency of the [coordinate] space waves to be proportional to the energy of the system in a manner which is not markedly artificial'.[90] A footnote contrasted this with his earlier attempt to justify the proportionality: 'In [my first paper] this relation appeared merely as an approximate equation, derived from pure speculation'.[91] Given his preference for the later justification, it is hard to imagine that Schrödinger would have bothered with the 'artificial' justification in the first paper if he were already aware of the basis for the proportionality in Hamilton's analogy. At the very least, we would expect him to have mentioned the more satisfactory approach. Instead he excused the approximation in the first paper as one which 'is wholly avoided when the relativistic theory is developed. . .'[92]

It would be unjustified to conclude, however, that Hamilton's analogy played no role at all in Schrödinger's discovery. Fues has asserted:

According to my subjective conviction, Schrödinger *perhaps* knew of the optical-mechanical analogy when [the first paper] was written, but he did not use it in the development of his wave equation.[93]

[89]*Abhandlungen*, pp. 13–14; *Collected Works*, pp. 9–10.
[90]*Ibid.*, pp. 25 and 19.
[91]*Ibid.*, pp. 25 and 19.
[92]*Ibid.*, pp. 14 and 9.
[93]E. Fues to T. S. Kuhn, October 31, 1963: 'Nach meiner subjektiven Überzeugung hat Schrödinger bei Niederschrift der 1. Mitteilung die optisch-mechanische Analogie zwar *vielleicht* gekannt, aber nicht zur Aufstellung seiner Wellengleichung benützt'.

According to Fues it was only after the second paper had been completed that Hamilton's analogy became central to Schrödinger's discussions of wave mechanics. He argued:

> The optical-mechanical analogy then became so important for Schrödinger that he even asked me to lecture on it in Stuttgart. It was an enormous confirmation for Schrödinger of the soundness of his ideas. If he had possessed it from the first, then it is my opinion that he would have also begun with it.[94]

A distinction should be drawn, however, between the use of the analogy as a guide for the detailed construction of waves in coordinate space, and its more general, motivating function. In the latter role, it was used by Schrödinger to support two arguments. First, Schrödinger argued that the non-classical behavior of matter inside the atom could be attributed to a breakdown of Hamilton's analogy due to a lack of well defined particle paths and rays in this region. This suggested that one follow the classical procedure for such a situation and develop a genuine wave theory of atomic structure. The assumption that the analogy did hold as long as rays were well defined allowed Schrödinger to argue further, that a wave theory of matter could be correct even though experimental evidence seemed to support a particle mechanics, since classical mechanics would still hold in all but extremely small regions. Hamilton's analogy provided both an inducement to look for a wave theory of matter and an assurance that this was a reasonable line of research. It is quite possible that both arguments played a part in Schrödinger's discovery of wave mechanics. The second had already been outlined in de Broglie's doctoral dissertation, the work in which Schrödinger had found the concept of matter waves. Schrödinger was almost certainly aware of the argument and probably recognized its importance. The first argument turned on supposing that electron orbits and rays were not well defined near the nucleus, a supposition that might have influenced and certainly would have been suggested by the shift from a construction approach to the search for a wave equation. Schrödinger's new excitement over Hamilton's analogy upon completing his second paper could be attributed to a realization that the analogy not only provided a general motivation, but that it could also provide a detailed construction which determined the wave parameters. These possibilities are put forward not as evidence that the analogy was a motivating force in the route to wave mechanics, but to show that this remains an open question.

---

[94]*Ibid.:* 'Die optisch-mechanische Analogie wurde Schrödinger ja dann so wichtig, dass er mich bat, sie in Stuttgart auch vorzutragen. Das war eine ungeheure Bestätigung der Gesundheit seiner Gedanken für Schrödinger. Hätte er sie im Voraus besessen, so hätte er meiner Meinung nach auch damit angefangen'.

Afterword

The wave model that inspired the search for a wave equation did not survive long after the discovery of that equation. For Schrödinger it was more than a heuristic device, however. The first successful application of the wave model was the result of his attempt to take into account molecular interactions whose nature was as yet unknown. Schrödinger identified the use of this model with the assumption that matter really did consist of waves, and turned immediately to address problems that arose when the wave picture was taken seriously. Without solutions to these problems, but with confidence that the matter wave concept did work in at least one area, and also with the insight that the rays and particle paths of de Broglie's approach were not needed to make use of the concept, Schrödinger turned to look for a wave theory of atomic structure. It is true that initially Schrödinger presented the wave equation as the product of formal restrictions on Hamilton's equation for particles. Several reasons for this uncharacteristic approach have been suggested. When a month later Schrödinger offered a second way of generating the wave equation, he took as the starting point his original wave picture of matter.

Schrödinger recognized that the wave equation could not be interpreted as a description of those original matter waves. In the final three papers of his series on wave mechanics, Schrödinger outlined a new vibrational interpretation, in which ψ^2 determined a continuous and often oscillating charge distribution surrounding the nucleus. This interpretation solved the problem of relating the multi-dimensional wave function to the separate oscillation amplitudes of several electrons, and guided Schrödinger in extending the theory to treat radiation intensities. As the number and variety of problems facing this interpretation grew to a critical mass in mid-1927, Schrödinger continued to insist that *some* physical picture of the processes underlying quantum mechanics be provided in order to complete the theory satisfactorily.[95] When it became clear in 1928 that most physicists had accepted the 'Copenhagen interpretation', Schrödinger abandoned his attempts to generate alternatives. He retained his conviction that a physical picture was necessary, however, and all but gave up his active participation in the further development of quantum theory. Almost all of his papers on quantum mechanics after 1928 were aimed at criticising the prevailing interpretation and, in the 1950s, at raising again the possibility of a wave interpretation.

[95]E. Mackinnon, 'The Rise and Fall of the Schrödinger Interpretation', *Foundations of Quantum Mechanics,* P. Suppes (ed.) (forthcoming, 1979), has argued that Schrödinger's tenacious search for an intuitive interpretation was motivated by a concern over priority. According to MacKinnon, Schrödinger feared that the discovery of matrix mechanics would overshadow his discovery of wave mechanics if quantum mechanics was not understood as a description of oscillatory processes. An alternative view is offered in L. Wessel's, 'Intellectual Sources of Schrödinger's Interpretations', also in *Foundations of Quantum Mechanics.*

340   *Studies in History and Philosophy of Science*

Acknowledgements — This paper is an extensively revised version of the first chapter of my dissertation. The results of Paul Hanle's recent work have been incorporated into this version; I thank him for generously supplying pre-publication copies of his papers and for his very helpful comments. I am grateful to Lewis Pyenson, Alberto Coffa and several anonymous referees for valuable suggestions on content and style. I am also indebted to Ronald Giere, Edward MacKinnon and Paul Teller for their comments and encouragement. This work was supported in part by a grant from the National Science Foundation.

The Discovery of Atomic Transmutation: Scientific Styles and Philosophies in France and Britain

By Marjorie Malley*

I T IS WELL KNOWN that in 1902–1903 two British scientists in Canada, Ernest Rutherford and Frederick Soddy, published the theory of atomic change popularly known as the transmutation theory.[1] Less widely known is the fact that Pierre Curie and his student André Debierne carried out researches contemporary with and parallel to those of Rutherford and Soddy. Yet even with the advantages of radium sources and an early start, the French researchers not only failed to arrive at the revolutionary discovery; they and their co-workers resisted the transmutation theory for several years.

The failure of the French to recognize transmutation and their delay in accepting it invites historical investigation. I propose that the key to this puzzle is contained in the divergent ways the French and British groups viewed and practiced science. The French, dominated by Pierre Curie, held a positivistic philosophy of science and were not willing to commit themselves to a specific theory of radioactivity. Both the positivism and Curie's preference for a thermodynamic model amenable to it impeded their progress. In contrast, Rutherford and Soddy were unselfconscious realists who chose material hypotheses, were unconcerned with philosophical subtleties, and were willing to pursue their specific hypothesis as far as it would go. Personal characteristics of the protagonists influenced their scientific styles, as well as the predominant national scientific proclivities. Competitiveness and pride also affected the outcome of this episode. As it turned out, the final arbiter, nature, had stacked the cards against the French.

EARLY RESEARCHES AND THE DISCOVERY OF TRANSMUTATION

When radioactivity was first discovered by Henri Becquerel in 1896,[2] most scientists viewed it as a form of secondary radiation. At first Becquerel thought his rays

---

*2017 North 6th Street Terrace, Blue Springs, Missouri 64015.

An earlier version of this article is in my dissertation, "From Hyperphosphorescence to Nuclear Decay: A History of the Early Years of Radioactivity, 1896–1914" (University of California, Berkeley, 1976). The research was funded in part by National Science Foundation Dissertation Research Grant GS–3247. I wish to express my appreciation to John L. Heilbron of the University of California at Berkeley for comments on the dissertation and to Henry Frankel of the University of Missouri, Kansas City, for suggestions on this paper; the Linda Hall Library, Kansas City, Missouri, the University of Kansas Libraries, and for access to manuscript materials, the Cambridge University Library, the Bodleian Library at Oxford, and the Bibliothèque Nationale.

[1]Ernest Rutherford and Frederick Soddy, "The Radioactivity of Thorium Compounds, I and II," *Transactions of the Chemical Society of London*, 1902, *81*:321–350, 837–860; "The Cause and Nature of Radioactivity, I and II," *Philosophical Magazine*, 1902, *4*:370–396, 569–585; "Radioactive Change," *Phil. Mag.*, 1903, *5*:576–591.

[2]Henri Becquerel, "Sur les radiations émises par phosphorescence," *Comptes Rendus de l'Académie des*

214   MARJORIE MALLEY

resembled visible light, but by 1899 the consensus was that Becquerel rays were a
mixture of primary and secondary X rays. A few physicists, notably Julius Elster and
Hans Geitel, J. J. Thomson, and Marie Curie speculated that radioactivity's energy
source lay within the atom, although the first three seem to have been thinking of
some process similar to ionization.[3] Most scientists initially saw radioactivity as just
one more of the epiphenomena observed in the wake of the 1895 discovery of
Röntgen rays.[4] Consequently radioactivity at first did not arouse much interest; even
Becquerel set it aside for eleven months to study optical phenomena. It took the
Curies' discovery in 1898 of two new energetic radioelements, polonium and radium,
to awaken the apathetic scientific community.[5]

Rutherford was one of a few physicists who became interested in radioactivity
before polonium and radium were discovered. Following the general view, he assumed in 1898 that the Becquerel rays were electromagnetic disturbances. The study
of radioactivity thus was an extension of his researches on X rays.

Starting in 1899, a series of anomalies which he encountered with thorium's
activity ultimately led him and Soddy to the theory of the exploding atom. The
resolution of this episode depended upon hypotheses. Although Rutherford was
careful to avoid much speculation in his publications and did not always make his
working hypotheses explicit, he always implicitly visualized possible explanations of
the phenomena. He did not trouble himself with the distinction between a heuristic
and an ontological hypothesis.[6] He chose hypotheses in which he also believed, in this
case that the thorium "emanation" and "excited activity" were material and particulate.

The first anomaly Rutherford confronted was a penetrating radiation from thorium which, unlike other radiations, seemed to depend upon air currents in the
laboratory.[7] To explain this strange behavior he postulated that, in addition to
Becquerel rays, thorium emitted highly penetrating particles. Wanting to avoid
committing himself on their nature, he termed the particles an emanation. But he
implicitly assumed that this emanation was material, and on the basis of his hypothesis he set out to determine what type of particles were involved. They could not be
dust from the thorium compound used, since they penetrated paper, metal, and
cotton wool and did not appear to affect a cloud chamber.[8] They could not be ions,
since an electric field did not influence them. Nor could they be a vapor from
thorium, since "the positive ion produced in a gas by the emanation possesses the
power of producing radioactivity in all substances on which it falls,"[9] a power which

Sciences, Paris, 1896, *122*:420–421.

[3]Johannes Elster and Hans Geitel, "Weitere Versuche an Becquerelstrahlen," *Annalen der Physik und
Chemie,* 1899, *69*:83–90; J. J. Thomson, "On the Diffuse Reflection of Röntgen Rays," *Proceedings of the
Cambridge Philosophical Society,* 1898, *9*:393–397; Marie Curie, "Les rayons de Becquerel et le polo-
nium," *Revue Générale des Sciences,* 1899, *10*:41–50.

[4]See Lawrence Badash, "Radioactivity before the Curies," *American Journal of Physics,* 1965,
33:128–135.

[5]*Ibid.*; Marjorie Malley, "The Discovery of the Beta Particle," *Am. J. Phys.,* 1971, *39*:1454–1460. My
count of the publications on radioactivity from 1896 to 1903 shows marked increases in the numbers of
both papers and contributors in 1899. There was an especially sharp rise in 1903, after the publication of
the Rutherford-Soddy theory.

[6]See Thomas S. Kuhn, *The Structure of Scientific Revolutions* (2nd ed., Chicago/London: University of
Chicago Press, 1970), p. 184.

[7]Ernest Rutherford and R. B. Owens, "Thorium and Uranium Radiation," *Transactions of the Royal
Society of Canada,* 1899, *2*:9–12; R. B. Owens, "Thorium Radiation," *Phil. Mag.,* 1899, *48*:360–387;
Ernest Rutherford, "A Radioactive Substance Emitted from Thorium Compounds," *Phil. Mag.,* 1900,
49:1–14.

[8]The specimen was enclosed in paper, which stopped the alpha particles.

[9]Rutherford, "A Radioactive Substance," p. 14.

thorium itself did not exhibit. The radioactivity so created he termed "excited activity," guessing that the emanation excited a secondary activity on the surface of other substances. But soon he found that this activity behaved like a layer of radioactive particles superimposed upon the surface excited. He could remove the activity from bodies made radioactive by scouring or with acids, and evaporation of the acid solutions left an active residue. Most significantly, the excited radioactivity was independent of the substance activated, always decaying exponentially at the same rate. It could also be concentrated on the negative electrode of his apparatus as though it were a positively charged body.[10]

By 1900 Rutherford had eliminated his first guess that the excited activity was a phosphorescence (secondary activity) produced by thorium. He supposed this activity came from emanation particles deposited upon neighboring bodies.[11] These particles had a tendency to collect positive charges from the surrounding air. In 1901 he tentatively adopted a suggestion by his friend and former teacher J. J. Thomson that the excited activity was produced by recoiling emanation atoms.[12] His particle hypothesis was strengthened when he showed that the carriers of excited activity traveled with about the same speed in air as positive ions.[13]

Rutherford still had to determine the chemical nature of the emanation particles, for which work he invited the collaboration of a gifted young chemist, Frederick Soddy. Soddy adopted Rutherford's belief that the thorium emanation was a gas, and when it refused to react with any of the reagents Soddy tried, he concluded that it belonged to the newly discovered inert gas family.[14] Upon this discovery Soddy reportedly blurted out: "Rutherford, this is transmutation: the thorium is disintegrating and transmuting itself into an argon gas."[15] Rutherford was still avoiding that surprising and possibly professionally hazardous conclusion.[16]

At that time scientists were uncertain whether all the newly discovered radioactive bodies were really new elements; some might merely be familiar elements which contained a trace radioactive impurity such as radium. After many attempts, Soddy and Rutherford thought they had separated such an active impurity from thorium, whose removal left the thorium considerably weakened in activity.[17] Thinking that the impurity was responsible for most if not all of thorium's activity, they labeled it "thorium X."[18]

[10]For a discussion of this effect see Thaddeus J. Trenn, "Rutherford and Recoil Atoms: The Metamorphosis and Success of a Once Stillborn Theory," *Historical Studies in the Physical Sciences*, 1975, *6*:513–547.

[11]Ernest Rutherford, "Radioactivity Produced in Substances by the Action of Thorium Compounds," *Phil. Mag.*, 1900, *49*:161–192.

[12]J. J. Thomson to Ernest Rutherford, Apr. 25, 1901, Rutherford Papers, T17, Cambridge; Ernest Rutherford, "Übertragung erregter Radioaktivität," *Physikalische Zeitschrift*, 1902, *3*:210–214; Trenn, "Rutherford and Recoil Atoms."

[13]Ernest Rutherford, "Transmission of Excited Radioactivity," *Bulletin of the American Physical Society*, 1901, *22*:37–43.

[14]Rutherford and Soddy, "The Radioactivity of Thorium Compounds, I"; personal records of Frederick Soddy, File I, Oxford. The gas was later named thoron, Rn^{220}.

[15]Muriel Howorth, *Pioneer Research on the Atom* (London: New World Publications, 1958), p. 90.

[16]The notion smacked of the old, discredited alchemy, which had recently seen a revival among nonscientific and fringe groups. See H. Carrington Bolton, "The Revival of Alchemy in France," *Chemical News*, 1898, *77*:69–70, 73–74, and "Recent Progress of Alchemy in America," *Chem. News*, 1897, *76*:61–62; Thaddeus Trenn, "The Justification of Transmutation: Speculations of Ramsay and Experiments of Rutherford," *Ambix*, 1974, *21*:53–77.

[17]Rutherford and Soddy, "The Radioactivity of Thorium Compounds, I."

[18]This name was chosen to indicate their substance's analogy to Crookes' uranium X, "the unknown substance in uranium." Becquerel made a similar separation independently. See William Crookes, "Radioactivity of Uranium," *Proceedings of the Royal Society of London*, A, 1900, *66*:409–422; Henri Becquerel, "Note sur le rayonnement de l'uranium," *Compt. Rend.*, 1900, *130*:1583–1585.

216   MARJORIE MALLEY

In December 1901 Soddy and Rutherford learned that uranium of reduced activity (similar to their inactive thorium) separated by Becquerel regained its activity with time.[19] Upon checking their thorium and thorium X preparations, they found that their weakened thorium had regained its activity, while their thorium X had become nearly inactive. A quick quantitative study showed that the decay and recovery curves of thorium and thorium X were complementary. Since they considered thorium X to be a particular material substance, not (as Curie would prefer) thorium activated by "induction" or energy transfer, their findings pointed to atomic transmutation: thorium was changing into thorium X at a constant rate![20] The emanation and excited activity would then also be material products in thorium's newly revealed decay series.

In the autumn of 1902 Rutherford extended his material interpretation of radioactivity by showing that the rays he had earlier named "alpha" were particulate, another product of the disintegrating atom.[21] This discovery enhanced the plausibility of the transformation theory, for the loss of these heavy particles, unlike the loss of radiation or electrons, would clearly alter the parent atom in a fundamental way. Thus he laid to rest the generally held X-ray theory of their nature which he had proposed in 1899.[22]

THE FRENCH NONDISCOVERY OF TRANSMUTATION

Shortly before Rutherford discovered thorium's excited activity the Curies found a similar phenomenon associated with radium.[23] At first, like Rutherford later, they thought the activity came from particles of radium dust or vapor. But soon, perhaps because of Pierre Curie's preference for explanations in terms of energy, they convinced themselves that this hypothesis was incorrect and that they had discovered a genuine "induced" radioactivity. The activity could be stimulated through a metal bottle closed with aluminum foil.[24] Furthermore, though their radium chloride was water soluble, they could not remove the activity by washing (they did not try to remove it by scouring, as Rutherford later did). And although radium chloride is not volatile, the induced radioactivity disappeared with time. The Curies obviously did not consider that the induced activity might be material, yet different from radium. They concluded that the activity was a secondary radiation provoked by Becquerel rays. It differed from secondary X rays by persisting after the exciting source had been removed. Upon learning of his colleagues' results, Becquerel also concluded that a type of phosphorescence was involved.[25]

Like Rutherford, the Curies found that the induced activity first increased to a maximum, then decayed according to an exponential curve. The activity was inde-

[19]William Crookes to Rutherford, Dec. 18, 1901, Rutherford Papers, C88, Cambridge, published in Arthur S. Eve, *Rutherford* (New York: Macmillan; Cambridge: Cambridge University Press, 1939), p. 79.
[20]Rutherford and Soddy, "The Radioactivity of Thorium Compounds, II" and "The Cause and Nature of Radioactivity." However, they had not entirely excluded electromagnetic induction: they evoked it to explain some irregularities in the decay and recovery curves.
[21]Ernest Rutherford, "The Magnetic and Electric Deviation of the Easily Absorbed Rays from Radium," *Phil. Mag.*, 1903, *5*:177–187. The Curies loaned him radium for this experiment.
[22]Ernest Rutherford, "Uranium Radiation and the Electrical Conduction Produced by It," *Phil. Mag.*, 1899, *47*:109–163; J. L. Heilbron, "The Scattering of α and β Particles and Rutherford's Atom," *Archive for History of Exact Sciences*, 1968, *4*:247–307.
[23]Pierre and Marie Curie, "Sur la radioactivité provoquée par les rayons de Becquerel," *Compt. Rend.*, 1899, *129*:714–716.
[24]For the Curies this precluded transport of matter. In his notes on one of Rutherford's 1902 papers Pierre Curie expressed amazement: "émanation traverse aluminum!" (Curie Papers, 10A2, Paris). Probably the inert gas radon penetrated more readily than other gases because of its lack of affinity for the foil.
[25]Henri Becquerel, a note in *Compt. Rend.*, 1899, *129*:716.

pendent of the substance upon which it was excited, but since secondary X rays were incompletely understood then, this finding did not shake their belief that the activity was a secondary radiation. The fact that induced activity decayed with time strengthened their conviction that they were not dealing with genuine, primary radioactivity.[26] They assumed true radioactivity continued at a constant level, because during the past three and a half years no one had observed any change in the activities of uranium, thorium, or radium.[27] From their viewpoint the induced activity was a relatively insignificant phenomenon, and they did not study it further at that time.

In 1901, after having read Rutherford's papers on thorium emanation and excited activity, Pierre Curie returned to radium-induced activity with the chemist André Debierne.[28] Ernst Dorn in Halle had recently found that radium also emitted an emanation.[29] With the evidence that Curie and Debierne quickly acquired they could have easily concluded that the radium emanation was a gas evolved by radium, which clearly was the hypothesis they were testing. Radium, whether solid or in solution, produced much more induced activity in a closed container than in the open air (where a gas could escape). The greater the surface area of solution in contact with free air, the greater the loss; by closing the container the activity could be recovered. No activity was induced outside of a closed glass vessel, but it could be produced in air outside of the direct path of the Becquerel rays (as a result of diffusion of the emanation). Marie and Pierre Curie had even observed in 1900 that sublimed pitchblende emitted a radioactive gas,[30] and Curie and Debierne conceded that the unwelcome conductivity of air in their laboratory was probably due to "the continuous formation of a radioactive gas."[31] Yet they rejected Rutherford's conclusion that thorium (and by implication radium) evolved a radioactive gas, because they thought he had strayed beyond the established facts by endowing his working hypothesis with the status of reality: "as one can easily conceive other satisfactory explanations it seems premature to us to adopt any theory whatsoever."[32] Their refusal to commit themselves to a specific hypothesis (because of Curie's positivism) or to attach much importance to the question of the nature of the emanation and excited activity (because Curie attributed them to radioactive "induction") stalemated their research.

Continuing their researches with radium emanation, Debierne and Curie found that induced activity was independent of the pressure and nature of the ambient gas except at very low pressures, when the activity mysteriously disappeared (Rutherford later traced this last effect to the pump's removing the emanation as fast as it was formed).[33] This independence of the gas made unlikely their previous suspicion that ordinary air became active and transferred radium's activity to other bodies. Soon they found that activity was transmitted in capillary tubes faster than they thought

[26]Pierre and Marie Curie, "Sur les corps radio-actifs," *Compt. Rend.*, 1902, *134*:85–87. Cf. Alfred Romer, *The Discovery of Radioactivity and Transmutation* (New York: Dover, 1964), p. 117.

[27]Their half-lives in years are 4.5×10^{10}, 1.39×10^{10}, and 1.62×10^3 (Samuel Glasstone, *Sourcebook on Atomic Energy*, New York: Van Nostrand, 1950, pp. 125–126).

[28]Pierre Curie and André Debierne, "Sur la radioactivité induite provoquée par les sels de radium," *Compt. Rend.*, 1901, *132*:548–551.

[29]Ernst Dorn, "Über die von radioaktiven Substanzen ausgesandte Emanation," *Abhandlungen der Naturforschenden Gesellschaft zu Halle*, 1900, *23*:3–15.

[30]Marie Curie, *Traité de radioactivité* (Paris: Gauthier-Villars, 1910), Vol. I, p. 198; Pierre and Marie Curie, "Les nouvelles substances radioactives et les rayons qu'elles émettent," *Rapports présentés au Congrès International de Physique* (Paris: Gauthier-Villars, 1900), Vol. III, pp. 79–114.

[31]Pierre Curie and André Debierne, "Sur la radio-activité induite et les gas activés par le radium," *Compt. Rend.*, 1901, *132*:768–770, p. 769.

[32]Curie and Debierne, "Sur la radioactivité induite provoquée par les sels de radium," p. 551. See n. 35 below.

[33]Ernest Rutherford, *Radio-activity* (Cambridge: Cambridge University Press, 1905), p. 226.

diffusion could occur,[34] and they dropped material explanations altogether. Instead, they turned to thermodynamics, which offered a descriptive mode adaptable to Curie's positivism:

> . . . one can suppose that each radium atom functions as a continuous and constant source of radioactive energy without it being necessary to specify besides where the energy comes from. The radioactive energy accumulated in a body tends to dissipate itself . . . 1) by radiation . . . 2) by conduction, that is to say by transmission by degrees to surrounding bodies by means of the intermediary of gas and liquids (induced radioactivity).[35]

If a body emitted heat continuously at a constant rate, its temperature would rise until thermal equilibrium was reached. Radioactive equilibrium could be described similarly, with intensity of radiation replacing temperature. Their hypothesis, descriptive, nonspecific, and positivistic, was framed in terms of energy[36] and postulated neither an energy source nor a transmission mechanism.

When Pierre Curie determined that the radium emanation's activity followed an exponential decay law, he observed that "The current experiments show that in the gas the energy is stored in a special form which dissipates itself according to an exponential law."[37] The emanation, he thought, could not be ordinary matter, for in addition to the unusual behaviors mentioned previously, it disappeared spontaneously in a sealed tube, it had no measurable weight, and its spectrum showed no new lines. Moreover, the activating power of radium solution did not behave like vapor tension.[38] Curie preferred the neutral statement that the emanation was "the energy emitted by radioactive bodies in the special form in which it is stored in the gas and in the vacuum."[39] The excited activity could be viewed as the energy under another form with a different decay constant. For Curie, thinking like the proponents of energetics, energy had become more real than matter. "I see energy," he told his colleague and former student, Paul Langevin.[40]

THE FRENCH RESPONSE TO THE TRANSMUTATION THEORY

The French reception of the Rutherford-Soddy theory, in contrast to the enthusiastic British response,[41] was predictably cool. Originally the Curies had preferred the view (proposed by Marie Curie in her first paper on radioactivity) that radioactivity's energy derived from an unknown external source, such as a cosmic radiation.[42] In her

[34]Curie and Debierne, "Sur la radio-activité induite et les gas activés par le radium." The unusual speed of transmission was probably due to radon's lack of chemical affinities.

[35]Pierre Curie and André Debierne, "Sur la radioactivité des sels de radium," *Compt. Rend.*, 1901, *133*:276–279, p. 277. In a note on p. 277 he enumerated the possible explanations of radioactivity given by Marie Curie (in "Les rayons de Becquerel et le polonium"): the energy could be previously stored; or it could derive from a modification of radium itself; from transformation of an unknown exterior radiation; or from ambient heat, contrary to Carnot's principle.

[36]Alfred Romer, "The Transformation Theory of Radioactivity," *Isis*, 1958, *49*:3–12, p. 9. Cf. Wilhelm Ostwald, *Die Energie* (Leipzig: Barth, 1908), Ch. 7; Mary Jo Nye, *Molecular Reality* (London: Macdonald; New York/Amsterdam: Elsevier, 1972), pp. 16–18.

[37]Pierre Curie, "Sur la constante de temps caractéristique de la disparition de la radioactivité induite par le radium dans une enceinte fermée," *Compt. Rend.*, 1902, *135*:857–859, p. 859.

[38]Pierre Curie and André Debierne, "Sur la radio-activité induite provoquée par des sels de radium," *Compt. Rend.*, 1901, *133*:931–934.

[39]Pierre Curie, "Sur la radioactivité induite et sur l'émanation du radium," *Compt. Rend.*, 1903, *136*:223–226, p. 226.

[40]Paul Langevin, "Pierre Curie," *Revue du Mois*, 1906, *2*:5–36.

[41]See Lawrence Badash, "How the 'Newer Alchemy' Was Received," *Scientific American*, 1966, *213*:88–95.

[42]Marie Curie, "Rayons émis par les composés de l'uranium et du thorium," *Compt. Rend.*, 1898, *126*:1101–1103.

THE DISCOVERY OF ATOMIC TRANSMUTATION   219

doctoral thesis (1903) Marie Curie remarked that none of the hypotheses proposed to explain radioactivity had been confirmed, without even mentioning Soddy and Rutherford.[43] Pierre Curie was skeptical of the transformation hypothesis, because his own researches had led him to doubt that the emanations and induced radioactivity were material. If it were necessary to form a hypothesis concerning the nature of the emanation's "support" (carrier), Curie would favor the theory that "the emanation does not have ordinary matter for its support, and there exist centers of condensation of energy located between the gas molecules and which can be carried along with it [the gas]."[44]

Pierre Curie's own experimental researches gradually forced him to alter his views. In 1902 he found that radium's activity remained constant between $-180°$ and $+450°$, thus excluding the possibility of an ordinary chemical change.[45] In March 1903 he showed, in conjunction with his assistant Albert Laborde, that radium continually maintained itself at a higher temperature than its surroundings.[46] According to their calculations, 1 gram atom of radium would generate 22,500 small calories per hour. This was of the same order of magnitude as the heat evolved in the most vigorous exothermal chemical reaction known, and Curie and Laborde conceded that this energy might arise from subatomic transformation. But, they maintained, the energy could still derive from an outside source—even though they had found that radium emitted the same amount of heat when they immersed it in an ice calorimeter. In June 1903 Curie and his *préparateur* Jacques Danne confirmed Rutherford and Soddy's contention that radium emanation diffused and condensed like an ordinary gas.[47]

The *tour de force* which finally convinced the informed skeptics came in August 1903 when Soddy (now in London) and Sir William Ramsay showed spectroscopically that helium was generated by radium emanation.[48] Ramsay and his colleague John Norman Collie found that radium emanation also produced new spectral lines,[49] thus removing one of Pierre Curie's chief objections to regarding the emanation as a new material gas. In the revised edition of her thesis published in 1904 Marie Curie acknowledged that "the most recent researches are favorable to the hypothesis of an atomic transformation of radium."[50] "The spontaneous production of helium in a sealed tube which contains radium is clearly a new fact of fundamental importance," observed Pierre Curie. "Helium could be, according to these results, one of the disintegration products of radium."[51] In January 1904 Curie and the chemist Sir James Dewar at the Royal Institution confirmed that radium produced helium.[52] Since the emanation and helium were now both patently material, Curie's hypothesis

[43]Marie Curie, "Radio-active Substances," *Chem. News*, 1903, *88*:271–272.

[44]Pierre Curie, "Sur la radioactivité induite et sur l'émanation du radium," p. 226.

[45]Pierre Curie, "Conductibilité des diélectriques liquides sous l'influence des rayons du radium et des rayons de Röntgen," *Compt. Rend.*, 1902, *134*:420–423; "Sur la mesure absolue du temps," *Bulletin des Seances Société Française de Physique*, 1902:60.

[46]Pierre Curie and Albert Laborde, "Sur la chaleur dégagée spontanément par les sels de radium," *Compt. Rend.*, 1903, *136*:673–675.

[47]Pierre Curie and Jacques Danne, "Sur l'émanation du radium et son coefficient de diffusion dans l'air," *Compt. Rend.*, 1903, *136*:1314–1316.

[48]William Ramsay and Frederick Soddy, "Experiments in Radioactivity, and the Production of Helium from Radium," *Proc. Roy. Soc. London*, A, 1903, *72*:204–207.

[49]William Ramsay and J. N. Collie, "The Spectrum of the Radium Emanation," *Proc. Roy. Soc. London*, A, 1904, *73*:470–476.

[50]Marie Curie, *Recherches sur les substances radioactives* (Paris: Gauthier-Villars, 1904), p. 150.

[51]Pierre Curie, "Recherches récentes sur la radioactivité," *Journal de Chimie Physique*, 1903, *1*:409–449, pp. 446–447.

[52]Pierre Curie and James Dewar, "Examen des gas occlus ou dégagés par le bromure de radium," *Compt. Rend.*, 1904, *138*:190–192.

220   MARJORIE MALLEY

of special energy forms became superfluous, and in his next publication he assumed that all the decay products were material.[53] In 1904–1905 Curie taught the Rutherford-Soddy theory in his lecture course at the Sorbonne.[54]

In 1903, under the influence of Pierre Curie's positivism, Becquerel concluded that a number of general hypotheses were feasible, in spite of his earlier thoughts that the emanations were gases and that radioactivity resulted from atomic disintegration.[55] In his last publication which treated radioactivity, Becquerel still declined to commit himself, stating that "it seems to have its origin in an extraordinarily slow decay of matter. Up to now it has not been determined whether . . . it is only a case of chemical changes in the arrangement of atoms, or rather a transformation in the atom itself."[56]

At first André Debierne echoed his colleagues, maintaining that it was impossible to choose between the various theories of radioactivity. But he admitted that Ramsay and Soddy's experiments almost required a transmutation hypothesis. Debierne apparently preferred a theory in which radium stored energy from an unknown external radiation and then used this energy to provoke the transformation of neighboring bodies.[57] This view enabled him to explain the puzzling observation that radium seemed to produce no emanation in a high vacuum. But instead of finding further evidence to support his theory, Debierne made a discovery which strengthened Rutherford and Soddy's position: emanation from actinium also produced helium.[58] By this time (1905) several German chemists had confirmed its production from radium.[59] Late in 1905 Debierne wrote that "Mr. Rutherford has given a complete theory of the formation of these substances which are produced through induced radioactivity which is generally accepted today."[60] The next year Georges Sagnac published the (negative) results of studies made in Pierre Curie's laboratory to test whether gravity supplied the energy for radioactivity.[61] Although Sagnac had tested for a possible external energy source, he did not doubt the reality of transmutation. The French were the last serious investigators of radioactivity to accept the Rutherford-Soddy theory, but doubts lingered on among less informed scientists for many years.[62]

PHILOSOPHIES AND STYLES OF SCIENCE

Pierre Curie's positivistic philosophy of science was expressed succinctly in the manifesto of scientific method published by the Curies in January 1902, while Soddy

[53]Pierre Curie and Jacques Danne, "Sur la disparition de la radioactivité induite par le radium sur les corps solides," *Compt. Rend.,* 1904, *138*:683–686.

[54]Curie Papers, Paris.

[55]Henri Becquerel, *Recherches sur une propriété nouvelle de la matière* (Paris: Firmin-Didot, 1903), pp. 329–336; "Sur la radio-activité secondaire," *Compt. Rend.,* 1901, *132*:734–739; "Sur la radio-activité de l'uranium," *Compt. Rend.,* 1901, *133*:977–980.

[56]Henri Becquerel, "Betrachtungen über eine moderne Theorie der Materie," *Jahrbuch der Radioaktivität und Elektronik*, 1907, *4*:361–369, pp. 363–364. Becquerel died in 1908.

[57]André Debierne, "Le radium et la radio-activité. Deuxième partie," *Rev. Gén. Sci.*, 1904, *15*:60–71, p. 70.

[58]André Debierne, "Sur les gaz produits par l'actinium," *Compt. Rend.,* 1905, *141*:383–385.

[59]Friedrich Giesel, "Ueber einen einfachen Nachweis von Helium aus Radiumbromid," *Berichte der Deutschen Chemischen Gesellschaft*, 1905, *38*:2299–2300; Franz Himstedt and Georg Meyer, "Über die Bildung von Helium aus der Radiumemanation," *Annalen der Physik*, 1904, *15*:184–192; 1905, *17*:1005–1008.

[60]André Debierne, "Über einige Eigenschaften des Actiniums," *Phys. Z.*, 1906, *7*:14–16, p. 14.

[61]Georges Sagnac, "Une relation possible entre la radioactivité et la gravitation," *Journal de Physique Théorique et Appliquée*, 1906, *5*:455–462. He began these researches in 1902.

[62]Chemists in particular objected to the new "chemistry of phantoms." See Arthur Smithells, Presidential Address to Section B, *Reports of the British Association for the Advancement of Science*, 1907, *77*:469–479, p. 477; Marie Curie, "Les radio-éléments et leur classification," *Rev. Mois*, 1914, *18*:5–41.

and Rutherford were doing their crucial experiments with thorium and thorium X. Becquerel had just advanced a speculative theory of atomic disintegration to explain the genesis of the emanation and the induced radioactivity, and the Curies responded with an admonishment:

> In the study of unknown phenomena, one can make very general hypotheses and advance step by step with the assistance of experience. This methodical and sure progress is necessarily slow. One can, on the contrary, make bold hypotheses, where one specifies the mechanism of the phenomena; this manner of proceeding has the advantage of suggesting certain experiments and above all by facilitating reasoning by rendering it less abstract by the use of an image. On the other hand, one cannot hope to imagine thus *a priori* a complex theory in accord with experience. Precise hypotheses assuredly contain some error along with some truth; the latter, if it exists, forms only part of a more general proposition to which it will be necessary to return some day.[63]

This pattern, adapted to a thermodynamic model, was followed by Pierre Curie in his studies of radioactivity.

Both the pattern and the model reflected Curie's scientific milieu. As a Frenchman he had imbibed the current continental positivism, which called for neutral correlative descriptions of phenomena; whereas his British colleagues had learned a more mechanistic approach and, as a whole, were surprisingly ignorant of the continental trends. Thermodynamics, that highly abstract science which exemplified Ernst Mach's dictum that "All science has for its goal to replace experience with the shortest possible intellectual operations,"[64] was congenial to the continental scene. It was a master of thermodynamics and a contemporary of Curie, Pierre Duhem, who perhaps most vividly characterized the differences between "typical" French and "typical" British science.[65] The continental view was that specific hypotheses can be useful, but they are readily disposable. The general framework of a theory, based upon experience, is essential and will remain: it is invariant, universal, and therefore objective.

Using what his friend Jean Perrin called the "intelligence of analogies,"[66] Curie had compared radioactivity to thermal phenomena. Energy replaced matter as the substratum for his explanations of radium emanation and induced radioactivity. The last term suggested a wave phenomenon rather than a particulate carrier, a model conformable to energetics and less adaptable to British mechanism. His choice of energetics over material hypotheses, which was dictated by his interpretation of positivism, doomed his researches to sterility. Curie's "confidence and . . . skill in the use of those general principles of energetics [thermodynamics]"[67] enhanced the subjective plausibility of both his philosophy of science and his model for radioactivity, and suggests an additional motive for these choices. Like the ancient Greek philosophers who made becoming rather than being the essence of reality, Curie chose process over substance. Energetics is Heraclitean physics.[68]

[63]P. and M. Curie, "Sur les corps radio-actifs," p. 87.

[64]Abel Rey, *La philosophie moderne* (Paris: Flammarion, 1921) p. 123. Cf. Charles Gillispie, *The Edge of Objectivity* (Princeton: Princeton University Press, 1960, p. 498: "Indeed, a positivist in philosophy at the end of the century was almost certain to be an energeticist in physics."

[65]Pierre Duhem, *The Aim and Structure of Physical Theory* (Princeton: Princeton University Press, 1954); Nye, *Molecular Reality*. See also Abel Rey, *La théorie de la physique chez les physiciens contemporains* (Paris: Alcan, 1923).

[66]Nye, *Molecular Reality*, p. 131.

[67]Langevin, "Pierre Curie," p. 29.

[68]See Giorgio de Santillana, *The Origins of Scientific Thought* (Chicago: University of Chicago Press; Toronto: University of Toronto Press, 1961), for discussion of the dichotomous trends.

Curie's papers, characterized by what his student Laborde called "rigor" and "probity,"[69] manifest his philosophy of science. They are clear and precise, for he was methodical and did not waste words. Marie Curie wrote correctly that "though he never hesitated to make hypotheses, he never permitted their premature publication."[70] Unfortunately, in radioactivity he also refused to commit himself to a specific hypothesis, a fact which Rutherford later noted.[71]

Rutherford (and later Soddy) was less inhibited and rigid. After his initial flirtation with "excited" radioactivity, Rutherford clearly leaned toward material explanations of the new discoveries. Although he too took care to avoid committing himself prematurely, he based his research program on the assumptions that the thorium emanation was a gas and the excited activity a material deposit. Most likely, suspicions that some fundamental process was at hand kept Rutherford, and later Rutherford and Soddy, at their work. In contrast, Curie viewed induced radioactivity as an insignificant electromagnetic event, and so he delayed exploring it further for sixteen months. By this time Rutherford had published several papers on thorium emanation and excited activity and had acquired some scientific stature. Consequently Curie could no longer afford to ignore these phenomena.

The parallel researches of the French and British teams became competitive, with Curie and Debierne taking a defensive stance toward the bold material conclusions of Rutherford and Rutherford-Soddy. Rutherford expressed the mood in a letter to his mother at the beginning of 1902:

> I have to keep going, as there are always people on my track. I have to publish my present work as rapidly as possible in order to keep in the race. The best sprinters in this road of investigation are Becquerel and the Curies in Paris, who have done a great deal of very important work in the subject of radioactive bodies during the last few years.[72]

Contemporaries noticed the rivalry, and national chauvinism is especially visible in French historical accounts of radioactivity.[73] The undercurrent of tension was not alleviated by Rutherford's knack for using French discoveries to support his own theories. Even without the differences in scientific philosophies and styles, we might expect the French to have been less than enthusiastic about the Rutherford-Soddy theory of atomic change.

Pierre Curie's significance for the course of radioactivity research in France goes beyond his direct role in the transmutation researches. He was largely responsible for the prevailing scientific style during this period (1899–1903), and his ghost haunted the Paris laboratories long after his premature death in 1906. Curie's role becomes especially clear when we compare the papers which he wrote jointly with Marie Curie with those which they published singly or with other colleagues. As early as 1898 Marie Curie suggested that radioactivity might originate in a profound subatomic change.[74] She usually discussed basic questions in her own papers, and her discussions sometimes were frankly speculative. Pierre Curie, in contrast, preferred to

[69]Albert Laborde, *Pierre Curie dans son laboratoire* (Paris: Palais de la Découverte, Conférences, Series A, No. 220, March 1956), p. 12.

[70]Marie Curie, *Pierre Curie* (New York: Macmillan, 1923), pp. 135–136.

[71]See Robert Reid, *Marie Curie* (New York: Mentor, 1974), p. 99.

[72]Eve, *Rutherford*, pp. 80–81.

[73]See esp. Becquerel, *Recherches sur une propriété nouvelle de la matière*; André Debierne, "L'état actuel de nos connaissances sur la radio-activité," *Rev. Gén. Sci.*, 1908, *19*:691–700, 730–738; Gaston Dupuy, "Notice sur la vie et les travaux de André Debierne (1874–1949)," *Bulletin, Société Chimique de France*, 1950, *17*:1023–1026.

[74]M. Curie, "Les rayons de Becquerel et le polonium."

THE DISCOVERY OF ATOMIC TRANSMUTATION   223

avoid such excursions into uncertainty. No doubt the Curies influenced one another, but after Marie introduced her spouse to radioactivity, Pierre dominated, for she, like Becquerel, never developed her own hypotheses. It was Pierre's positivistic philosophy of science which they preached in their joint article on scientific methodology.[75] Later work in Marie Curie's laboratory not only avoided specific hypotheses, as Pierre would have wished; it eschewed basic theoretical questions. Busy work became the main pursuit in Paris until after World War I.[76]

CONCLUDING REMARKS

The parallel studies of radioactivity in France and Canada during 1899–1903 illustrate the obvious fact that one's science is intertwined with one's philosophy. The scientific milieu was uniform enough for Curie and Rutherford to ask similar questions using similar experimental methods (although Curie's choices were partially determined by Rutherford's); yet they came to very different conclusions, because they made opposite assumptions about the nature of the emanation and held different views of the function of scientific theory. Curie's positivism clearly did not facilitate progress in radioactivity. Not only did it make him excessively cautious; it predisposed him to an unproductive thermodynamic model. He supplied only the most general of organizing schemes—one which was not fruitful since it did not suggest further experiments. Nor was he much interested in verifying it, for as a positivist he regarded models as relatively unimportant. In the face of the experimental evidence favoring material hypotheses Curie's stand appears obstinate, even irrational. He seems to have been motivated as much by aloofness and pride as by commitment to a scientific philosophy of noncommitment.

Since only a handful of persons devoted themselves to this nascent science, individual personalities wielded great influence. Pierre Curie inhibited the growth of original ideas in France, even indirectly after his death. In Britain Rutherford was the prime mover behind radioactivity research for most of his life. His outgoing, boisterous personality contrasted with Curie's quiet, reserved, independent, and introspective spirit, and the personality difference was reflected in their scientific styles, which to some extent meshed with the corresponding national styles. Curie's theory was hesitant, conservative, vague, and abstract; Rutherford's hypotheses were bold, ultimately revolutionary, specific, and concrete. Scientific faith was an important ingredient in the outcome of this episode. Rutherford's belief in the validity of his material hypotheses provided an impetus which Curie's noncommitment precluded. The personal factors would have favored the British team in any case, but nature too was working against the French. Unfortunately for them the study of radioactivity, an atomic phenomenon, progressed rapidly with British material and particulate hypotheses, but languished under the continental energetics which had succeeded so admirably with heat.

[75] See O. A. Starosel'skaia-Nikitina, *Istoriia radioaktivnosti i vozniknoveniia iadernoi fiziki* (Moscow: Izdatel'stvo Akademii Nauk SSSR, 1963), Ch. 4.

[76] Lawrence Badash uses this term in his review of Reid, *Marie Curie* in *Isis*, 1975, *66*:566–568.

AMERICAN JOURNAL OF PHYSICS   VOLUME 37, NUMBER 10   OCTOBER 1969

Einstein and the "Crucial" Experiment

GERALD HOLTON
Department of Physics, Harvard University, Cambridge, Massachusetts 02138
(Received 17 June 1969)

This paper was presented on 5 February 1969 at a joint symposium of the AAPT and APS on the history and philosophy of physics, held during the AAPT–APS annual meeting.

When asked to speak at this session, I thought it might interest you if I chose a debated episode in the history of recent science that can be illuminated by documents found in the Einstein Archive at Princeton. Over the past few years, I have helped to supervise the cataloging of the papers and correspondence there, and some of you may remember that from time to time I have reported at these meetings what has turned up.

Today I shall use some documentary materials to examine the widespread opinion that it was a crucial *experiment* that led Einstein to formulate the special relativity theory.

Our discussion will start innocently enough, but by the end I hope to have raised some rather unsettling questions, for example these:

What shall the historian of science do when he does not have "all the facts"?

EINSTEIN AND THE "CRUCIAL" EXPERIMENT   969

Is there something built into the educational process itself that tends to make us tell a wrong story about scientific developments?

Is our vision of the role of theory in physics today to some degree distorted by an outdated or erroneous philosophy of science?

We start with a letter dated 2 February 1954, about a year before Albert Einstein's death, from Davenport of the Department of History of Monmouth College, Illinois, who wrote to Einstein that he was looking into evidence that Michelson had "influenced your thinking and perhaps helped you to work out your theory of relativity." Not being a scientist, Davenport asked for "a brief statement in nontechnical terms, indicating how Michelson helped to pave the way, if he did, for your theory."

Einstein answered very soon after receipt, on 9 February 1954 (according to the copy of his hitherto unpublished reply in the files of the Einstein Archive of the Institute for Advanced Study at Princeton). Of course, Einstein had frequently been asked about Michelson's influence, particularly during the previous few years (e.g., by Shankland and Balazs). Having perhaps pondered again over the answers he had given earlier, Einstein now seemed remarkably willing to respond to a stranger. He gave not only a detailed reply, but even volunteered his offer to allow the letter to be published, and invited further correspondence[1]:

Dear Mr. Davenport:

Before Michelson's work it was already known that within the limits of the precision of the experiments there was no influence of the state of motion of the coordinate system on the phenomena, resp. their laws. H. A. Lorentz has shown that this can be understood on the basis of his formulation of Maxwell's theory for all cases where the second power of the velocity of the system could be neglected (effects of the first order).

According to the status of the theory it was, however, natural to expect that this independence would not hold for effects of second and higher

orders. To have shown that such expected effect of the second order was *de facto* absent in one decisive case was Michelson's greatest merit. This work of Michelson, equally great through the bold and clear formulation of the problem as through the ingenious way by which he reached the very great required precision of measurement, is his immortal contribution to scientific knowledge. This contribution was a new strong argument for the non-existence of "absolute motion," resp. the principle of special relativity which, since Newton, was never doubted in Mechanics but *seemed* incompatible with electro-dynamics.

In my own development Michelson's result has not had a considerable influence. I even do not remember if I knew of it at all when I wrote my first paper on the subject (1905). The explanation is that I was, for general reasons, firmly convinced that there does not exist absolute motion and my problem was only how this could be reconciled with our knowledge of electro-dynamics. One can therefore understand why in my personal struggle Michelson's experiment played no role or at least no decisive role.

You have my permission to quote this letter. I am also willing to give you further explanations if required.

Sincerely yours,
Albert Einstein.

It is a thoughtfully composed reply, and the last letter of Einstein I have been able to find on this subject. We may in fact regard it as an excellent summary of what one can learn from a study of the many other first-hand documents in the Archive dealing with the question. In particular, the text of the letter shows the need for four kinds of sharp differentiations: (1) between the effect the experiment had on the development of physics, and the effect it may have had on the development of Einstein's own thought, his "personal struggle;" (2) between the beauty of the immortal experiment, and its subsidiary place in theory; (3) between the statements Einstein made in direct response to repeated requests to deal with the possible genetic role of the Michelson experiment, and the rather different statement he made whenever he volunteered any comments concerning the genesis of relativity (in which case Einstein almost always spoke only about the experiment of Fizeau and the aberration measurements, insofar as he spoke of measurements at all); and (4) between the large interest the whole question seemed to have held for many people, and the small interest it held for Einstein, who

[1] I thank the Estate of Albert Einstein, and particularly Miss Helen Dukas, for permission to publish this and the other cited documents from the Archive.

This article constitutes a summary of portions of a detailed study scheduled to be published in the summer 1969 issue of *Isis*.

970   GERALD HOLTON

neither could recall whether he really knew of the experiment in 1905 nor seemed particularly disturbed by this fact.

Today I shall not be concerned with providing further support for Einstein's retrospective evaluation (that Michelson's result "has not had a considerable influence" on Einstein's development of the special theory of relativity, that "it played no role or at least no decisive role"). Such a task requires a close reading of the evidence, mostly long available, as found in Einstein's first paper on relativity (in which Michelson's experiment is not specifically mentioned), and in Einstein's other voluminous work, comments and letters. (An essay of this kind will appear elsewhere). In a nutshell, Lorentz's electrodynamics appeared to Einstein to contain too many *ad hoc* features—among them, as one of several factors, the patent artificiality of its explanation of the ether-drift experiments. What we shall ask here is, first of all, this: If we accept Einstein's response, *why* is it that the published evaluations by others, with very few exceptions, are so very different? After all, it has been the overwhelming preponderance of opinion of scientists and others over the last half century that Michelson's was the fundamental experiment that led Einstein directly to his relativity theory. Most of the popular literature and many scholarly discussions concur with the characterization given in the caption under

FIG. 1. "Albert Abraham Michelson, first American scientist to win the Nobel prize in physics, made the measurements on which are based Einstein's Special Theory of Relativity. Michelson is shown here in his laboratory at the University of Chicago." Photograph and caption as published in *Your Career in Optics* (The Optical Society of America, New York, 1965), p. 3. Reproduced by permission.

Michelson's picture in a publication of the Optical Society of America: Michelson "made the measurements on which are based Einstein's Special Theory of Relativity."

A more detailed account of the experimental origins of relativity theory, coming from an authoritative source, and printed at an unusual occasion, but otherwise quite representative of recent and current thought, is found in Millikan's essay entitled "Albert Einstein on his Seventieth Birthday." It was the lead article in a special issue in Einstein's honor of the Rev. Mod. Phys. **21**, (1949), and the early parts are worth quoting at some length:

The year 1905 was a notable year in that at the age of 26, Einstein published in that year's issue of the *Annalen der Physik* three brief but remarkable papers which have had very important bearings upon my own work as a physicist throughout my whole life. These three papers were on the following subjects: (1) the special theory of relativity; (2) the Brownian movements; and (3) photoelectric stopping potentials.

Every one of these three papers represents new and far-reaching *generalizations* of immense importance. For the first and second of these the stage had already been set, and the *experimental* foundations on which all sound generalizations must rest had already been built. In the case of relativity, the prime experimental builder had been my own chief at the University of Chicago, Albert A. Michelson, who made his first experiment on aether-drift at Berlin in 1881 . . .. The special theory of relativity may be looked upon as starting essentially in a generalization from Michelson's experiment. And here is where Einstein's characteristic boldness of approach came in, for the distinguishing feature of modern scientific thought lies in the fact that it begins by discarding all *a priori* conceptions about the nature of reality—or about the ultimate nature of the universe—such as had characterized practically all Greek philosophy and all medieval thinking as well, and takes instead, as its starting point, well-authenticated, carefully tested *experimental* facts, no matter whether these facts seem at the moment to be reasonable or not. In a word, modern science is essentially empirical . . .

But this experiment, after it had been performed with such extraordinary skill and refinement by Michelson and Morley, yielded with great definiteness the answer that there is . . . no observable velocity of the earth with respect to the aether. That unreasonable, apparently inexplicable experimental fact was very bothersome to 19th century physics, and so for almost twenty years after this fact came to light physicists wandered in the wilderness in the disheartening effort to make it seem reasonable. Then Einstein called out to us all, "Let

EINSTEIN AND THE ''CRUCIAL'' EXPERIMENT   971

us merely accept this as an established experimental fact and from there proceed to work out its inevitable consequences,'' and he went at that task himself with an energy and a capacity which very few people on earth possess. Thus was born the special theory of relativity. (Italics in original.)

I. THE STORY IN DIDACTIC WRITINGS

Many other prominent scientists have proposed quite similar scenarios. So it is not surprising that authors of science textbooks generally mirror, or at least do not contradict, the mythology current among the foremost research scientists.

On the question under study here, the textbooks are virtually unanimous on the history they imply. Selecting practically at random from recent physics texts on your own shelf, you will find essentially the same story as in Millikan's article.

This version is not limited to recent American textbooks, nor indeed to writers who themselves did not participate in the development of the field in its early days. Thus, Max von Laue published one of the first serious textbooks on relativity (*Das Relativitätsprinzip*) in 1911; the ''Michelson experiment'' is described early: ''. . . the experiment became, as it were, the fundamental experiment for the relativity theory. . . .''

But things get more puzzling when we find that Einstein has left in some of his readers a similar impression in at least two of his own publications. One is his early ''gemeinverständliche'' (generally understandable, popularized) book, *Über die spezielle und die allgemeine Relativitätstheorie* (1917). Here we find a route which was to become so familar in almost all subsequent text presentations:

> . . . for a long time the efforts of physicists were devoted to attempts to detect the existence of an aether-drift at the earth's surface. In one of the most notable of these attempts Michelson devised a method which appears as though it must be decisive . . . . But the experiment gave a negative result—a fact very perplexing to physicists. Lorentz and FitzGerald rescued the theory from this difficulty by assuming that the motion of the body relative to the aether produces a contraction of the body in the direction of motion . . . . But on the basis of the theory of relativity, the method of interpretation is incomparably more satisfactory.

While no genetic relation is presented here, it is almost irresistible to read one into the text.

The second of these early publications by Einstein is his contribution on ''Relativity Theory'' to a collection of 36 essays by some of the foremost physicists, intended to convey the ''state of physics in our time.'' [E. Warburg, Ed., *Die Physik* (B. G. Teubner, Leipzig, 1915)].

Einstein begins as follows: ''It is hardly possible to form an independent judgment of the justification of the theory of relativity, if one does not have some acquaintance with the experiences and thought processes which preceded it. Hence, these must be discussed first'' (p. 703). There follows a discussion of the Fizeau experiment, leading to Lorentz's theory based on the hypothesis of the stagnant ether. Despite its successes, ''the theory had *one* aspect which could not help but make physicists suspicious'' (p. 705): it seemed to contradict the relativity principle, valid in mechanics and ''as far as our experience reaches, generally'' beyond mechanics also. According to it, all inertial systems are equally justified. But not so in Lorentz's theory: A system at rest with respect to the ether has special properties; for example, with respect to this system, the light velocity is constant.

Einstein pinpoints the difficulty: ''The successes of Lorentz's theory were so significant that the physicists would have abandoned the principle of relativity without qualms, had it not been for the availability of an important experimental result, of which we now must speak, namely Michelson's experiment.'' There follows a description of the experiment, and of the contraction hypothesis invoked by Lorentz and FitzGerald. Einstein adds to it sharply, ''This manner of theoretically trying to do justice to experiments with negative result through *ad hoc* contrived hypotheses is highly unsatisfactory'' (p. 707). The preferable conception is to hold on to the relativity principle, and to accept the impossibility-in-principle of discovering relative motion.

But how is one to make the principle of constancy of light velocity and the principle of relativity compatible after all? ''Whoever has deeply toiled with attempts to replace Lorentz's theory by another one that takes account of the experimental facts will agree that this way of beginning appears to be quite hopeless at the present state of our knowledge'' (pp. 707–708). Rather, one can attain compatibility of the two apparently contradictory principles through a

972   GERALD HOLTON

reformulation of the conception of space and time, and by abandoning the ether.

The rest of Einstein's short essay is concerned with the introduction of the relativity of simultaneity and of time, the transformation equations, and the length measurement of a rod moving with respect to the observer. "One sees that the above-mentioned hypothesis of H. A. Lorentz and FitzGerald for the explanation of the Michelson experiment results as a consequence of the relativity theory" (p. 712). But this result does not seem to be worthy of listing as one of the achievements of the relativity theory a little later: "We will now briefly enumerate the individual results achieved so far for which we have to thank the relativity theory" (p. 712). The list, as of 1915, was not long: "a simple theory of the Doppler Effects, of aberration, of the Fizeau experiment"; applicability of Maxwell's equations to the electrodynamics of moving bodies, and in particular to the motion of electrons (cathode rays, β-rays) "without invoking special hypotheses"; and "the most important result," the relation between mass and energy, although for that there was then no direct experimental confirmation.

Now it is essential to see that both of these publications by Einstein are frankly *didactic*. The essay of 1915 for example, is plainly not meant as an historic account of the road Einstein himself followed. The whole essay is introduced as dealing with the "justification" of the theory of relativity, not with the genesis. It is "the physicists," not Einstein himself, of whom the author says that they would have abandoned the principle of relativity without qualms if it had not been for the Michelson experiment.

Yet without having actually said anything here about his own historic route, Einstein's singling out of the Michelson experiment in his didactic writings cannot have failed to reinforce later, second-hand, didactic writings by others—even after the publication of Einstein's very different and frankly historical accounts, for example in the important interviews which R. S. Shankland published in the Amer. J. Phys. **31**, 47–57 (1963).

We get here a hint of a first answer to our question: the reason why there has been so much agreement on the supposed genetic role of Michelson's experiment is the frequently made confusion between science in the sense of the private ac-

tivity, the "personal struggle," and science in the sense of a developed field, a public institution.

If one takes the trouble to read the available documents carefully, one discovers a clear distinction: whenever Einstein writes or speaks on the origins of relativity in the passive voice, in answer to a direct question or in response to an obligation such as at a public lecture, he notes the importance the Michelson experiment had for the further development and acceptance of the theory by *other* physicists. But when he writes or speaks of the influence of the experiment on himself, explicitly and in first person singular, Einstein says in all letters that the effect on himself was "negligible," "indirect," "rather indirect," "not decisive," or (according to his letter to B. Jaffe, published in 1944) at most was of "considerable influence upon my work insofar as it strengthened my conviction concerning the validity" of relativity theory, adding at once "I was pretty much convinced of the validity of the principle before I did know this experiment and its results." We must therefore distinguish between Einstein's evaluations of different effects upon public science and private science. The failure to make such distinctions is among the most insidious defects of our instruction, as I shall have occasion to elaborate below.

II. WHAT DID EINSTEIN SAY TO MICHELSON?

At this point you might well confront me with one well-known document that flatly contradicts everything I have said so far. It is an account of Einstein's speech given in 1931 on the occasion of his visit to Pasadena, California, when, for the first and last time, he came face to face with Michelson. The occasion must have been moving. Michelson, 27 years his senior, was much admired by Einstein from a distance. Shankland later reported that Einstein particularly appreciated "Michelson's artistic sense and approach to science, especially his feeling for symmetry and form. Einstein smiled with pleasure as he recalled Michelson's artistic nature—here there was a kindred bond."

But Michelson was known to be no friend of relativity, the destroyer of the ether. Like so many others, Michelson was convinced that his own ill-fated experiments were the basis for the theory. Einstein reminisced later that Michelson "told me more than once that he did not like the

EINSTEIN AND THE "CRUCIAL" EXPERIMENT   973

theories that had followed from his work," and that he had told Einstein he was a little sorry that his own work started this "monster."

Michelson was now 79 years old, weak after a serious stroke that had first forced him to his sickbed two years earlier. The picture taken on that occasion shows the frail old man, standing next to Einstein with his usual erect dignity on this last public appearance; but he was marked for death that came three months later.

Among others present at a grand dinner in the new Athenaeum in Pasadena on 15 January 1931 were distinguished physicists and astronomers: W. S. Adams, W. W. Campbell, G. E. Hale, E. P. Hubble, C. E. St. John, R. A. Millikan, R. C. Tolman, as well as Mrs. Einstein and 200 members of the California Institute Associates. Millikan set the stage with some remarks on what he saw to be the characteristic features of modern scientific thought. It is, in fact, largely the very same material Millikan was to republish 18 years later as part of his introduction for the Einstein issue of the Reviews of Modern Physics. But after the sentence, "Thus was born the special theory of relativity," Millikan went on to say in 1931: "I now wish to introduce the man who laid its experimental foundations, Professor Albert A. Michelson . . .."

Michelson kept his short response in the channel laid out for him:

> I consider it particularly fortunate for myself to be able to express to Dr. Einstein my appreciation of the honor and distinction he has conferred upon me for the result which he so generously attributes to the experiments made half a century ago in connection with Professor Morley, and which he is so generous as to acknowledge as being a contribution on the experimental side which led to his famous theory of relativity.

In fact, Einstein had not yet responded, if the published record of the meeting is a guide [Science **73**, 375–381 (1931)]. Millikan next called on Campbell of the splendid group of experimental astronomers, saying, "I am herewith assigning him the task of sketching the development of the experimental credentials of the general theory of relativity." Campbell recounted the success of the three chief tests, in which the California astronomers had played leading roles.

Millikan next started to introduce Einstein, but prefaced it with a last reinforcement of the philosophical message that he had been building

up. Pursuing his theme with tenacity, Millikan now referred to his own "experimental verification" of predictions contained in the early papers of Einstein. Seen from his perspective, Millikan's evaluation of Einstein's paper on the quantization of light energy (1905) was not surprising:

> The extraordinary penetration and boldness which Einstein showed in 1905 in accepting a new group of experimental facts and following them to what seemed to him to be their inevitable consequences, whether they were reasonable or not as gauged by the conceptions prevalent at the time, has never been more strikingly demonstrated.

At last, the stage was set, the expectations fully aroused for Einstein's response. What happened now—or rather, what is supposed to have happened—is widely known from the account given in Michelson's only biography available so far, that of B. Jaffe [*Michelson and the Speed of Light* (Doubleday & Co., Inc., Garden City, N. Y., 1960)]. Jaffe writes:

> "Einstein made a little speech. Seated near him were Michelson, Millikan, Hale, and other eminent men of science. 'I have come among men,' began Einstein, 'who for many years have been true comrades with me in my labors.' Then, turning to the measurer of light, he continued, 'You, my honored Dr. Michelson, began with this work when I was only a little youngster, hardly three feet high. It was you who led the physicists into new paths, and through your marvelous experimental work paved the way for the development of the Theory of Relativity. You uncovered an insidious defect in the ether theory of light, as it then existed, and stimulated the ideas of H. A. Lorentz and FitzGerald, out of which the Special Theory of Relativity developed. Without your work this theory would today be scarcely more than an interesting speculation; it was your verifications which first set the theory on a real basis.'
>
> Michelson was deeply moved. There could be no higher praise for any man." (p. 168)

This was indeed the kind of response expected from the preparatory speeches; Jaffe gives a natural and clear-cut answer to the question of the possible genetic connection between Michelson's experiment and Einstein's work. He says on another page of the book simply: "In 1931, just before the death of Michelson, Einstein publicly attributed his theory to the experiment of Michelson." (p. 101)

Reading Jaffe's passage carefully we need not go so far. Michelson "stimulating the ideas" of Lorentz and FitzGerald, out of which in turn the

974   G E R A L D   H O L T O N

special theory of relativity "developed," is not a scenario in contradiction with the likely chain of events: The then-current Lorentz–FitzGerald contraction–explanation of the Michelson experiment, as found in the works of 1892 and 1895 of Lorentz which we know Einstein had read, by its unappealing *ad hoc* character, compromised further the ether-committed theory of electrodynamics that Einstein already knew for many other reasons to be inadequate. But the remarks "without your work ... it was your verification ..." sound indeed like a personal acknowledgement to Michelson, a public attribution of the kind that Jaffe clearly saw in it. And in that case we must confess, as Kepler put it half-way through the *Astronomia Nova*, "Dear Reader, Our hypothesis goes up in smoke."

But there is another explanation. There exists Einstein's German manuscript of his talk, as well as translations of it that were published in 1931 and 1949. From these we can see that Jaffe's widely read version of Einstein's talk has fallen into the trap that had unwittingly been set up for Einstein. Missing from Jaffe's version is a heading a little sentence, and a long ending from Einstein's talk. They make a lot of difference.

The talk starts with "Liebe Freunde!" It is, of course, addressed to the whole company. And just between the last two sentences quoted by Jaffe, we find there was another sentence that switches the discussion away from Michelson and special relativity, and toward the assembled astronomers and general relativity: "You uncovered an insidious defect in the ether theory of light, as it then existed, and stimulated the ideas of H. A. Lorentz and FitzGerald, out of which the special theory of relativity developed. *These in turn led the way to the general theory of relativity, and to the theory of gravitation.* Without your work this theory would today be scarcely more than an interesting speculation; it was your verifications which first set the theory on a real basis." (Italics supplied.) Then follows immediately an acknowledgement of experimental contributions by the California astronomers that had "furnished the real basis for the [general] theory," those by Campbell, St. Johns, Adams, and Hubble.

What emerges is still a fine compliment to Michelson. Yet even standing before him, and under the accumulated pressures of the dramatic affair, Einstein agreed neither with Millikan's

nor with Michelson's version of the genetic connection (nor, of course, with Jaffe's). He did not avail himself of the occasion to say straight out what everyone seemed to have come to hear him say, for example: "Michelson's is the crucial experiment that was the basis for my own work." He rather seems to see Michelson as one of the figures on the continuous, long way leading to relativity theory.

As to the unfortunate omission of the sentence in Jaffe's otherwise very useful book, one knows how such things happen at the most awkward point. The significance lies in this: Mistakes favor the prepared mind. And worse, through no fault of Jaffe's, his evaluation had been repeatedly republished by others who, apparently without scholarly examination of the available original text, have found comfort in his evaluation for their own purpose of forging a tight genetic link from Michelson to Einstein.

III. REASONS FOR LINKING EXPERIMENT AND THEORY

The birth of a new theory as the response to a puzzling empirical finding: this is the message clearly stated by eminent scientists such as Millikan, gladly adopted by textbooks, and found in avowedly historical accounts, too. Why is this sequence so seductive? There are half a dozen reasons. One is that almost every science textbook of necessity places a high value on clear, unambiguous, inductive reasoning. The norm of rationalism in the classroom would seem to be threatened if the text were to allow that a correct inductive generalization may be made without unambiguous experimental evidence. Hence, the likelihood is *a priori* great that any pedagogic presentation of any scientific subject will suggest a clear genetic link from experiment to theory.

Moreover, in a textbook or a didactic essay where a large amount of ground has to be covered, it is likely (for reasons of space if for no other) that the author will select one suitable experiment that can be convincingly presented, rather than a number of different experiments that in historical fact may have been equally good or better candidates.

But in the particular case of the growth of relativity theory, the author of a didactic account has an added incentive to foreshorten the period of doubt in the scientific community that actually

EINSTEIN AND THE "CRUCIAL" EXPERIMENT   975

followed upon Einstein's 1905 publication. A student can be expected to accept more easily a theory as non-common-sensical as Einstein's if he can be shown that Einstein, or at least Einstein's contemporaries, became convinced on the basis of some clearcut experiment.

Hence little is said in such books about the sometimes dramatic battles that in fact were required for the gradual acceptance of a new theory. That lack fits in well with another moralizing function of scientific instruction, namely to underplay the scientist's private and emotional involvements in the pursuit of his own scientific work, or in the evaluation of others, to avoid seemingly to place value on his making premature commitments before "all the evidence is in," and in all these ways to introduce the student to what the textbook author usually, perhaps unconsciously, conceives to be the accepted public norms of professional behavior.

Texts probably cannot and certainly do not want to deal with the private aspect of doing science, an aspect which can be so different from scientist to scientist, and one that is so far from being fully understood in any case. It seems simpler to deal with the public aspect of science on which there is (though perhaps falsely) some consensus—the more so, in this case, as Einstein himself, in his didactic publications of 1915 and 1917, had written only of the public aspects of the rise of relativity theory. Therefore, the elements that hold a historian's attention, the elements that carry the possibility for a classic case study of the difference between private and public science—or for that matter of the relative roles of theory and experiment in modern scientific innovation, or of the quasi-aesthetic criteria for a decision between rival conceptual systems embracing the same "facts" in different ways—all these give way in textbooks to other, simpler purposes. It is just a by-product of this attitude that has caused Einstein's supposed use of the Michelson result to have become a fixed part of the folkloric consensus about the history of science, a story as widely known and believed as the story of the falling apple in Newton's garden and of the two weights dropped from the Leaning Tower in Galileo's Pisa—two other cases in which experimental fact is supposed to have provided the genetic occasion for synthetic theory.

There are yet other special circumstances that have made it almost irresistible for pedagogic accounts to give a place of importance to the Michelson experiment. It is not just any experiment, but rather one of the most fascinating experiments in the history of physics. Its fascination has been felt equally by text writers and research physicists, and derives from its beauty and its mystery. Despite the central position of the question of ether drift in late nineteenth-century physics, nobody before Michelson was able to imagine and construct an apparatus to measure the second-order effect of the presumed ether drift. The interferometer was a lovely thing. Invented by the 28-year-old Michelson in response to a challenge by Maxwell, it was capable of revealing an effect of the order of one part in ten billion. It is to this day one of the most sensitive scientific instruments, and the experiment is one that carried precision to the extreme limit. Einstein himself later paid warm and sincere tribute to Michelson's experimental genius and artistic sense, and he added, "many negative results are not highly important, but the Michelson experiment gave a truly great result which everyone should understand."

The interpretation of the experiment, however, initially presented mysterious difficulty on two levels. One was understanding the way the apparatus works in the context of ether theory, regardless of the meaning of its outcome. On this score, one finds nowadays everywhere some adequate though quite oversimplified outlines of the experiment. But Michelson himself, on presenting an account of his first experiment to the Académie des Sciences in 1882, acknowledged he had made an error in his earlier report of 1881 and had neglected the effect of the earth motion on the path of light in the interferometer arm at right angles to the motion. Potier, who had pointed out the error to Michelson in 1882, was in error also. A debate continued for over 30 years on the question how the moving reflector affects the angle of reflection.

But beyond that, the findings obtained with the instrument were enormously puzzling to everyone at the time, and for many remained so for a long time afterwards. A significant displacement of the interference fringes had been expected, owing to the presumed effect of the ether on the motion of light, but virtually none was obtained. The glorious device had yielded a puzzling, disap-

976 G E R A L D H O L T O N

pointing, even incomprehensible result in the context of the then-current theory. Michelson himself called his experiment a "failure"; the repeatedly obtained null or nearly null results were contrary to all his expectations. And contrary to the stereotype that the true scientist accepts the experimental test that falsifies a theory, he refused to grant the importance of his own result, saying, "Since the result of the original experiment was negative, the problem is still demanding a solution." He even tried to console himself with the remarkable observation that "the experiment is to me historically rich because it was for the solution of this problem that the interferometer was devised. I think it will be admitted that the problem, by leading to the invention of the interferometer, more than compensated for the fact that this particular experiment gave a negative result."

Others were just as mystified and displeased. Lorentz wrote to Rayleigh on 18 August 1892:

> I am utterly at a loss to clear away this contradiction, and yet I believe if we were to abandon Fresnel's theory [of the ether], we should have no adequate theory at all.... Can there be some point in the theory of Mr. Michelson's experiment which has as yet been overlooked?

Lord Kelvin could not reconcile himself to the negative findings into the 1900's. Rayleigh, who, like Kelvin, had encouraged Michelson to repeat his first experiment, confessed he found the null result obtained by Michelson and Morley in Cleveland to be "a real disappointment."

As L. S. Swenson has pointed out, Michelson and Morley were so discouraged by the "null" result of their experiment in 1887 that they disregarded the promise in their original paper that the measurements, which they had taken during only 6 hours within one 5-day period, "will therefore be repeated at intervals of 3 months, and thus all uncertainty will be avoided." Instead, Michelson stopped their work on this experiment, and turned to the new use of the interferometer for measuring lengths. (As it turned out, it was a wise move. That was the work that led to his Nobel Prize award.) In short, to everyone's surprise, including Michelson's, the experiment had turned out to be one of "test," not merely of "application," to use the terminology of Duhem. It is therefore rather ironic that neither Michelson nor Einstein, in their different ways, considered the

famous experiment as decisive for himself, not to speak of "crucial." However, for most ether theoreticians it became a crucial experiment *malgré lui* in the only valid sense of the term, namely, as the pivotal occasion causing a significant part of the scientific community to reexamine its previously held basic convictions.

If the result of the Michelson experiment was a mystery for a long time, the relativity theory for its part was even more mysterious at its announcement in 1905, and remained so for some time afterwards. It took a few years before one could say that even among German scientists there was a preponderance of opinion in favor of it. The turning point came perhaps with the publication of Minkowski's address on "Space and Time" in 1909. Among physicists outside Germany the lag was much longer, and among most teachers, students, and the general public, the lag was longer still.

It now seems almost inevitable that in this situation there occurred a natural act of symbiosis, especially in the didactic literature. It was the marriage of the puzzling Michelson experiments and the all-but-incomprehensible relativity theory. The undoubted result of Michelson's experiments could be thought to provide an experimental basis for the understanding of relativity theory which otherwise seemed contrary to common sense itself. And the relativity theory in turn could provide an explanation for Michelson's experimental result in a manner not as "artificial" or *ad hoc* as reliance on the supposed Lorentz–FitzGerald contraction was widely felt to be.

This strategy was the more necessary for proponents of relativity theory because, after its publication, no new experimental results came forth for many years which could be used to "verify" this theory in the way most physicists are used to looking for verification. On the contrary, as I have indicated elsewhere, the very first response within the scientific community to Einstein's relativity paper of 1905, and in the same journal in which he had published it, was a categorical experimental disproof of the relativity theory by W. Kaufmann. It took until 1915 for Kaufmann's experimental equipment to be shown to have been defective, and for unambiguously favorable results to be established. Max Planck noted in 1907 that Michelson was then still regarded as provider of the only experimental support. The physicist W. Wien was not convinced of the theory until 1909; and then it was not by any clearcut evidence from experiment, but on more nearly aesthetic grounds, in words which Einstein must have appreciated and which to this day are applicable:

> What speaks for it [the theory] most of all, however, is the inner consistency which makes it possible to lay a foundation having no self-contradictions, one that applies to the totality of physical appearances, although thereby the customary conceptions experience a transformation.

What finally made special relativity theory a widely accepted basic part of physics throughout the scientific community were developments far from the scope of Einstein's 1905 paper itself. Foremost among these developments were the *experimental* successes such as the eclipse expedition of 1919, the use of relativistic calculations to explain the fine structure of spectral lines, the Compton effect, etc. In the meantime, for the interested public and for many physicists, Michelson's result was the crutch that supported relativity theory, particularly in the face of its challenging paradoxes and iconoclastic demands.

To summarize, the particular missions of pedagogic accounts backed up by the popular writings of distinguished experimental physicists, and the particular history surrounding the acceptance of the Michelson results and the Einstein publication, together provided two sets of pressures that tended to the same end, namely to proclaim the existence of a direct genetic link between Michelson's and Einstein's work.

I believe these pressures can also be traced to a common root, namely, to a certain philosophical view concerning science as a whole. This view was supported by a vocal group of philosophers in the U. S. and Europe, and was widely current after the victories of the empiricist schools around the turn of the century.

IV. THE DOCTRINE OF EXPERIMENTICISM

There exists a view of science at the extreme edge of the time-honored tradition of empiricism which, for want of a better name, will here be called *experimenticism*. It is best recognized by the unquestioned priority assigned to experiments and experimental data in the analysis of how scientists do their own work and how their work is incorporated into the public enterprise of science. A few examples will suffice to indicate the per-

vasiveness of this attitude. With specific reference to relativity theory, it is well illustrated by Ernst Mach's disciple Joseph Petzoldt, the moving spirit behind the Gesellschaft für positivistische Philosophie of Berlin and its journal, Zeitschrift für positivistische Philosophie. As lead article of its inaugural issue (Vol. 1, 1913), he printed the text of the speech he had delivered at the opening session of the Gesellschaft on 11 November 1912. With relativity theory, he said there had come "the victory over the metaphysics of absolutes in the conceptions of space and time," and a "fusion of mathematics and natural science which at last and finally shall lead beyond the old rationalistic, Platonic–Kantian prejudice." But the fixed hinge on which this desired turn of events moved was, again, the Michelson experiment:

> Clarity of thinking is inseparable from knowledge of a sufficient number of individual cases for each of the concepts used in investigation. Therefore, the chief requirement of positivistic philosophy: greatest respect for the facts. The newest phase of theoretical physics gives us an exemplary case. There, one does not hesitate, *for the sake of a single experiment*, to undertake a complete reconstruction. The Michelson experiment is the cause and chief support of this reconstruction, namely the electrodynamic theory of relativity. To do justice to this experiment, one has no scruples to submit the foundation of theoretical physics as it has hitherto existed, namely Newtonian mechanics, to a profound transformation. (Italics supplied)

The real enemy and the full ambitions were both revealed again in the next volume (1914), where Petzoldt wrote, "Lorentz's theory is at its conceptual center pure metaphysics, nothing else than Schelling's or Hegel's *Naturphilosophie*." Again, the Michelson experiment, as the one and only experiment cited, is given the credit for ushering in the new era: "... the Einsteinian theory is entirely tied to the result of the Michelson experiment, and can be derived from it." Einstein himself "from the beginning conceived of the Michelson experiment relativistically. We are dealing here with a principle, a foremost postulate, a particular way of understanding the facts of physics, a view of nature, and finally a Weltanschauung .... The sequence Berkeley–Hume–Mach shows us our direction and puts into our hands the epistemological standard."

A few years before, Michelson had been awarded the 1907 Nobel Prize in Physics "for his optical precision instruments and the search which he has

978   GERALD HOLTON

carried out with their help in the fields of precision metrology and spectroscopy." The relativity theory was, of course, still far too new and was regarded as too speculative to be mentioned in the citations or responses; indeed, at the time Petzoldt was writing his eulogies to it, the theory had become too speculative for Mach himself. (The Nobel Prize Committee did not award Einstein's prize until 1922, and then, as Einstein was specifically reminded by the Committee, it was for the experimentally well-confirmed work on the photoelectric effect). In any case, theory was not of interest on that day in 1907. As the presentation address by K. B. Hasselberg showed, the award to Michelson was clearly motivated by the same experimenticist philosophy of science of which we have had ample indications. Hasselberg said,

> As for physics, it has developed remarkably as a precision science, in such a way that we can justifiably claim that the majority of all the greatest discoveries in physics are very largely based on the high degree of accuracy which can now be obtained in measurements made during the study of physical phenomena. (Accuracy of measurement) is the very root, the essential condition, of our penetration deeper into the laws of physics—*our only way to new discoveries*. It is an advance of this kind which the Academy wishes to recognize with the Nobel Prize for Physics this year. (Italics supplied).

Somehow, everyone managed to keep a decorous silence on the experiment which Petzoldt and others of his persuasion were soon to hail as the crucial turning point for physics and for Weltanschauung. Nobody referred here to Michelson's ether-drift experiments—neither the Swedish hosts, nor Michelson himself in his responding lecture ("Recent Advances in Spectroscopy"). It was as embarrassing an experiment now for experimenticists with ether-theoretic presuppositions as it was welcome later for experimenticists with relativistic presuppositions.

What about the philosophers of science? In their work, the discussion of relativity theory is quite frequently found linked tightly to the Michelson experiment, though rarely more enthusiastically than in one of the essays, entitled "The Philosophical Dialectic of the Concepts of Relativity," collected in honor of Einstein in Schilpp's volume:

> As we know, as has been repeated a thousand times, relativity was born of an epistemological shock; it was born of the "failure" of the Michelson experiment . . .. To paraphrase Kant, we might say that the Michelson experiment roused classical mechanics from its dogmatic slumber . . .. Is so little required to "shake" the universe of spatiality? Can a single experiment of the twentieth century [*sic*] annihilate—a Sartrian would say *neantiser*—two or three centuries of rational thought? Yes, a single decimal sufficed, as our poet Henri de Regnier would say, to "make all nature sing." [P. Schilpp Ed., *Albert Einstein, Philosopher–Scientist* (Library of Living Philosophers, Evanston, Illinois, 1949), pp. 566–568.]

Einstein chose not to respond to this apotheosis of the Michelson experiment in his replies to critics published at the end of the same volume. But he makes a lengthy and subtly devastating reply to another essay in this collection, that by Hans Reichenbach (on "The Philosophical Significance of the Theory of Relativity"), which has the same kind of experimenticist basis.

We must remember that over the years Reichenbach was one of the most persistent and interesting philosophical analysts of the epistemological implications of relativity, and he published several attempts to cast the theory into axiometric form. But Reichenbach's empiricist conviction never flagged. For example, he wrote that Einstein's work "was suggested by closest adherence to experimental facts . . .. Einstein built his theory on an extraordinary confidence in the exactitude of the art of experimentation." [H. Reickenbach, *From Copernicus to Einstein* (Philosophical Library, New York, 1942), p. 51.] The only specific historic experiment Reichenbach associates with the genesis of Einstein's theory is, of course, the Michelson experiment; for example, "the theory of relativity makes an assertion about the behavior of rigid rods similar to that about the behavior of clocks . . .. This assertion of the theory of relativity is based mainly on the Michelson experiment." [*The Philosophy of Space and Time* (Dover Publishing Co., New York 1957); a translation by M. Reichenbach and John Freund of H. Reichenbach, *Die Philosophie der Raum-Zeit Lehre*, Walter de Gruyter & Co., Berlin, (1928), p. 195.] In fact Reichenbach and his faithful followers prefer to give a crucial place to the Michelson experiment in the supposed development rather merely showing the usefulness of the relativity theory. Thus Reichenbach claims, "It would be mistaken to argue that Einstein's theory gives an explanation of Michelson's experiment, since it does not do so. Michelson's experiment is simply taken over as an axiom." (*op. cit.* p. 201).

In Reichenbach's essay in Schilpp's collection of 1949 honoring Einstein, he reverts to the same points (e.g. p. 301), but they are only preludes to the conclusion that "it is the philosophy of empiricism, therefore, into which Einstein's relativity principle belongs . . .. In spite of the enormous mathematical apparatus, Einstein's theory of space and time is the triumph of such a radical empiricism in a field which had always been regarded as a reservation for the discoveries of pure reason," (pp. 309–310).

In his reply to this essay at the end of the volume, Einstein devoted most of his attention to a denial of this claim. He preferred to hold fast to the basic conceptual distinction between "sense impressions" and "mere ideas"—despite the expected reproach "that, in doing so, we are guilty of the metaphysical 'original sin.'" (p. 673). Einstein pleads that one must also accept features not only of empiricism but also of rationalism, indeed that a "wavering between these extremes appears to be unavoidable." (p. 680). He adopts the role of the "non-positivist" in an imaginary dialogue with Reichenbach, and urges the useful lesson of Kant that there are concepts "which play a dominating role in our thinking, and which, nevertheless, cannot be deduced by means of a logical process from the empirically given (—a fact which several empiricists recognize, it is true, but seem always again to forget)." (p. 678).

The difference is, of course, one of relative scientific taste or style. To Reichenbach, the interest in a scientific theory does not reside in the details of its historical development in the work of an actual person. As Reichenbach said honestly, "The philosopher of science is not much interested in the thought processes which lead to scientific discoveries; he looks for a logical analysis of the completed theory, including the relationships establishing its validity. That is, he is not interested in the context of discovery, but in the context of justification." (*Ibid.*, p. 292).

Unfortunately, however, Reichenbach and his followers have not always remembered his laudable attempt to distinguish clearly between private and public science, nor have they always adhered to his wise disclaimer of interest in the thought processes leading to the discovery. The desire to see a theory as a logically complete structure arising from empirical observations brings them, after all, to assume presumed historical sequences on the road that led to the discovery. Thus, implicit history is produced after all, e.g., that "Einstein incorporated its [the Michelson experiment's] null result as a physical axiom in his light principle," and similar attempts at "the unravelling of the history" of relativity theory.

When such attempts are then faced with some of the direct evidence against the priority and importance of the Michelson experiment in Einstein's thinking, the response is this: without the genetic role of this particular experiment, an understanding of the discovery of the theory would become "quite problematic," and one would be left "puzzled concerning the logical, as distinct from psychological grounds which would then originally have motivated Einstein to have confidence in the principle of relativity without the partial support of the Michelson–Morley experiment . . .." [A. Grünbaum, *Philosophical Problems of Space and Time* (Alfred A. Knopf, Inc., New York, 1963), p. 381.] Not surprisingly, Jaffe's authority is invoked at that very point to the effect that, "In 1931, just before the death of Michelson, Einstein publicly attributed his theory to the experiment of Michelson." (*Ibid.*)

To Einstein himself, the possibility of a lack of sufficiently secure "logical" grounds in the original motivation was not a puzzle. It was a well recognized fact of actual scientific work. Indeed, he went out of his way to make this point. For example, when asked to provide a statement honoring Michelson at the Centenary celebration on 19 December 1952, Einstein stressed that "The influence of the crucial Michelson–Morley experiment upon my own efforts has been rather indirect," and he ended with a paragraph that is as significant as it may have appeared gratuitous on that occasion:

> There is, of course, no logical way leading to the establishment of a theory, but only groping constructive attempts controlled by careful consideration of factual knowledge.

This is entirely in accord with the honest self-appraisal of the experience many a creative scientist has had in his own work. Einstein was perhaps more forthright in this confession, one that is so far from the still widely current myths which present scientific research as the inexorable result of the pursuit of logically sound conclusions from experimentally indubitable premises. The truth, alas, is different.

Einstein freely explained that matters are not so simple, e.g., in speaking to Shankland about the origins of his own work of 1905, Shankland reported [Amer. J. Phys. **31**, 48 (1963)]:

> This led him to comment at some length on the nature of mental processes in that they do not seem at all to move step by step to a solution, and he emphasized how devious a route our minds take through a problem . . . . It is only at the last that order seems at all possible in a problem.

Similarly, in commenting on the correct view a historian should take of the work of physicists, Einstein told him: "The struggle with their problems, their trying everything to find a solution which came at last often by very indirect means, is the correct picture." (*Ibid.*, p. 50). Contrary to the expectations of systematizers, axiomatizers, text writers, and others who, as we have seen, yearn for linearized sequences both in scientific work itself and in accounts given of such work, Einstein opposes with the gentle warning that there is no straightforward logical way, at least not to a theory of such magnitude as that contained in his work of 1905.

It was, after all, a warning Einstein had made repeatedly—from about 1918 on, and more emphatically from about the early 1930's. Examples may be found in his essay for Max Planck in 1918: "There is no logical way to the discovery of these elementary laws. There is only the way of intuition" (based on *Einfühlung* in experience). It is repeated in his Herbert Spencer lecture of 1933 (concerning the "purely fictitious character of the fundamentals of scientific theory"), and in his Autobiographical Notes written in 1946 ["A theory can be tested by experience, but there is no way from experience to the setting up of a theory" (p. 89)]. Again, when J. Hadamard asked Einstein for a self-analysis of his thought processes, Einstein replied

> The words or the language, as they are written or spoken, do not seem to play any role in my mechanism of thought. The psychical entities which seem to serve as elements in thought are certain signs and more or less clear images which can be "voluntarily" reproduced and combined.

in believing that there can be an essential abyss between experience and logically structured theory, and in believing also in the related distinction between "sense impressions" on the one hand and "mere ideas" on the other, Einstein separated himself from most of the prominent philosophies of science of his time. That he did not do this lightly we know from many evidences, including the frequency with which he kept coming back to reiterate these points over the years.

V. A LESSON FOR THE HISTORIAN OF SCIENCE

Historians are quite used to finding large discrepancies between the documentable history of science on the one hand, and, on the other, the popular history found in texts and in the writings of some eminent scientists and some philosophical analysts. We have looked here at a quite limited case, but there were some more widely applicable conclusions. Above all, we are forced to ask anew what are the most appropriate styles and functions of historical scholarship today, particularly against the background of the prevailing experimenticist doctrine. On this point, Einstein's own opinion is illuminating. Shankland had asked Einstein during their first conversation in 1950 whether "he felt that writing out the history of the Michelson–Morley experiment would be worthwhile":

> He said, 'Yes, by all means, but you must write it as Mach wrote his *Science of Mechanics*.' Then he gave me his ideas on historical writing of science. 'Nearly all historians of science are philologists and do not comprehend what physicists were aiming at, how they thought and wrestled with their problems. Even most of the work on Galileo is poorly done.' A means of writing must be found which conveys the thought processes that lead to discoveries. Physicists have been of little help in this, because most of them have no 'historical sense.' Mach's *Science of Mechanics*, however, he considered one of the truly great books and a model for scientific historical writing. He said, 'Mach did not *know* the real facts of how the early workers considered their problems,' but Einstein felt that Mach had sufficient insight so that what he said is very likely correct anyway. The struggle with their problems, their trying everything to find a solution which came at last often by very indirect means, is the correct picture.' (*op. cit.*, p. 50).

In discussing the approach of "nearly all historians" (perhaps somewhat too brusquely), Einstein accentuates the need for historical work

EINSTEIN AND THE ''CRUCIAL'' EXPERIMENT   981

to deal with the private phase of scientific effort—how a man thinks and wrestles with a problem. In discussing the physicists themselves (perhaps also somewhat too brusquely), Einstein accentuates the need for a particular kind of historical sense, one that largely intuits how a scientist may have proceeded, even in the absence of "the real facts" about the creative phase. It is a challenging statement, nothing less than a recommendation to adopt for research in the history of science a lesson Einstein had learned from his research in physics: *Just as for doing physics itself*, Einstein here advises the historian of science to leap across the unavoidable gap between the necessarily too limited "facts" and the mental construct that must be formed to handle the facts. And in such an historical study, *as in physics itself*, the solution comes often "by very indirect means," and the best outcome one can hope for is not certainty but only a probability of being "correct anyway."

One can well agree with this call for new ways of writing about the thought processes that led to major discoveries, without having to agree at this

FIG. 2. Einstein on bicycle in front of Hale's Solar Laboratory during his 1931 visit to Pasadena. (Courtesy of California Institute of Technology.)

late date with the particular model of Mach's *Science of Mechanics*. The most obvious difficulty with following Einstein's advice is of course the unspecifiability of "sufficient insight." Another is that any study of the processes of discovery—that evanescent, partly unconscious, unobserved, unverbalized, unreconstructable activity—is by definition going to yield a report with apparently vague and contradictory elements. Yet another is that the invitation to leap courageously may cause even some of the most pertinent and easily available documents (historical "facts") to be overlooked. And a fourth trouble is that there are some problems which now seem largely unsolvable by any method, and may remain so for a long time: the problem of genius, of reasons for thematic and aesthetic choices, of interaction between private and public science, not to speak of the problem of induction.

Ernst Mach himself would perhaps have objected to Einstein's characterization of his work on the history of science, laudatory though it was intended to be. But Einstein was right nevertheless in ascribing the Mach, and recommending to others, an unconventional method, despite the difficulties and dangers it may pose. For in this way one can at least hope to penetrate beyond the more pedestrian or trivial aspects of an historic case of such magnitude, to recognize more fully the feat of intellectual daring and superb taste that was needed to create the theory.

Of course, experiments are essential for the progress of science. Of course, the chain from a puzzling new experiment to a theoretical scheme that explains it is the more usual process, particularly in the everyday accomplishments of most scientists. Of course, experiments influenced also the developing thought processes of the young Einstein struggling with the problem of understanding electrodynamics in a new way, to get at the "heart of the matter." Of course, Michelson's experiment played an indirect role in this, if only because Einstein found one inadequacy in H. A. Lorentz's theory of electrodynamics to be that "it was leading to an interpretation of the result of the Michelson–Morley experiment which seemed to be artificial," as Einstein wrote in his Michelson Centenary message.

And yet, the experimenticist fallacy of imposing a logical textbook sequence, rigorously from experiment to theory, must be resisted. Not only

982   G E R A L D   H O L T O N

is it false to the actual development of historic cases of thought processes that may have led to major scientific discoveries. Not only might the doctrine, if taken seriously, inhibit creative work in science. But worse, by drawing attention primarily to the externally visible clay that provides factual support and operational usefulness for the developed theory, it does not do adequate justice to the full grandeur of the theory. The basic achievement of Einstein's theory was not to preserve hallowed traditional concepts or mechanisms; it was not to produce a logically tightly structured sequence of thoughts; it was not to build on a beautiful and pedagogically persuasive experiment. Rather, the basic achievement of the theory was that even at the cost of sacrificing all these, it gave us a new unity in the explanation of nature.

On Einstein's Invention of Special Relativity

Arthur I. Miller

University of Lowell & Harvard University

Albert Einstein distinguished sharply between the context of discovery and the context of invention. For example, in a letter of 6 January 1948 to Michele Besso, Einstein wrote:

> Mach's weakness, as I see it, lies in the fact that he believed more or less strongly, that science consists merely of putting experimental results in order; that is, he did not recognize the free constructive element in the creation of a concept. He thought that somehow theories arise by means of *discovery* [durch Entdeckung] and not by means of *invention* [nicht durch Erfindung]. (1972, p. 391, italics in original).

By "invention" Einstein meant the mind's ability to leap across what he took to be the essential abyss between perceptions and data on the one side in order to create concepts and axioms on the other. Although sometimes Einstein interchanged the terms discovery and invention, he deemed invention to be the route of creative scientific thinking. Today I shall develop that to a good approximation this motif describes Einstein's invention of special relativity in contrast to the efforts of Max Abraham, H.A. Lorentz and Henri Poincaré, among others, who sought to discover an electromagnetic basis for physical theory.[1]

One of the problems in dealing with the emergence of special relativity is the dearth of extant archival documentation from the period 1901-1905. The scenario that I shall present here is woven from conjectures that receive support over the widest possible number of archival, primary and secondary sources.[2] It is apropos to quote from Einstein's own view of what constitutes worthwhile history of science — namely, that history of science should "convey the thought processes that led to discoveries" and thus focus on how physicists "struggle with their problems, their trying to find a solution which came at last often by very indirect means, is the correct picture." The historian of science, continued Einstein, just like the scientist, can

PSA 1982, Volume 2, pp. 377-402
Copyright © 1983 by the Philosophy of Science Association

378

construct a scenario that is at best "only very likely correct."[3]

The three main paths to special relativity are philosophy, physics and technology. Before 1905 many German scientists and engineers considered these paths to intersect, but Einstein combined them in a unique way. First I shall review these paths as they confronted Einstein in 1905 through works with which he was acquainted. This will enable us to glimpse the dynamics of thinking that led Einstein to realize that the contemporaneous notion of time had to be re-examined, which led him to invent the concept of the relativity of simultaneity, and then to propose the axioms of special relativity. This sequence squares with Einstein's own words to the effect that his creative thinking occurred in a predominantly nonformal manner, usually with the aid of visual thinking, and axiomatization followed.[4] From analysis of his published papers during 1901-1907, and his correspondence, there emerges a portrait of someone who was aware of developments in philosophy, technology and physics. Biographies that depict him as virtually cut off from the world of science do an injustice to his achievements in 1905. In fact, his awareness of contemporary research renders all the more dazzling how he opted for a course of action so different from that of other physicists in 1905.

1. Philosophy

By the end of the 19th century Mach's criticisms in his 1883 *Science of Mechanics* of Newtonian absolute space and time as idle metaphysical conceptions had been heeded by most major physicists (e.g., Mach 1883, pp. 273, 279-284). On the other hand, Mach's empiricistic emphasis did not deter philosopher-scientists such as Heinrich Hertz and Henri Poincaré from exploring the Kantian notion of a priori organizing principles. Compared with Mach these men placed a strong premium on the primacy of the imagination and the deep meaning of mathematics. Before 1905 Einstein had read at least the introduction to Hertz's 1894 *Principles of Mechanics* (Sauter 1965). Although Einstein, as did others, disagreed with Hertz's implementation of his program, by 1905 Einstein may well have been impressed with the power of an approach to physical theory that emphasized axioms and mental pictures over empirical data.

During 1902-1904 Einstein had read Poincaré's 1902 book *Science and Hypothesis* (Seelig 1954, p. 69). In this display of intellectual virtuosity Poincaré developed arithmetic, geometry, classical mechanics and electromagnetic theory within a neo-Kantian framework that emphasized the close connection between the construction of pre-scientific and scientific knowledge.[5] Although the connection in the hands of Poincaré turned out to be too close for Einstein, variations on this theme would pervade his scientific work. In fact, it is reasonable to conjecture that Poincaré's writings impressed Einstein no less than did Mach's "incorruptable skepticism" (Einstein 1946, p. 21). (See Miller 1981b and in press for evidence to support this conjecture.)

I turn next to survey those experimental data that we can suggest with some certainty Einstein was aware of in 1905, and how they were explained by current physics.

2. Data

    In the 1905 special relativity paper entitled, "On the Electrodyna-
mics of Moving Bodies," Einstein referred to "unsuccessful attempts to
discover any motion of the earth relatively to the 'light medium'."
(Einstein 1905, p. 392).[6] He gave no explicit citations to these
ether-drift experiments; in fact, there are no literature citations
anywhere in the paper. Then he specialized to the class of experiments
accurate to the "first order of small quantities" — that is, accurate
to order (v/c) where v is the velocity of the earth relative to the
ether and c is the velocity of light in the free ether that is measured
by an observer at rest in the ether. (Einstein 1905, p. 392). For
example, Martinus Hoek's (1868) experiment is an ether-drift experiment
of first-order accuracy. Among the wider class of "unsuccessful attempts" was the second-order experiment of Albert A. Michelson and
Edward W. Morley (1887). No serious historian has argued that Einstein
was unaware of this experiment before 1905   (Holton 1969).   Then there
were the second-order null experiments of Lord Rayleigh (1902) and
Dewitt B. Brace (1904) to detect double refraction in isotropic crystals.
Rayleigh and Brace's experiments were analyzed in Max Abraham's widely-
cited paper of 1904 in the *Annalen der Physik* entitled, "On the Theory
of Radiation and of Radiation Pressure." Since it is inconceivable
that someone would publish in a journal to which he had no access, we
can safely conjecture that Einstein had at least perused Abraham's 1904
paper. Another group of first-order experiments are positive experiments that did not attempt to detect the earth's motion through the
ether — for example, observations of stellar aberration and experiments to verify Fresnel's dragging coefficient that were performed by
Hippolyte Fizeau (1851), and Michelson and Morley (1886). In these
experiments v was the velocity of ponderable matter relative to the
laboratory.

    Einstein's emphasis in the special relativity paper on the first-
order experiments supports his later comments that stellar aberration
and Fizeau's experiment had been the most influential of the often-
cited empirical data to his thinking toward the special relativity
theory. In a number of places Einstein later recalled of these data,
"They were enough."[7] Their explanation, as we shall discuss in a moment, turned about a quantity that also explained systematically the
failure of first-order ether-drift experiments — namely, Lorentz's
local time coordinate. In historical context, Einstein's predilection
for first-order data is not surprising. Others also emphasized these
data over the 1887 Michelson-Morley data. For example, Lorentz especially framed his electromagnetic theory of 1892 to systematically
include Fresnel's dragging coefficient. Second-order data are nowhere
mentioned in his seminal opus, *Maxwell's Electromagnetic Theory and Its
Application to Moving Bodies* (1892a). In a short sequel publication
(1892b) Lorentz proposed the ad hoc hypothesis of contraction to explain the irksome 1887 Michelson-Morley experiment. Another example
is Max Abraham's widely-read book *Theory of Electricity* (1904c), where

380

he considered Fizeau's 1851 experiment to be critical for deciding between Lorentz's and Hertz's theories of the electrodynamics of moving
bodies (pp. 435-436). Almost certainly Einstein had read Abraham's
1904 book, and we know from a letter of 19 February 1955 to his bio-
grapher Carl Seelig (Born 1969, p. 104) that before 1905 Einstein had
read Lorentz's 1892 presentation of his new version of Maxwell's theory
as well as Lorentz's monograph entitled, *Treatise on Electrical and
Magnetic Phenomena in Moving Bodies* (1895). In this treatise Lorentz
reviewed available first-order experiments and explained them in a more
systematic manner than he had in 1892.

3. H.A. Lorentz's Electromagnetic Theory

Since the results of Einstein's analysis of Lorentz's electromagnetic theory played a key role in his thinking toward the special relativity theory, it is useful to analyze how Lorentz treated data on the
optics of moving bodies. Lorentz postulated that the sources of the
electric and magnetic fields are submicroscopic particles that move
about in an all-pervasive absolutely resting ether. The five fundamental equations of Lorentz's theory are:

$$\vec{\nabla} \times \vec{E} = -\frac{1}{c}\frac{\partial \vec{B}}{\partial t}$$

$$\vec{\nabla} \times \vec{B} = \frac{1}{c}\frac{\partial \vec{E}}{\partial t} + \frac{4\pi}{c}\rho\vec{v}$$

$$\vec{\nabla} \cdot \vec{E} = 4\pi\rho \qquad\qquad\qquad \textit{Maxwell-Lorentz Equations}$$

$$\vec{\nabla} \cdot \vec{B} = 0$$

$$\vec{F} = \rho\vec{E} + \rho(\vec{v}/c) \times \vec{B} \qquad \textit{Lorentz Force Equation}$$

where $\vec{E}$ and $\vec{B}$ are the electric and magnetic fields, respectively, and
ρ is the electron's volume density of charge. Since Lorentz's fundamental equations are written relative to a reference system at rest in
the ether, which we shall call S, then c is the velocity of light measured in S, and $\vec{v}$ is the electron's velocity relative to S. Lorentz's
electromagnetic field equations possess the property expected of a
wave theory of light; namely, that relative to a reference system that
is fixed in the ether, the velocity of light is independent of the
source's motion and is a determined constant c. But this may not
necessarily be the result of measuring the velocity of light in an
inertial reference system. Therefore, the reference systems in the
ether are preferred reference systems. But experiments had not revealed any effect of the earth's motion through the ether on optical
or electromagnetic phenomena.

The situation concerning the velocity of light was as follows:
Newtonian mechanics postulates that the velocity of light emitted from
a moving source should differ from the velocity of the light emitted
from a source at rest by the amount of the source's velocity; consequently, the velocity of light c' from a source moving with velocity v

381

is given by Newton's law for the addition of velocities, $\vec{c}' = \vec{c} + \vec{v}$.
On the other hand, according to the wave theory of light, the quantity
c' measured by an observer at rest in the ether is c' = c, and Lorentz's
equations agreed with this requirement. But the effect of the ether on
measurements of the velocity of light done on the moving earth was expected to yield a result in agreement with Newtonian mechanics, i.e.,
observers on the moving earth should measure the velocity c' = c + v,
where now c' is the velocity of light relative to the earth and v is
the ether's velocity relative to the earth. However, experiments accurate to second-order in (v/c) led to the result c' = c. To this order
of accuracy, optical and electromagnetic phenomena occurred on the moving
earth as if the earth were at rest in the ether. Therefore, to second-
order accuracy in (v/c), Newtonian mechanics and electromagnetism are
inconsistent with optical phenomena occurring in inertial reference
systems.

In the 1895 treatise Lorentz responded systematically to the failure
of the first-order ether-drift experiments to detect any effects of the
earth's motion on optical phenomena as follows. For regions of the
ether that were free of matter, or within neutral matter that is neither magnetic nor dielectric, the Lorentz equations for the electric
field $\vec{E}$ and magnetic field $\vec{B}$ in the ether-fixed reference system is the
set of Equations (S) in Figure 1.

Lorentz's Theorem of Corresponding States

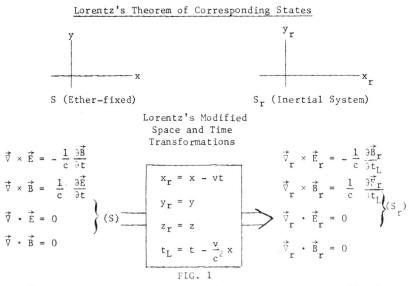

FIG. 1

Lorentz's modified space and time transformations contain the "local
time coordinate" t_L and Lorentz referred to the electromagnetic field
quantities $\vec{E}_r = \vec{E} + \vec{v}/c \times \vec{B}$ and $\vec{B}_r = \vec{B} - \vec{v}/c \times \vec{E}$ as "new" vectors.

382

Applying the modified space and time transformations to the Eqs. (S),
Lorentz obtained their analogues in the inertial reference system S_r.
We can appreciate Lorentz's achievement at a glance because to first
order in the quantity (v/c), the Lorentz equations have the same form
in the inertial system S_r as in the ether-fixed system S, and thus the
same physical laws pertain to both these reference systems; in other
words, to this order of accuracy optical experiments could not reveal
the motion of the system S_r. Lorentz called this stunning and desira-
ble result the "theorem of corresponding states" (p. 84); it rested on
the hypothesis of the mathematical "local time coordinate" t_L. The real
or physical time was still the one from the transformations of classical mechanics that leave Newton's second law covariant and to which I
shall refer as the Galilean transformations. Thus, although the physical time of electromagnetic theory was not absolute in the meaning of
Newton (1687, p. 6), i.e., of flowing without reference to anything
external, there was an absoluteness associated with the time in Lorentz's
theory because there was no reason to believe that the time in different
reference systems should differ, in agreement with our perceptions.

In summary, to order (v/c) the velocity of light in S_r was the same
as in S, i.e., c' = c, and so to this order of accuracy Lorentz's theorem of corresponding states removed the inconsistency between Newton's
prediction and that of electromagnetic theory, in favor of electromagnetic theory.

In the final chapter of the 1895 treatise Lorentz presented the ad
hoc hypothesis of contraction. This blemish on Lorentz's theory was
emphasized in several of Poincaré's philosophic-scientific criticisms
of the state of physical theory, e.g., in *Science and Hypothesis* (1902,
p. 172). Nevertheless, Poincaré was impressed with the theorem of corresponding states because absolute motion had no place in his philosophy or physics, a point that he addressed with great vigor in *Science
and Hypothesis*.

4. The Frontier of the Physics of 1905: The Electromagnetic World-
    Picture

Owing to the wide range of impressive successes of Lorentz's theory
in 1900 Wilhelm Wien suggested the "possibility of an electromagnetic
foundation for mechanics" (1900, p. 97), i.e., pursuance of an electromagnetic world-picture based on Lorentz's electromagnetic theory, instead of the relatively unsuccessful inverse research effort called the
mechanical world-picture. A far-reaching implication of the electromagnetic world-picture was that the electron's mass originated in its
own electromagnetic field, and should therefore be velocity dependent.
During 1901-1902 this implication was verified by Walter Kaufmann (1901,
1902a, 1902b). Theoretical developments of Kaufmann's data were rapid.
During 1902-1903 Kaufmann's colleague at Göttingen, Max Abraham, formulated a theory of a rigid sphere electron that agreed with Kaufmann's
latest data (1902a, 1902b, 1903). However, Abraham's theory offered no
explanation for the Michelson-Morley experiment and disagreed with the
data of Rayleigh and Brace.

Prompted by new second-order data and by Kaufmann's measurements, as well as Poincaré's criticisms, Lorentz (1904b) proposed his own theory of a deformable electron that also agreed with Kaufmann's data. In this theory the contraction hypothesis was deemed no longer to be ad hoc because it became one of several hypotheses that could explain more than one experiment accurate to second order in v/c — that is, in addition to Michelson and Morley the experiments of Rayleigh, Brace and of F.T. Trouton and H.R. Noble (1903). Poincaré agreed (1904). But in 1904 Abraham immediately countered with a severe criticism of Lorentz's theory (1904b): Lorentz's deformable electron was unstable because it could explode under the enormous repulsive forces between its constituent parts.

Einstein's predictions in the special relativity paper for the transverse and longitudinal masses for the electron are ample evidence that he was aware of Kaufmann's data and Abraham's elegant attempts in 1902-1903 to explain them (1905, p. 414). In fact, Abraham's most complete exposition of his theory of the electron was published in the *Annalen der Physik* in 1903. But before 1905 Einstein had neither seen Lorentz's 1904 paper on his new theory of the electron nor was he aware of Poincaré's elegant (1905) version of Lorentz's theory in which Poincaré ensured the stability of Lorentz's electron.[8]

To summarize: By 1905 physicists believed that fundamental physical theory was proceeding in the correct direction. The most successful theory was Lorentz's. It was what Einstein would refer to as a "constructive theory" (1919) because it explained such effects as the directly unobservable contraction of length, the observed variation of mass with velocity, and the fact that the measured velocity of light always turned out to be the same as if the earth were at rest in the ether, all to be caused by the interaction of constituent electrons with the ether.

The stage was set for a great new era in science to emerge from what everyone considered to be the cutting edge of scientific research — namely, high-velocity phenomena. But as we shall see this turned out not to be the case. We move next to the technological path, which is an area where basic problems were deemed unimportant for progress on the frontier of physics — the area of electromagnetic induction. In German-speaking countries problems in this area interfaced technology and basic research; they received a particularly interesting treatment because, as the intellectual historian J.T. Merz has written, the "German man of science was a philosopher." (Merz, Vol. 1, p. 215).

5. Electromagnetic Induction, Kant, Electrons, and Relativity

Michael Faraday's law of electromagnetic induction states that the rate at which the lines of force are cut determines the strength of the current induced; furthermore, the direction and magnitude of the induced current depend on only the relative velocity between the circuit and magnet. But Faraday's interpretation of electromagnetic induction differed when circuit and magnet were rotating relative to each other.

384

An apparatus of the sort in Figure 2 was important to Faraday because by 1851 he had convinced himself that lines of force participated in the magnet's linear motion, but not in its rotation.[9]

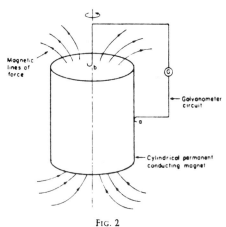

FIG. 2
An example of a unipolar dynamo.

In Figure 2 the wire loop makes sliding contact with the rotating magnet's periphery and touches one of the magnet's poles; hence, the sort of electromagnetic dynamo in Figure 2 became known as a unipolar dynamo. Faraday's interpretation of unipolar induction is: if the loop rotates counterclockwise, a current appears in it owing to its cutting the lines of force. If the loop remains at rest, and the magnet rotates clockwise, a current of the same magnitude and direction appears in the loop; the current was thought to originate in the magnet owing to the magnet's rotating through its own lines of force. When magnet and loop turn together in the same direction, there is no net current in the loop because the loop's current is cancelled by the magnet's internal current. Thus, two different explanations were required for the current induced in the loop, depending on whether the loop or magnet rotated. Clearly, the experimental data are more easily understood in terms of relative motion between the wire loop and the magnet with its co-moving lines of force.

Representing the magnetic field by lines of force found fertile ground in the German-speaking countries because the Kantian philosophical background of German scientists and engineers led them to consider lines of force to be a fundamental *Anschauung* (Miller 1981a). In this context *Anschauung* refers to the intuition through pictures formed in the mind's eye from previous visualizations of physical processes in the world of perceptions; *Anschauung* is superior to viewing merely with senses. In short, lines of force were seen everywhere. There ensued a controversy in the German-speaking scientific-engineering community on

the merits of the *Anschauungen* of Faraday versus the rotating-line view;
experiments were offered to distinguish between them. One engineer emphasized that this controversy was "not an academic moot point" for two
chief reasons (Weber 1895, p. 514): (1) There was intense research and
development toward unipolar direct-current and alternating-current gen-
erators owing to the relative simplicity of these machines vis-à-vis
multipolar machines. (2) Calculating the effect of a rotating magnet
on stationary armature coils required knowing whether the magnetic field
lines remained stationary; this problem appeared also in multipolar
machines with stationary field magnets. So seriously did this controversy rage that despite available technology in Germany the first large-
scale unipolar dynamo was built elsewhere.

With confidence I can conjecture that Einstein encountered this controversy during his independent and classroom studies at the ETH,
Zurich, and then while a Patent Clerk. Among the evidence to support
this conjecture is that while at the ETH Einstein had read August
Föppl's 1894 book, *Introduction to Maxwell's Theory of Electricity* (Holton (1967-68). There Föppl discussed unipolar induction and frankly confessed that fundamental conceptual, and perhaps also quantitative problems, remained in electromagnetic induction (Chapter 5). But the
relativity paper itself contains irrefutable evidence that Einstein had
thought about unipolar induction, because in the paper's Section 6 he
audaciously dismissed as "meaningless" problems concerning the seat of
emf in "unipolar machines." (Einstein 1905, p. 406). Einstein had come
to realize that the crux of the conceptual problems in unipolar induction could be resolved through studying the simple case of a permanent
magnet and conducting loop in relative inertial motion. But we are
getting ahead of ourselves.

On the theoretical side of the problem of electromagnetic induction,
Föppl's 1894 book and the 1904 edition rewritten by Abraham (1904c)
emphasized that the laws of mechanics and electromagnetism are used to
explain electromagnetic induction as follows (pp. 398-409): (1) The
laws of mechanics are used to discuss the motion of circuit and magnet;
(2) the current in the loop depends on only the relative velocity between loop and magnet, and this agrees with the principle of relative
motion from mechanics. This principle asserts that physical phenomena
occur as a result of the relative motion of material bodies, and that
the laws of mechanics are the same for all inertial systems. Abraham
stressed that experiments supported the principle of relative motion
for electrodynamics only to first order in (v/c) (p. 404).

In a widely-cited review paper entitled *Extensions of Maxwell's
Theory. Theory of Electrons* (1904a), Lorentz moved toward extending his
electromagnetic theory to take account of the bulk magnetic and dielectric properties of moving bodies. Lorentz's extension required his
postulating three different sorts of electrons: conduction electrons
responsible for electric current; polarization electrons that produce
dielectric properties; and magnetization electrons to explain magnetic
properties of matter. Once again the local time coordinate played a
key role. For example, using the local time coordinate Lorentz calcu-

386

lated the electric field due to the moving magnet, and its effect on a
stationary current loop — that is, a constructive explanation for
electromagnetic induction (pp. 235-239). Lorentz himself, however,
found his calculation puzzling because it meant that a permanent conducting magnet, supposedly comprised only of conduction and magnetization electrons, developed dielectric properties (see also Lorentz 1910,
p. 225).

Besides appending hypotheses of additional electrons to an already
overburdened superstructure, Lorentz by the end of 1904 was still unable
to explain the stability of his electron (1905, p. 100). Einstein
recalled in 1946 that he had been aware of this shortcoming which he
took to be a "fundamental crisis" (p. 37). He probably learned of this
defect in Lorentz's theory from Abraham's 1904 paper in the *Physikalische Zeitschrift* where Abraham first proposed it (1904b). Then there
was Max Planck's 1900 explanation for the radiation from hot bodies
(Planck 1900, 1901), which brings us to the final ingredient necessary
for describing Einstein's unique view of the physics of 1905 — problems
concerning radiation.

6. Radiation

Planck's radiation law shocked Einstein. For not only did it violate
classical electromagnetism and mechanics, but Einstein could not even
adapt classical physics to it, as he recalled (1946, p. 53). To Einstein alone Planck's law was a "second fundamental crisis" in addition
to the instability of Lorentz's electron (1946, p. 37). He chose to
accept Planck's law and to see what general conclusions could be inferred concerning the electromagnetic and mechanical world-pictures. It
is reasonable to conjecture that Planck's law piqued Einstein's interest
in problems concerning radiation. Evidence in support of this conjecture and of Einstein's 1946 recollections can be found in his 1907 paper
"On the Consequences of the Principle of Relativity for the Inertia of
Energy" (1907a).[10] To the best of my knowledge this paper contains Einstein's earliest published views on the physics of 1905. Einstein wrote
that his investigations of the structure of light revealed that classical electromagnetism failed in volumes of the order of the electron's.
Thus, the electromagnetic world-picture could not succeed.[11] His Brownian motion investigations had yielded a similar result for mechanics;
hence, exit any possibility for a mechanical world-picture. From these
startling results Einstein concluded that the laws of mechanics, thermodynamics and electromagnetism can be employed in space and time regions
large enough so that fluctuations can be neglected. This is precisely
what he did in the special relativity paper. But prior to this point,
wrote Einstein in 1946, he "despaired of the possibility of discovering
the true laws by means of constructive efforts based on known facts."
(1946, p. 53).

In summary, by the end of 1904 Einstein was aware of most of the key
"known facts", e.g., the ether-drift experiments and Kaufmann's data, of
philosophical currents (of which I shall have more to say in a moment),
of fundamental problems in electromagnetic induction, and had concluded,

387

as he said later, that the physicists of 1905 were theorizing "out of
[their] depth." (1923, p. 484).

7. Mental Pictures and Concepts

    At this juncture the three paths to special relativity merged in
Einstein's predilection for visual thinking. It is safe to conjecture
that the value of this sort of thinking for a fundamental analysis of
science had been underscored for Einstein chiefly by Ludwig Boltzmann
and Hermann von Helmholtz.[12] Two of Einstein's more reliable biogra-
phers, Philipp Frank and Carl Seelig, wrote of Einstein's having read
these master philosopher-scientists on his own at the ETH (Frank 1947,
p. 38; Seelig 1954, p. 30). Einstein read others of Helmholtz's
"papers and lectures" at the informal study group called the "Olympia
Academy" of which he was a charter member along with Conrad Habicht and
Maurice Solovine (Einstein 1956, p. viii).

    In his 1897 *Lectures on the Principles of Mechanics*, Boltzmann sug-
gested that "unclarities in the principles of mechanics [derive from]
not starting at once with hypothetical mental pictures but trying to
link up with experience from the outset." (Boltzmann 1897a, p. 225).
Elsewhere in 1897 Boltzmann delved into an analysis of perceptions in
order to set the notion of useful mental pictures as best as he could.
He defined useful mental pictures to be common to various perceptual
complexes or groups of phenomena, thereby providing an understanding of
the complexes. Boltzmann referred to useful mental pictures as "concepts." (Boltzmann 1897b, p. 59).

    In 1894 Helmholtz had given a similar definition of a concept in an
essay where he discussed the important role of mental pictures and non-
verbal thinking in the creative process: "...the memory images of pure
sense impressions can also be used as elements in combinations of ideas,
where it is not necessary or even possible to describe those impressions
in words and thus to grasp them conceptually...the idea of the stereo-
metric form of a material object plays the role of a concept formed on
the basis of the combination of an extended series of sensuous intuition
images." (Helmholtz 1894, p. 507).

    In the special relativity paper the measuring rods and inertial re-
ference systems of the thought experimenter play the role of concepts
in the meaning of Boltzmann and Helmholtz. The relativistic notion of
simultaneity and the two principles of relativity are concepts that
function also as a priori organizing notions in a sense analogous to
that of Hertz and Poincaré.[13] Three decades later Einstein would define
a concept in much the same way as did Boltzmann and Helmholtz — namely,
as a "memory-picture" that serves as an ordering element for perceptions
(Einstein 1946, p. 7). Needless to say, I am not claiming that Einstein consciously took his notions of "concept" and of the dynamics of
thinking from these philosophers. Rather the relationship of Einstein
to Boltzmann and Helmholtz, among the other philosophers and scientists
that he had read before 1905, was best described by Frank, who wrote
that from their works Einstein "learned how one builds up the mathema-

388

tical framework and then with its help constructs the edifice of physics." (1947, p. 38). It is to how Einstein built up the edifice of
physics that I turn next.

8. The Trend Toward Axiomatization

As Wien put it in 1900, Hertz's program was diametrically opposite
to the goal of an electromagnetic world-picture. Yet Wien hoped that
in due course the new research effort would measure up to the logical
structure of Hertz's mechanics. Abraham wrote similarly of Hertz in
the *Annalen* paper of 1903 in which he presented his electron theory.
Wien's suggestion for an electromagnetic world-picture appeared in the
1900 Lorentz *Festschrift* and also in the *Annalen* of 1901. We can conjecture that before 1905 Einstein had read the *Annalen* version of Wien's
paper. The trend toward axiomatics was in the air in 1905. It was to
be achieved by degrees, after all extant data were analyzed. Yet owing
to his research on fluctuation phenomena, by 1905 Einstein knew that
the electromagnetic world-picture could not succeed.

That notions of axiomatization were mentioned at key places in the
scientific literature permits us to begin to fill in Einstein's 1946
recollection of a turning point in his thinking toward special relativity. Rather tersely in his "Autobiographical Notes" he wrote that in
the midst of the "despair" of 1904 he decided to try his hand at a
purely axiomatic theory that could offer a "universal formal principle
...the example I saw before me was thermodynamics." (1946, p. 53). A
problem ripe for historical analysis is how Einstein realized that the
principles of thermodynamics were exemplars for the double-edged sword
that he would use in his Gordian resolution of problems confronting the
physics of 1905 — that is, his realization that the laws of thermodynamics could be wielded as restrictive principles that *also* make no
assumptions on the constitution of matter. In the "Autobiographical
Notes" he mentioned only their usefulness as restrictive principles. A
possible reply to this problem is: Does not everyone know these two
properties of the laws of thermodynamics? Although this reply may be
the case, it omits assessing the problem in historical context. For
implicitly knowing the double-edged power of thermodynamics is quite
different from applying it to the physics of 1905. I add that the
straight-forward reply is also uninteresting to the historian who seeks
to place Einstein in the matrix of science in 1905.

A conjecture that may be closer to the mark is that in Abraham's
1904 *Annalen* paper "On the Theory of Radiation and of Radiation Pressure," Einstein saw the laws of thermodynamics applied specifically
because they make no assumptions on the constitution of matter. Abraham resorted to this strategy because he refused to permit the data of
Rayleigh and Brace to falsify his theory of the electron. After acknowledging this disagreement between data and theory Abraham went on to
emphasize the need for further research on the optics of moving bodies.
His route was to investigate the thermodynamics of radiation in order
to analyze a quantity indigenous to every theory of the electron —
namely, the Poynting vector, all the while working within a theory that

389

made no assumptions on the constitution of matter.

Seventy-odd pages after Abraham's 1904 paper in the *Annalen* is Einstein's paper in which he calculated that black-body radiation could exhibit observable fluctuations (1904). Then, as I conjectured earlier (with support, in part, from Einstein's own testimony in 1907), Einstein went on in unpublished calculations to conclude that classical electromagnetism failed in spatial regions of the size of the electron. Thus, Abraham's elegant applications of the thermodynamics of radiation may well have struck a responsive chord in someone who "despaired of discovering the true laws [through] constructive efforts based on known facts."

9. Einstein's Method for Axiomatization

Einstein's next step was to decide what was to be axiomatized. He recalled that he found the clue in a thought experiment over which he had been pondering since 1895, an experiment that permitted him to supplement the "known facts." (1946, p. 53).[14]

The essence of this experiment is: (1) Current physics asserts that an observer who is moving alongside a light wave whose source is in the ether should be able to discern the effects of his motion by, for example, measuring the velocity of light — that is, this observer can perform all possible ether-drift experiments. (2) But to Einstein it was "intuitively clear" that the laws of optics could not depend on the state of the observer's motion (1946, p. 53). Statements (1) and (2) are mutually contradictory, and to Einstein this thought experiment contained a paradox.

In their own way, Lorentz and Poincaré were also attempting to resolve this paradox, one degree of accuracy in (v/c) at a time that is, by proposing hypotheses such as the local time and the ad hoc contraction of moving bodies. Einstein, on the other hand, would realize that the key to the paradox lay, as he said later, in the "axiom of the absolute character of time, viz., of simultaneity [which] unrecognizedly was anchored in the unconscious." (Einstein 1946, p. 53).

The scientific-philosophic-technological route toward this realization we can conjecture was as follows. Of all the ether-drift experiments that had been performed, those accurate to first-order in (v/c) were explainable by Lorentz's theorem of corresponding states. Lorentz's theorem was based on a set of modified space and time transformations that included the mathematical local-time coordinate (see Figure 3). Although Lorentz's equations remained unchanged under the modified space and time transformations, this was not the case for the equations of mechanics. Thus, according to Lorentz's modified transformations, the equations of mechanics were not the same in every inertial reference system. But this result violated Newton's exact principle of relative motion (Corollary V from the *Principia*, 1687, p. 20), which states that no mechanical experiment could reveal an inertial system's motion. The mathematical statement of Newton's principle of relative motion lies in

390

Galilean Transformations		*Lorentz's Modified Space and Time Transformations*
$x_r = x - vt$		$x_r = x - vt$
$y_r = y$		$y_r = y$
$z_r = z$		$z_r = z$
$t_r = t$		$t_L = t - (vx/c^2)$
(a)		(b)

FIG. 3

(a) The coordinates (x_r, y_r, z_r, t_r) and (x, y, z, t) refer to the two *inertial* reference systems S_r and S.

(b) The coordinates $(x_r, y_r, z_r, t_r = t)$ and (x, y, z, t) refer to the inertial reference system S_r and to the ether-fixed reference system S; and t_L is the mathematical "local time coordinate."

*   *   *

the Galilean transformations of Figure 3(a), where for mechanics the reference systems S_r and S are both inertial reference systems. Consequently, the transformation rules for the laws of mechanics and of electromagnetism depended on two different notions of time — one physical and the other mathematical — contrary to Lorentz's goal of reduction. Thus whereas most scientists in 1905 considered the inexactness in electromagnetism of the principle of relative motion from mechanics as basic to the tension between these disciplines,[15] Einstein delved deeper and found that current physics rendered mechanics and electromagnetism incompatible.

The texts of Föppl and Abraham emphasized to Einstein the intimate connection between mechanics and electromagnetism for interpreting electromagnetic induction. Faraday's law was basic also to Lorentz's electromagnetic theory where the local time made it possible to calculate the moving magnet's effect on either a resting conductor or the open circuit in unipolar induction. But, as we recall, this calculation rested on certain special assumptions on the constitution of matter. Moreover, Lorentz's electromagnetic theory explained electromagnetic induction in two different ways, depending on whether the conductor or the magnet was moving, even though the physically measurable effect was a function of only their relative velocity. For Einstein this was, as he wrote in 1905, an asymmetry that was "not inherent in the phenomena." (1905, p. 392). The new ingredient in Einstein's notion of symmetry, compared to those found in the contemporaneous scientific literature, was that Einstein's was essentially aesthetic and concerned physical processes and not mathematics.[16] Furthermore, symmetry considerations enabled Einstein to further supplement the "known facts." In 1919 Einstein recalled that he found the asymmetry in the interpretation of electromagnetic induction to be so "unbearable" that it forced him to focus on the necessity for an equivalence of viewpoints between observers on the wire loop and on the magnet, rather than on the source of the moving magnet's electric field as had Lorentz (Holton 1971-1972, p. 364).

10. Einstein's Invention of the Special Theory of Relativity

All of the ingredients were now present for Einstein to invent special relativity. As we saw, the Galilean transformations could not explain the velocity of light measurements. So Einstein assumed that the modified transformations of Lorentz with their apparently ubiquitous local time coordinate would have to play a role in relating phenomena between reference systems. From Lorentz's modified transformations in Figure 3(b), Einstein could have deduced a new result for the addition of velocities:[17] $w_r = (w - v)/(1 - vw/c^2)$, where w_r is the velocity of a moving point relative to S_r and w is its velocity relative to S. For the case of $w = c$, then $w_r = c' = c$, instead of Newton's addition law of velocities, $c' = c - v$. Thus Einstein could have realized that, to first order in (v/c), the addition law for velocities from Lorentz's modified transformations produced a result that agreed with the intuition of his thought experimenter. Since Lorentz's modified transformations differed from the Galilean transformations only in the local time coordinate, Einstein recalled in a review paper of 1907 asking himself whether the local time might be *the* time (1907b, p. 413)? But this step required asserting that the times in inertial reference systems differed because the local time coordinate depends on their relative velocity. Yet the absoluteness of time had always been accepted. Furthermore, the thought experimenter's intuition demanded an examination of the mathematical relation between Newton's principle of relative motion and Lorentz's theorem of corresponding states: after all, the spatial coordinates of Lorentz's modified transformation equations of 1895 were mathematically the same as the ones in the Galilean transformations, and the local time coordinate had been invented for use in electromagnetic theory. Einstein's imposing a Newtonian unity on Lorentz's modified transformation equations meant also his asserting the equivalence of the reference systems S and S_r. This was a big step, for it meant rejecting Lorentz's ether, and with it the dynamical interpretations of an enormously successful and, for the most part, satisfying theory. To his surprise, Einstein had found that the notion of time was both the central point and the Achilles heel of the electrodynamics of moving bodies.

For aid in analyzing the nature of time, Einstein recalled that he benefited from the "critical reasoning [in] David Hume's and Ernst Mach's philosophical writings." (Einstein 1946, p. 53). In the 1948 letter to Besso from which I quoted earlier, Einstein wrote that "Hume had a greater effect on me" than Mach. In his 1944 paper entitled, "Remarks on Bertrand Russell's Theory of Knowledge," Einstein provided a discussion of Hume's insights that enable me to offer the following conjecture. [We know that Einstein had read Hume's *Treatise of Human Nature* during 1902-1904 at the informal study group called the "Olympia Academy" (Einstein 1956, p. viii; Seelig 1954, p. 69).] My conjecture is that Hume's analysis of sense-perceptions offered strong evidence that exact laws of nature could not be induced from empirical data.

392

And Hume's analyses of the limits imposed by sense-perceptions on notions of causality and of time enabled Einstein to realize that the
high value of the velocity of light, compared with the other velocities we encounter daily, had prevented our appreciating that "the absolute character of time, viz., of simultaneity, unrecognizedly was anchored in the unconscious." (Einstein 1946, p. 53). Poincaré's preg-
nant statement in *Science and Hypothesis* to the effect that we have no
direct intuition of the simultaneity of two distant events may also
have been helpful here (p. 90).

But Einstein's fundamental analysis of physical theory went far beyond science as it is normally conceived: from an analysis of physics
and technology per se as regards, for example, electromagnetic induction, into an analysis of sensations, and then much as Boltzmann had
suggested into the "mode of our own thinking." (Boltzmann 1899, p.
104). Einstein concluded that the customary sensation-based notions
of time and simultaneity resulted in a physics burdened with asymmetries, hypotheses concerning the constitution of matter, unobservable
quantities, and ad hoc hypotheses. Thus prevented from lapsing into a
dogmatic slumber, Einstein found suggestive a combination of axiomatics
with a neo-Kantian view that was predicated on the usefulness of organizing principles such as the second law of thermodynamics. From the
eclectic philosophical view that had been tempered by over four decades
of scientific research, in 1944 Einstein wrote that in his opinion
Kant's reply to Hume's crushing message was that in thinking one must
use "concepts to which there is no access from the materials of sensory
experience, if the situation is viewed from the logical point of view."
(1944, p. 287).

And to a good approximation this is what Einstein had done in 1905,
based in part on a nascent form of his mature philosophical position.
For example, he enlarged Newton's principle of relativity to include
Lorentz's theory even though existing data did not indicate to be
the case. Einstein's masterstroke was to accomplish this extension of
Newton's principle of relativity by linking the low-velocity experiment
of magnet and conductor in relative inertial motion, that he analyzed
in the first paragraph of the paper, to the unnamed ether drift experiments. He reasoned that electromagnetic induction depends on the laws
of mechanics and electromagnetism, which covers optics too; then he
"conjectured" that Newton's principle of relativity covers these three
disciplines to order (v/c). (1905, p. 392). He boldly continued by
raising this widened principle of relativity to a postulate or axiom
to be applied in its fullest thermodynamic sense. Then he went on to
propose another postulate that is basic to every wave theory of light:
in Lorentz's theory this postulate asserts that in the ether-fixed
system S the velocity of light is independent of the source's motion
and is always c. In special relativity which does not contain Lorentz's
ether, this is axiomatic in every inertial reference system although
empirical data gave support to only $(v/c)^2$.[18]

On these two axioms Einstein built a theory of pristine beauty that
accounts for but does not explain phenomena, and it makes no assump-

393

tions on the constitution of matter. Like thermodynamics it was a theory of principle (Einstein 1919). In the special relativity paper he deduced from the two axioms the relativity of time and of simultaneity. Thus, the order of development of these concepts in the published paper is the reverse of their invention.

11. Concluding Remarks

We can depict Einstein's approach to axiomatization in the special relativity theory as a hybrid version of the views of Boltzmann who emphasized mental pictures, but who was not daring enough in raising concepts to axiomatic status owing to his anti-Kantian stance; of Hertz's brilliant use of axioms as organizing principles, but within a scheme that was inapplicable to real mechanical phenomena; of Poincaré's far-reaching neo-Kantian organizing principles, but which placed too much emphasis on perception and empirical data; of Mach, who, with Poincaré, presented to Einstein paths not to follow without care — that is, not to emphasize perceptions; of Wien's suggestion of axiomatics as a goal; and of Abraham's 1904 paper that suggested an approach both to fundamentals and to not permitting data to decide the issue, a lesson soon to be useful to Einstein.

All of this enabled Einstein to cull away nonessentials such as those arising from excessive preoccupation with ether-drift experiments and to move boldly counter to the prevailing currents of theoretical physics by resolving problems in a Gordian manner — by inventing a view of physics in which certain problems do not occur, a view in which the 1895 paradox became mere fiction. He accomplished this feat by realizing the necessity for a demarcation between data and the mental constructs that are concepts or axioms, in order to pluck out of the air the version of space-time that was out there beyond our perceptions.

Today I have tried to demonstrate how complex is the mosaic from which special relativity emerges. Further work remains to be done to better understand how this scenario was played out against the backdrop of one man's view of philosophy, physics and technology in 1905.

394

Notes

[1] See Miller (1981b) for details concerning the theoretical and experimental physics of the 19th and early 20th centuries that were germane to
the emergence of the special theory of relativity. The book's bibliography contains an extensive listing of secondary sources.

[2] The problem of source materials for the emergence of special relativity is analyzed in my (1981b, esp. Chapter 1). See also Holton (1960).

[3] Report of an interview with Einstein on 4 February 1950 (Shankland
1963, p. 48).

[4] Curiously enough, in their contribution to this session, John Earman,
Clark Glymour and Robert Rynaciewicz completely deemphasize the philosophical component in Einstein's thinking toward special relativity.
Instead these three philosophers of science depict Einstein as having
discovered special relativity as a result of manipulating primes and
exchanging v with -v in certain transformation rules for the electric
and magnetic fields from Lorentz's electromagnetic theory. The result
of their speculations is a caricature of Einstein that is antithetical
to Einstein's own description of his creative thinking and to his own
emphasis on the importance of philosophical considerations toward inventing special relativity.

[5] By neo-Kantian I mean that Poincaré based his philosophical view on
two synthetic a priori organizing principles, namely, the principle of
mathematical induction (1902, p. 13) and the notion of continuous groups
of transformation (1902, p. 70). For further discussion of Poincaré's
philosophy of science see Miller (1981b and in press).

[6] For the purpose of serious historical analysis I had to retranslate
Einstein's relativity paper from the *Annalen* version for my book (1981b),
where the retranslation appears in the Appendix (pages 391-415). The
hitherto most frequently quoted English translation is by W. Perrett and
G.B. Jeffery in the Dover reprint volume, The Principle of Relativity.
But that contains some substantive mistranslations, infelicities, and
outdated Britishisms. For example, Einstein's second principle of the
relativity theory is mistranslated as: "Any ray of light moves in the
'stationary' system of co-ordinates with the determined velocity c,
whether the ray be emitted by a stationary or by a moving body." The
correct translation is: "Any ray of light moves in the 'resting' coordinate system with the definite velocity c, *which is independent* of
whether the ray was emitted by a resting or by a moving body" (italics
added to indicate a key phrase that was omitted in the Dover translation).
The Dover translation was made from a retypeset version of Einstein's
relativity paper that had appeared in a Teubner reprint volume, thereby
adding to the misprints in the original *Annalen* version. In addition,
the Dover translation does not distinguish between Einstein's footnotes
and those added to the Teubner edition by Arnold Sommerfeld. [The history of the Dover volume is in my 1981b, pages 391-392.] This state of

395

affairs is an example of the pitfalls inherent in using a translation that was not made from the original paper and of the importance of going back to the original papers.

Hereinafter all citations to Einstein's relativity paper are to the translation in my book.

[7]Report of an interview with Einstein of 4 February 1950 (Shankland 1963, p. 48). Further evidence from Einstein's own writing (e.g., Einstein 1907b, p. 413) is discussed in Miller (1981b, Chapter 3).

[8]Needless to say, one can be aware of the results of a paper without having actually seen it. Thus, it is pertinent to mention that Wilhelm Wien (1904) summarized certain results in Lorentz's new theory of the electron and Emil Cohn (1904) discussed Lorentz's (1904b) space and time transformations. Although Einstein probably read Wien's 1904 *Annalen* paper, there is no extant documentation that he had seen the more detailed paper of Cohn that contained the exact Lorentz transformations. [See Miller (1981b) for discussion of the papers of Wien and Cohn.]

[9]Faraday (1852, pages 336-337, §§3088-3090). See Miller (1981a) for discussion of Faraday's experiments with rotating magnets and for analysis of problems concerning unipolar induction. Miller (1981b, esp. Chapter 3) focuses on the importance of unipolar induction to Einstein's thinking toward the special theory of relativity.

[10]I have analyzed this important paper in Miller (1981b).

[11]See Miller (1981b, Chapter 2) for historical evidence to support the conjecture that in 1904 Einstein had deduced the wave-particle duality for light, which he did not publish until 1909 (Einstein 1909a, 1909b).

[12]See my (1981b, 1983, in press) for a development of this conjecture for Boltzmann and Helmholtz. The importance of visual thinking for Einstein was first emphasized by Holton (1971-1972).

[13]See my (1981b and in press) for a comparative analysis of the philosophies of science of Poincaré and Einstein.

[14]Einstein's stunning use in thought experiments of images constructed from objects that had actually been perceived is an example of his utilizing the notion of *Anschauung* toward inventing the special theory of relativity. [See my (1983, in press) for further development.]

[15]This problem is connected with the inability of mechanics to explain the measured velocity of light and to whether the principle of action and reaction from mechanics is applicable to electromagnetic theory. [See Abraham (1904c).] For example, whereas Hertz's theory is compatible with both principles of mechanics it fails to explain Fizeau's 1851 experiment. By 1904 mechanical models of the ether were not seriously discussed because most physicists deemed them to be unfruitful.

396

[16]See my (1981b), especially Chapters 1 and 3, for a discussion of the notions of symmetry used by Abraham, Lorentz and Poincaré.

[17]This result turned out to be valid to all orders in (v/c). It is noteworthy that the new addition law for velocities is not necessarily a relativistic result. In fact, the new addition law is valid in every theory of a deformable electron that transforms according to the generalized Lorentz transformations:

$$x' = \ell\gamma(x - vt)$$
$$y' = \ell y$$
$$z' = \ell z$$
$$t' = \ell\gamma(t - \frac{v}{c^2} x)$$

where $\gamma = (1 - v^2/c^2)^{-\frac{1}{2}}$ and ℓ is a function of the magnitude of the relative velocity v between the two reference systems related by the transformations. For Lorentz's theory of the electron and for special relativity $\ell = 1$. For Paul Langevin's theory of a deformable electron $\ell = \gamma^{-1/3}$. Thus, Langevin's electron when in motion undergoes a contraction in the direction of its motion and an expansion in the transverse direction so that its volume remains unchanged. But the same velocity addition law holds because ℓ cancels when the quantity dx/dt is calculated. Poincaré (1906) investigated the class of electron theories that transform according to the generalized Lorentz transformations where $\ell \neq 1$ (see Miller 1973, 1981b esp. Chapter 1).

[18]Einstein went on to remove the asymmetries inherent in the contemporaneous treatment of electromagnetic induction in Section 6 of the relativity paper. Incidentally, contrary to Earman, Glymour and Rynaciewicz, Part II of the relativity paper (comprised of Sections 6, 7, 8, 9 and 10) is not a virtual potpourri of topics in dire need of an editor. Rather Part II is a quite "logical" application of the kinematics that Einstein developed in Part I: Sections 6, 7 and 8 discuss problems that concern electromagnetic fields and not their sources — for example, the relativity of the electric and magnetic fields, stellar aberration, Doppler's principle and radiation pressure; Section 9 introduces as a source a macroscopic charged body that is nondielectric, nonmagnetic and nonconducting; and Section 10 specializes the source of Section 9 to the submicroscopic electron with no assumptions on its structure. Nor, contrary to Earman, Glymour and Rynaciewicz, is there any extant archival evidence to the effect that Einstein wrote Part II before Part I, or that he realized the relativity of simultaneity as a result of deliberations on calculational details concerning topics in Part II.

397

<u>References</u>

Abraham, Max. (1902a). "Dynamik des Elektrons." <u>Göttinger Nachrich-</u>
<u>ten</u> 20-41.

------------. (1902b). "Principien der Dynamik des Elektrons."
<u>Physikalische Zeitschrift</u> 4: 57-63.

------------. (1903). "Principien der Dynamik des Elektrons." <u>Annalen</u>
<u>der Physik</u> 10: 105-179.

------------. (1904a). "Zur Theorie der Strahlung und des Strahlungs-
druckes." <u>Annalen der Physik</u> 14: 236-287.

------------. (1904b). "Die Grundhypothesen der Elektronentheorie."
<u>Physikalische Zeitschrift</u> 5: 576-579.

------------. (1904c). <u>Theorie der Elektrizität: Einführung in die</u>
<u>Maxwellsche Theorie der Elektrizität.</u> Leipzig: Teubner.
(Revision of Föppl (1894).)

Boltzmann, Ludwig. (1897a). <u>Vorlesungen über die Principe der Mecha-</u>
<u>nik. I. Theil.</u> Leipzig: Barth. (The Preface and §§1-12 are
reprinted in Boltzmann (1974). Pages 223-254.)

-----------------. (1897b). "Über die Frage nach der objectiven
Existenz der Vorgänge in der unbelebten Natur." <u>Sitzungsberichte</u>
<u>der mathematisch-naturwissenschaftlichen classe der kaiserlichen</u>
<u>Akademie der Wissenschaften. Wien</u> 106(Abt. 2a): 83-109. (As
reprinted as "On the Question of the Objective Existence of
Processes in Inanimate Nature." In Boltzmann (1974). Pages
57-76.)

-----------------. (1899). "Über die Grundprinzipien und Grundgleich-
ungen der Mechanik." In <u>Clark University 1889-1899, decennial</u>
<u>celebration.</u> Worcester, Mass.: Clark University. Pages 261-309.
(As reprinted as "On the Fundamental Principles and Equations of
Mechanics." In Boltzmann (1974). Pages 101-128.)

-----------------. (1974). <u>Theoretical Physics and Philosophical</u>
<u>Problems.</u> (ed.) B. McGuiness, (trans.) P. Foulkes. Boston:
Reidel. (This contains English versions of selected essays from
Boltzmann's, <u>Populäre Schriften.</u> Leipzig: Barth, 1905 as well as
others of Boltzmann's writings.)

Born, Max. (1969). <u>Physics in my Generation.</u> New York: Springer-
Verlag.

Brace, Dewitt B. (1904). "On Double Refraction in Matter moving
through the Aether." <u>Philosophical Magazine</u> 7: 317-329.

Cohn, Emil. (1904). "Zur Elektrodynamik bewegter Systeme." <u>Berlin</u>
<u>Berichte</u> 40: 1294-1303.

398

Einstein, Albert. (1904). "Allgemeine molekulare Theorie der Wärme."
    Annalen der Physik 14: 354-362.

----------------. (1905). "Zur Elektrodynamik bewegter Körper."
    Annalen der Physik 17: 891-921. (Translated in Miller (1981b).
    Pages 392-415.)

----------------. (1907a). "Die vom Relativitätsprinzip geforderte
    Trägheit der Energie." Annalen der Physik 23: 371-384.

----------------. (1907b). "Relativitätsprinzip und die aus demselben
    gezogenen Folgerungen." Jahrbuch der Radioaktivität und Elektro-
    nik 4: 411-462.

----------------. (1909a). "Zum gegenwärtigen Stande des Strahlungs-
    problems." Physikalische Zeitschrift 10: 185-193.

----------------. (1909b). "Enwicklung unserer Anschauungen uber das
    Wesen und die Konstitution der Strahlung." Physikalische Zeit-
    schrift 10: 817-825.

----------------. (1919). "What is the Theory of Relativity." The
    London Times November 28, 1919. (Versions appear in A. Einstein,
    Ideas and Opinions. New York: Bonanza Books, n.d. Pages
    227-232; A. Einstein, Out of My Later Years. Totowa, New Jersey:
    Littlefield Adams and Co., 1967. Pages 54-57.)

----------------. (1923). "Grundgedanken und Probleme der Relativitats-
    theorie." Stockholm: Imprimerie royale (An address delivered
    before the Nordische Naturforscherversammlung, Göteborg, July 11,
    1923.) (As reprinted as "Fundamental Ideas and Problems of the
    Theory of Relativity." In Nobel Lectures, Physics: 1901- 1921.
    New York: Elsevier, 1967. Pages 479-490.)

----------------. (1944). "Remarks on Bertrand Russell's Theory of
    Knowledge." In The Philosophy of Bertrand Russell. (The Library
    of Living Philosophers. Volume V.) Edited by P.A. Schilpp.
    Evanston: The Library of Living Philosophers. Pages 277-291.

----------------. (1946). "Autobiographical Notes." In Albert
    Einstein: Philosopher-Scientist. (The Library of Living Philos-
    ophers. Volume VII.) Edited by P.A. Schilpp. Evanston: The
    Library of Living Philosophers, 1949. Pages 2-94.

----------------. (1956). Lettres à Maurice Solovine. (ed.) M.
    Solovine. Paris: Gauthier-Villars. (Translated into French by
    M. Solovine who also wrote an introduction.)

----------------. (1972). Albert Einstein -- Michele Besso: Corres-
    pondance 1903-1955. Paris: Hermann. (Translated into French by
    P. Speziali who also supplied notes and an introduction.)

399

Faraday, Michael. (1852). "On Lines of Magnetic Force; their definite character; and their distribution within a magnet and through space." Philosophical Transactions of the Royal Society of London 142: 25-56. (As reprinted in Experimental Researches in Electricity. Volume 3. New York: Dover, 1965. Pages 328-370.)

Fizeau, Hippolyte. (1851). "Sur les hypothesès relatives à l'éther lumineaux, et sur une expérience quî parait démontrer que le mouvement des corps change la vitesse avec laquelle la lumière se propage dans leur intérieur." Comptes rendus hebdomadaires des séances de l'académie des sciences 33: 349-355.

Föppl, August. (1894). Einführung in die Maxwell'sche Theorie der Elektricität. Leipzig: Teubner.

Frank, Philipp. (1947). Einstein: His Life and Times. New York: Knopf.

Helmholtz, Hermann von. (1894). "Über den Ursprung der richtigen Deutung unserer Sinneseindrucke." Zeitschrift für Psychologie und Physiologie der Sinnesorgane VII: 81-96. (As reprinted as "The Origin and Correct Interpretation of our Sense Impressions." In Selected Writings of Hermann von Helmholtz. (ed.) (trans.) R. Kahl. Middletown, CT: Wesleyan University Press, 1971. Pages 501-572.)

Hertz, Heinrich. (1894). Die Prinzipien der Mechanik in neuen zusammenhange. (Gesammelte Werke. bd. III.) Edited by P. Lenard. Leipzig: J.A. Barth. (As reprinted as The Principles of Mechanics. (trans.) D.E. Jones and J.T. Walley. New York: Dover, 1956.)

Hoek, Martinus. (1868). "Détermination de la vitesse avec laquelle est entrainée une onde traversant un milieu en mouvement." Archives Néelandaises des Sciences Exactes et Naturelles 3: 180-185.

Holton, Gerald. (1960). "On the Origins of the Special Theory of Relativity." American Journal of Physics 28: 627-636. (As reprinted in Holton (1973). Pages 165-183.)

--------------. (1967-1968). "Influences on Einstein's Early Work." The American Scholar 37: 59-79. (As reprinted in Holton (1973). Pages 197-217.)

--------------. (1969). "Einstein, Michelson, and the 'Crucial' Experiment." Isis 60: 133-197. (As reprinted in Holton (1973). Pages 261-352.)

--------------. (1971-1972). "On Trying to Understand Scientific Genius." The American Scholar 41: 95-110. (As reprinted in Holton (1973). Pages 353-380.)

400

--------------. (1973). Thematic Origins of Scientific Thought:
    Kepler to Einstein. Cambridge, MA: Harvard University Press.

Kaufmann, Walter. (1901). "Die magnetische und electrische Ablenk-
    barkeit der Becquerelstrahlen und die scheinbare Masse der Elek-
    tronen." Göttinger Nachrichten 143-155.

--------------. (1902a). "Über die elektromagnetische Masse des
    Elektrons." Göttinger Nachrichten 291-303.

--------------. (1902b). "Die elektromagnetische Masse des Elek-
    trons." Physikalische Zeitschrift 4: 54-57.

Lorentz, H.A. (1892a). "La théorie électromagnétique de Maxwell et
    son application aux corps mouvants." Archives Néelandaises des
    Sciences Exactes et Naturelles 25: 363-551. (As reprinted in
    Lorentz (1935-1939), Volume 2. Pages 164-343.)

------------. (1892b). "The relative Motion of the Earth and the
    Ether. Koninklijke Akademie van wetaschappen te Amsterdam 1: 74.
    (As reprinted in Lorentz (1935-1939), Volume 4. Pages 219-233.)

------------. (1895). Versuch einer Theorie der elektrischen und
    optischen Erscheinungen in bewegten Körpern. Leiden: Brill.
    (As reprinted in Lorentz (1935-1939), Volume 5. Pages 1-137.)

------------. (1904a). "Weiterbildung der Maxwellschen Theorie. Elek-
    tronentheorie." Encyklopädie der mathematischen Wissenschaften
    14: 145-288.

------------. (1904b). "Electromagnetic Phenomena in a System Moving
    with any Velocity Less than that of Light." Proceedings of the
    Academy of Sciences of Amsterdam 6: 809-831. (As reprinted in
    Lorentz (1935-1939), Volume 5. Pages 172-197.)

------------. (1905). "Ergebnisse und Probleme der Elektronentheorie."
    Elektrotechnischen Verein zu Berlin, 1904. Berlin: Springer-
    Verlag. Pages 76-124. (As reprinted in Lorentz (1935-1939),
    Volume 8. Pages 79-124.)

------------. (1910). "Alte und neue Fragen der Physik." Physika-
    lische Zeitschrift 11: 1234-1257. (As reprinted in Lorentz
    (1935-1939), Volume 7. Pages 205-257.)

------------; Einstein, A.; Minkowski, H.; and Weyl, H. (1923). The
    Principle of Relativity: A Collection of Original Memoirs on the
    Special and General Theories of Relativity. (trans.) W. Perrett
    and G.B. Jeffery. London: Metheuen. (As reprinted New York:
    Dover, 1952.)

------------. (1935-1939). Collected Papers. 9 vols. The Hague:
    Nijhoff.

401

Mach, Ernst. (1883). Die Mechanik in ihrer Entwicklung historisch-kritisch dargestellt. Leipzig: F.A. Brockhaus. (As reprinted from the 9th German edition as The Science of Mechanics: A Critical and Historical Account of Its Development. (trans.) T.J. McCormack. La Salle, IL: Open Court, 1960.)

Merz, John T. (1904-1912). A History of European Scientific Thought in the Nineteenth Century. 2 vols. Edinburgh: Blackwood. (As reprinted New York: Dover, 1965.)

Michelson, Albert A. and Morley, Edward. (1886). "Influence of Motion of the Medium on the Velocity of Light." American Journal of Science 31: 377-386.

--. (1887). "On the Relative Motion of the Earth and the Luminiferous Ether." American Journal of Science 34: 333-345.

Miller, Arthur I. (1973). "A Study of Henri Poincaré's 'Sur la Dynamique de l'Electron'." Archive for History of Exact Sciences 10: 207-328.

-----------------. (1981a). "Unipolar Induction: A Case Study of the Interaction Between Science and Technology." Annals of Science 38: 155-189.

-----------------. (1981b). Albert Einstein's Special Theory of Relativity: Emergence (1905) and Early Interpretation (1905-1911). Reading, MA: Addison-Wesley.

-----------------. (1983). "On the Origins, Methods, and Legacy of Ludwig Boltzmann's Mechanics." Scheduled to appear in Proceedings of the International Conference on Ludwig Boltzmann. Braunschweig: Vieweg.

-----------------. (in press). On the Nature of Scientific Discovery. Boston: Birkhäuser.

Newton, Isaac. (1687). Philosophiae Naturalis Principia Mathematica. London: Royal Society. (As reprinted as Sir Isaac Newton's Mathematical Principles of Natural Philosophy and his System of the World. 2 vols. (trans.) A. Motte, revised by F. Cajori. Berkeley: University of California Press, 1973.)

Planck, Max. (1900). "Zur Theorie des Gesetzes der Energieverteilung im Normalspektrum." Verhandlungen der Deutschen Physikalischen Gesellschaft 2: 237-245.

-----------. (1901). "Über das Gesetz der Energieeverteilung in Normalspektrum." Annalen der Physik 4: 553-563.

402

Poincaré, Henri. (1902). <u>La Science et l'Hypothese.</u> Paris:
    Flammarion. (As reprinted as <u>Science and Hypothesis.</u> (trans.)
    W.J. Greenstreet. New York: Dover, 1952.)

----------------. (1904). "L'état actuel et l'avenir de la Physique
    mathématique." <u>Bulletin des Sciences mathématiques et astrono-
    miques</u> 28: 302-324. (As reprinted in <u>The Value of Science.</u>
    (trans.) G.B. Halsted. New York: Dover, 1958. Pages 91-111.
    This is a translation of Poincaré's, <u>La Valeur de la Science.</u>
    Paris: Flammarion, 1905.)

----------------. (1905). "Sur la dynamique de l'électron." <u>Comptes
    rendus hebdomadaires des seances de l'academie des sciences</u> 140:
    1504-1508. (As reprinted in <u>Oeuvres de Henri Poincare.</u> Paris:
    Gauthier-Villars, 1934-1953, Volume 9. Pages 489-493.)

----------------. (1906). "Sur la dynamique de l'electron." <u>Rendiconti
    del Circolo matematico di Palermo</u> 21: 129-175. (As reprinted in
    <u>Oeuvres de Henri Poincare.</u> Paris: Gauthier-Villars, 1934-1953,
    Volume 9. Pages 494-550.)

Lord Rayleigh. (1902). "Does Motion through the Aether case Double
    Refraction?" <u>Philosophical Magazine</u> 4: 678-683.

Sauter, Joseph. (1965). <u>Erinnerungen an Albert Einstein.</u> This pham-
    phlet (unpaginated) was published in 1965 by the Patent Office in
    Bern, and contains documents pertaining to Einstein's years at
    that office as well as a note by Sauter.

Seelig, Carl. (1954). <u>Albert Einstein: Eine dokumentarische Bio-
    graphie.</u> Zürich: Europa-Verlag.

Shankland, Robert S. (1963). "Conversations with Albert Einstein."
    <u>American Journal of Physics</u> 31: 47-57.

Trouton, F.T. and Nobel, H.R. (1903). "The Mechanical Forces Acting
    on a Charged Electric Condenser moving through Space." <u>Philoso-
    phical Transactions of the Royal Society London</u> 202: 165-181.

Weber, C.L. (1895). "Über unipolare Induktion." <u>Elektrotechnische
    Zeitschrift</u> 16: 513-514.

Wien, Wilhelm. (1900). "Über die Möglichkeit einer elektromagnetischen
    Begründung der Mechanik." In <u>Recueil de travaux offerts par
    les auteurs à H.A. Lorentz.</u> The Hague: Nijhoff. Pages 96-107.
    (As reprinted in <u>Annalen der Physik</u> (1901) 5: 501-513.)

-------------. (1904). "Erwiderung auf die Kritik des Hrn. Abraham."
    <u>Annalen der Physik</u> 14: 635-637.